Two-Dimensional Materials: From Synthesis to Applications

Two-Dimensional Materials: From Synthesis to Applications

Guest Editors

Sake Wang
Nguyen Tuan Hung
Minglei Sun

Basel • Beijing • Wuhan • Barcelona • Belgrade • Novi Sad • Cluj • Manchester

Guest Editors

Sake Wang
College of Science
Jinling Institute of Technology
Nanjing
China

Nguyen Tuan Hung
Frontier Research Institute for
Interdisciplinary Sciences
Tohoku University
Sendai
Japan

Minglei Sun
Materials Science &
Engineering
University of Texas at Dallas
Richardson
United States

Editorial Office
MDPI AG
Grosspeteranlage 5
4052 Basel, Switzerland

This is a reprint of the Special Issue, published open access by the journal *Molecules* (ISSN 1420-3049), freely accessible at: www.mdpi.com/journal/molecules/special_issues/0PF00PDM2E.

For citation purposes, cite each article independently as indicated on the article page online and using the guide below:

Lastname, A.A.; Lastname, B.B. Article Title. *Journal Name* **Year**, *Volume Number*, Page Range.

ISBN 978-3-7258-3450-1 (Hbk)
ISBN 978-3-7258-3449-5 (PDF)
https://doi.org/10.3390/books978-3-7258-3449-5

© 2025 by the authors. Articles in this book are Open Access and distributed under the Creative Commons Attribution (CC BY) license. The book as a whole is distributed by MDPI under the terms and conditions of the Creative Commons Attribution-NonCommercial-NoDerivs (CC BY-NC-ND) license (https://creativecommons.org/licenses/by-nc-nd/4.0/).

Contents

About the Editors vii

Preface ix

Sake Wang, Nguyen Tuan Hung and Minglei Sun
Two-Dimensional Materials: From Synthesis to Applications
Reprinted from: *Molecules* **2025**, *30*, 741, https://doi.org/10.3390/molecules30030741 1

Vytautas Stankus, Andrius Vasiliauskas, Asta Guobienė, Mindaugas Andrulevičius and Šarūnas Meškinis
Synthesis and Characterization of Boron Nitride Thin Films Deposited by High-Power Impulse Reactive Magnetron Sputtering
Reprinted from: *Molecules* **2024**, *29*, 5247, https://doi.org/10.3390/molecules29225247 7

Jun Xing, Hongxin Wang and Fei Yan
Carbon Nitride Nanosheets as an Adhesive Layer for Stable Growth of Vertically-Ordered Mesoporous Silica Film on a Glassy Carbon Electrode and Their Application for CA15-3 Immunosensor
Reprinted from: *Molecules* **2024**, *29*, 4334, https://doi.org/10.3390/molecules29184334 24

Mariel Amparo Fernandez Aramayo, Rafael Ferreira Fernandes, Matheus Santos Dias, Stella Bozzo, David Steinberg and Marcos Rocha Diniz da Silva et al.
Eco-Friendly Waterborne Polyurethane Coating Modified with Ethylenediamine-Functionalized Graphene Oxide for Enhanced Anticorrosion Performance
Reprinted from: *Molecules* **2024**, *29*, 4163, https://doi.org/10.3390/molecules29174163 36

Wanhai Liu, Fuyan Wu, Zao Yi, Yongjian Tang, Yougen Yi and Pinghui Wu et al.
Broadband Solar Absorber and Thermal Emitter Based on Single-Layer Molybdenum Disulfide
Reprinted from: *Molecules* **2024**, *29*, 4515, https://doi.org/10.3390/molecules29184515 55

Guangzhao Wang, Wenjie Xie, Sandong Guo, Junli Chang, Ying Chen and Xiaojiang Long et al.
Two-Dimensional GeC/MXY (M = Zr, Hf; X, Y = S, Se) Heterojunctions Used as Highly Efficient Overall Water-Splitting Photocatalysts
Reprinted from: *Molecules* **2024**, *29*, 2793, https://doi.org/10.3390/molecules29122793 70

Erik Biehler, Qui Quach and Tarek M. Abdel-Fattah
Gold Nanoparticle Mesoporous Carbon Composite as Catalyst for Hydrogen Evolution Reaction
Reprinted from: *Molecules* **2024**, *29*, 3707, https://doi.org/10.3390/molecules29153707 86

Maryam Yaldagard and Michael Arkas
Enhanced Mass Activity and Durability of Bimetallic Pt-Pd Nanoparticles on Sulfated-Zirconia-Doped Graphene Nanoplates for Oxygen Reduction Reaction in Proton Exchange Membrane Fuel Cell Applications
Reprinted from: *Molecules* **2024**, *29*, 2129, https://doi.org/10.3390/molecules29092129 97

Dobrina Ivanova, Hristo Kolev, Bozhidar I. Stefanov and Nina Kaneva
Enhanced Tribodegradation of a Tetracycline Antibiotic by Rare-Earth-Modified Zinc Oxide
Reprinted from: *Molecules* **2024**, *29*, 3913, https://doi.org/10.3390/molecules29163913 125

Rodolfo Fernández-Martínez, Isabel Ortiz, M. Belén Gómez-Mancebo, Lorena Alcaraz, Manuel Fernández and Félix A. López et al.
Transformation of Graphite Recovered from Batteries into Functionalized Graphene-Based Sorbents and Application to Gas Desulfurization
Reprinted from: *Molecules* **2024**, *29*, 3577, https://doi.org/10.3390/molecules29153577 . . . . . . **141**

Ana Carolina de Jesus Oliveira, Camilla Alves Pereira Rodrigues, Maria Carolina de Almeida, Eliane Teixeira Mársico, Paulo Sérgio Scalize and Tatianne Ferreira de Oliveira et al.
Ethylene Elimination Using Activated Carbons Obtained from Baru (*Dipteryx alata* vog.) Waste and Impregnated with Copper Oxide
Reprinted from: *Molecules* **2024**, *29*, 2717, https://doi.org/10.3390/molecules29122717 . . . . . . **158**

Grazia Giuseppina Politano
Optical Properties of Graphene Nanoplatelets on Amorphous Germanium Substrates
Reprinted from: *Molecules* **2024**, *29*, 4089, https://doi.org/10.3390/molecules29174089 . . . . . . **174**

Ruba Mohammad Alauwaji, Hassen Dakhlaoui, Eman Algraphy, Fatih Ungan and Bryan M. Wong
Binding Energies and Optical Properties of Power-Exponential and Modified Gaussian Quantum Dots
Reprinted from: *Molecules* **2024**, *29*, 3052, https://doi.org/10.3390/molecules29133052 . . . . . . **183**

He Huang, Fan He, Qiya Liu, You Yu and Min Zhang
Magnetic Exchange Mechanism and Quantized Anomalous Hall Effect in Bi_2Se_3 Film with a $CrWI_6$ Monolayer
Reprinted from: *Molecules* **2024**, *29*, 4101, https://doi.org/10.3390/molecules29174101 . . . . . . **195**

Yuriko Uetake and Hiroyuki Takemura
Complex Formation of Ag^+ and Li^+ with Host Molecules Modeled on Intercalation of Graphite
Reprinted from: *Molecules* **2024**, *29*, 3987, https://doi.org/10.3390/molecules29173987 . . . . . . **206**

Azamat Mukhametov, Insaf Samikov, Elena A. Korznikova and Andrey A. Kistanov
Density Functional Theory-Based Indicators to Estimate the Corrosion Potentials of Zinc Alloys in Chlorine-, Oxidizing-, and Sulfur-Harsh Environments
Reprinted from: *Molecules* **2024**, *29*, 3790, https://doi.org/10.3390/molecules29163790 . . . . . . **214**

Jun Wang, Andong Wang, Jiayuan Liu, Qiang Niu, Yijia Zhang and Ping Liu et al.
Polyethyleneimine Modified Two-Dimensional GO/MXene Composite Membranes with Enhanced Mg^{2+}/Li^+ Separation Performance for Salt Lake Brine
Reprinted from: *Molecules* **2024**, *29*, 4326, https://doi.org/10.3390/molecules29184326 . . . . . . **223**

Xiaoping Yan, Jinguo Wang, Chao Chen, Kai Zheng, Pengfei Zhang and Chao Shen
Remote Sulfonylation of Anilines with Sodium Sulfinates Using Biomass-Derived Copper Catalyst
Reprinted from: *Molecules* **2024**, *29*, 4815, https://doi.org/10.3390/molecules29204815 . . . . . . **236**

About the Editors

Sake Wang

Sake Wang earned his PhD in physics at Southeast University, China, in 2016. During this period, he was awarded the national scholarship for doctoral students. Since 2021, he has been an Associate Professor at the Jinling Institute of Technology, China. He was a Visiting Scientist at Tohoku University, Japan, from 2019 to 2021. His current interests focus on theoretical studies of spin and valley transport, as well as valley-optoelectronic devices in two-dimensional materials. He is a PI of the National Science Foundation for Young Scientists of China and the Natural Science Foundation of Jiangsu Province. He has published 71 papers with more than 3,900 citations, and he published 29 of these papers as the first author or corresponding author. Four of his first-authored and corresponding-authored papers are in the top 1% of ESI highly cited papers. He was ranked in the World's Top 2% most-cited scientists in 2023 and 2024 by Stanford University. In addition, he has served as an Associate Editor of the *Journal of Superconductivity and Novel Magnetism* (Springer Publishing) since 2020 and as a guest editor of the *Journal of Physics D: Applied Physics* (IOP Publishing) since 2023, as well as having been an outstanding reviewer of three SCI journals.

Nguyen Tuan Hung

Nguyen Tuan Hung received his Ph.D. in Physics and Interdepartmental Doctoral Degree from Tohoku University in March 2019. Since April 2019, he has been an assistant professor at the Frontier Research Institute for Interdisciplinary Sciences (FRIS), Tohoku University. He has been visiting scholars at the Chinese Academy of Sciences (2017) under his MD Program and the Massachusetts Institute of Technology (2023–2024) under the Researcher, Young Leaders Overseas Program. In addition, he received the Aoba Society Prize for the Promotion of Science from Tohoku University (2017), Research Fellowships for Young Scientists from the Japan Society for the Promotion of Science (2018), and Prominent Research Fellow from Tohoku University (2021–2025).

Minglei Sun

Dr. Minglei Sun is a research scientist at the Department of Materials Science and Engineering at the University of Texas in Dallas. His research focuses on first-principles studies of nanomaterials and their applications, particularly in energy conversion and storage. He is actively engaged in exploring the theoretical design and optimization of advanced materials for sustainable energy solutions and electronic devices. Currently, he is working on Bayesian optimization and nonlinear physics.

Preface

The field of 2D materials has rapidly gained momentum in recent years due to its transformative potential across a wide spectrum of scientific and technological domains. This Special Issue (SI) titled "Two-Dimensional Materials: From Synthesis to Applications" brings together 17 meticulously selected papers that exemplify cutting-edge research in this exciting field. By presenting advancements that span from innovative synthesis techniques to practical applications, this compendium serves as a testament to the versatility and promise of 2D materials.

The scope of this SI is broad, encompassing fundamental studies, novel methodologies, and practical implementations. Topics range from the synthesis and characterization of materials such as boron nitride and molybdenum disulfide to their deployment in areas as diverse as optoelectronics, catalysis, and environmental remediation. These contributions collectively highlight the multidisciplinary nature of 2D materials research and underscore its relevance to both academic and industrial communities.

The motivation for compiling this SI stems from the need to provide a platform where researchers can explore the synthesis, properties, and applications of 2D materials in a cohesive and comprehensive manner. Our aim is to stimulate further innovations by fostering collaboration and dialogue among researchers from various disciplines. We hope that this reprint will not only serve as a valuable resource for experts in the field but also inspire newcomers to delve into the fascinating world of 2D materials.

We extend our heartfelt gratitude to the authors who have contributed their exceptional work to this issue. The papers featured herein represent the collective effort of researchers worldwide, i.e., from Brazil, Bulgaria, Chile, China, Greece, Iran, Italy, Japan, Lithuania, Russia, Saudi Arabia, Singapore, Spain, Turkey, and the USA. They have dedicated themselves to advancing our understanding of 2D materials. We are also deeply appreciative of the reviewers whose meticulous evaluations have ensured the scientific rigor and quality of this SI.

Additionally, we would like to acknowledge the invaluable support of the editorial team, whose guidance and expertise have been instrumental in bringing this publication to fruition. Their unwavering commitment to excellence has made this endeavor a success.

This SI is addressed to researchers, scientists, and engineers in materials science, nanotechnology, physics, chemistry, and related disciplines. We are confident that it will serve as an essential reference for those seeking to explore the synthesis, properties, and applications of 2D materials and that it will spark new ideas and collaborations that propel the field forward.

We hope you find the articles as enlightening and inspiring as we did and that you contribute to the next edition of the SI "Two-Dimensional Materials: From Synthesis to Applications, 2nd Edition".

Sake Wang, Nguyen Tuan Hung, and Minglei Sun
Guest Editors

Editorial

Two-Dimensional Materials: From Synthesis to Applications

Sake Wang [1,*], Nguyen Tuan Hung [2] and Minglei Sun [3]

1 College of Science, Jinling Institute of Technology, 99 Hongjing Avenue, Nanjing 211169, China
2 Frontier Research Institute for Interdisciplinary Sciences, Tohoku University, Sendai 980-8578, Japan; nguyen.tuan.hung.e4@tohoku.ac.jp
3 NANOlab Center of Excellence and Department of Physics, University of Antwerp, Groenenborgerlaan 171, 2020 Antwerp, Belgium; minglei.sun@uantwerpen.be
* Correspondence: isaacwang@jit.edu.cn

1. Introduction

Two-dimensional materials have become a cornerstone of modern materials science, offering unique structural, electronic, optical, and mechanical properties [1]. This SI of *Molecules* brings together 17 diverse contributions, showcasing the latest advancements in 2D material synthesis, theoretical insights, and applications [2,3]. These papers collectively highlight the transformative potential of 2D materials in energy, environmental sustainability, catalysis, sensors, and more. This editorial provides a detailed analysis and categorization of these works, emphasizing their importance and future implications.

2. Synthesis and Characterization: Building the Foundation of 2D Materials

The synthesis of high-quality 2D materials is the first step in realizing their potential. This SI includes three contributions on optimizing synthesis techniques and characterizing the resulting materials to enhance their applicability.

1. Hexagonal boron nitride thin films.
   This study employed high-power impulse reactive magnetron sputtering to deposit hexagonal boron nitride (h-BN) thin films. By tuning deposition parameters such as temperature, nitrogen flow, and time, the team achieved control over crystallinity, surface roughness, and chemical composition. The results provide insights into optimizing h-BN films for electronic and optical devices, underscoring the importance of process parametrization. (Contribution 1).

2. Carbon nitride nanosheets for silica film adhesion.
   A critical challenge in fabricating VMSF is ensuring their stability on carbonaceous electrodes. This study demonstrated that carbon nitride nanosheets act as effective adhesive layers, improving the mechanical stability of VMSF. This enhancement enabled the development of robust immunosensors with broad detection ranges, highlighting the role of hybrid material interfaces. (Contribution 2).

3. GO-based anticorrosion coatings.
   Functionalizing GO with ethylenediamine introduced a sustainable way to enhance waterborne polyurethane coatings' anticorrosion performance. The low GO content effectively improved the coatings' barrier properties, offering environmentally friendly solutions for corrosion protection. This study illustrates the potential of GO-based additives for large-scale industrial applications. (Contribution 3).

Received: 21 January 2025
Accepted: 26 January 2025
Published: 6 February 2025

Citation: Wang, S.; Hung, N.T.; Sun, M. Two-Dimensional Materials: From Synthesis to Applications. *Molecules* 2025, 30, 741. https://doi.org/10.3390/molecules30030741

Copyright: © 2025 by the authors. Licensee MDPI, Basel, Switzerland. This article is an open access article distributed under the terms and conditions of the Creative Commons Attribution (CC BY) license (https://creativecommons.org/licenses/by/4.0/).

3. Energy Applications: Powering a Sustainable Future

Energy sustainability is one of the most critical challenges of our time. Four contributions in this SI highlight the role of 2D materials in advancing energy conversion and storage technologies.

1. MoS_2-based broadband solar absorbers.
   A single-layer molybdenum disulfide (MoS_2) was used to design a broadband solar absorber with exceptional absorption efficiency, maintaining over 95% absorption across a wide spectral range. The absorber's polarization insensitivity and thermal stability make it a promising candidate for photovoltaics and other energy conversion applications. (Contribution 4).
2. Photocatalytic water splitting with GeC/MXY heterojunctions.
   GeC/MXY (M = Zr, Hf; X, Y = S, Se) heterojunctions were explored for water splitting. These structures demonstrated strong light-harvesting capabilities, small bandgaps, and efficient charge carrier separation due to built-in electric fields at the heterointerface. This research emphasizes the potential of 2D materials for producing clean hydrogen energy. (Contribution 5).
3. Catalysts for hydrogen evolution reaction.
   A gold nanoparticle–mesoporous carbon composite was developed as a catalyst for hydrogen evolution. The material achieved high stability and low activation energy, making it an efficient solution for sustainable hydrogen production. This work exemplifies the importance of 2D material composites in addressing the growing energy demands. (Contribution 6).
4. Pt-Pd nanoparticles for fuel cells.
   Pt-Pd alloy nanoparticles anchored on graphene nanoplates demonstrated enhanced mass activity and durability in oxygen reduction reactions. These materials hold significant promise for advancing proton exchange membrane fuel cells, a key technology for clean energy systems. (Contribution 7).

4. Environmental and Catalytic Applications

The environmental impact of industrial processes and the need for efficient catalysis have spurred interest in 2D materials. This SI includes three contributions to addressing these challenges.

1. Tribocatalytic degradation of antibiotics.
   Rare-earth-modified zinc oxide powders were shown to degrade tetracycline antibiotics using triboelectric effects. This approach converts mechanical energy into catalytic activity, providing a sustainable method for water purification and pollution control. (Contribution 8).
2. Gas desulfurization with recycled graphite.
   Recycled graphite from spent Zn/C batteries was transformed into reduced GO-based sorbents. These materials demonstrated competitive desulfurization performance, offering a cost-effective and environmentally friendly alternative to commercial sorbents. (Contribution 9).
3. Ethylene adsorption for agricultural applications.
   Activated carbons derived from agricultural waste were impregnated with copper oxide, significantly improving ethylene adsorption. This innovation has practical applications in prolonging the post-harvest life of fruits and vegetables, showcasing the versatility of 2D materials in agriculture. (Contribution 10).

5. Sensors and Optoelectronics

The unique electronic and optical properties of 2D materials enable their application in advanced sensing and optoelectronic devices. This SI includes two contributions to the development and design of high-performance optoelectronic devices.

1. Graphene nanoplatelets for optoelectronics.
   Graphene nanoplatelets integrated with amorphous germanium substrates exhibited enhanced optical absorption and increased refractive indices. These findings pave the way for the development of high-performance optoelectronic devices, such as photodetectors and solar cells. (Contribution 11).
2. Modified quantum dots for optical devices.
   Theoretical studies on GaAs quantum dots with modified confining potentials revealed tunable optical absorption coefficients. This study contributes to the design of new optoelectronic devices leveraging inter-sub-band transitions. (Contribution 12).

6. Theoretical Insights and Fundamental Studies

Theoretical studies provide critical insights into the behavior and design of 2D materials, guiding experimental efforts. This SI includes three contributions to the DFT calculations for the 2D materials.

1. Topological insulators and magnetic proximity.
   $Bi_2Se_3/CrWI_6$ heterostructure was studied to achieve magnetic proximity-induced spin splitting, enabling the quantum anomalous Hall effect. This study advances the field of topological insulators, essential for quantum computing and spintronics. (Contribution 13).
2. Cation–π interactions in graphene.
   Pi-stacked host molecules were synthesized to study their interactions with metal cations. The results demonstrated enhanced binding in stacked configurations, informing the design of materials for ion intercalation and energy storage. (Contribution 14).
3. Corrosion potentials of zinc alloys.
   Using DFT simulations, researchers analyzed the corrosion behavior of zinc alloys in various harsh environments. The findings provide valuable indicators for developing biodegradable metals and corrosion-resistant coatings. (Contribution 15).

7. Advanced Functional Materials

Novel functional materials were developed for targeted applications, including ion separation and chemical catalysis. This SI consists of two contributions discussing these applications.

1. GO/MXene composite membranes for ion separation.
   Polyethyleneimine-coated GO/MXene membranes exhibited high efficiency in separating Mg^{2+} and Li^+ ions from salt lake brines. This technology addresses the critical need for lithium extraction in renewable energy systems. (Contribution 16).
2. Biomass-derived catalysts for sulfonylation.
   A biomass-derived copper catalyst enabled efficient sulfonylation of aniline derivatives, demonstrating a recyclable and sustainable approach to heterogeneous catalysis. (Contribution 17).

8. Conclusions

The 17 papers in this SI of *Molecules* exemplify the versatility and transformative potential of 2D materials. From cutting-edge synthesis techniques to innovative applications in energy, environment, and technology, these studies demonstrate the profound impact of 2D materials on addressing global challenges [4].

The integration of experimental advancements with theoretical insights will undoubtedly drive future breakthroughs. As researchers continue to explore the possibilities of 2D materials, their role in shaping sustainable and efficient technologies will only expand [5]. This collection not only highlights current progress, but also sets the stage for a future where 2D materials redefine the boundaries of science and engineering. We sincerely hope the inspirations will contribute to the next edition of this SI, "Two-Dimensional Materials: From Synthesis to Applications, Second Edition" [6].

Funding: S.W. was funded by the China Scholarship Council (No. 201908320001), the Natural Science Foundation of Jiangsu Province (No. BK20211002), and Qinglan Project of Jiangsu Province of China. N.T.H. was funded by financial support from the Frontier Research Institute for Interdisciplinary Sciences, Tohoku University, Japan. M.S. was supported by funding from Research Foundation-Flanders (FWO; No. 12A9923N).

Acknowledgments: The authors would like to thank all the staff in MDPI Publishing and the editors of *Molecules* for establishing and running this SI, as well as reviewers around the globe who spent their valuable time thoroughly reviewing and improving the articles published in this SI. We also feel grateful to all the authors from Brazil, Bulgaria, Chile, China, Greece, Iran, Italy, Japan, Lithuania, Russia, Saudi Arabia, Singapore, Spain, Turkey, and the USA for choosing this SI to publish their excellent science.

Conflicts of Interest: The authors declare no conflicts of interest.

Abbreviations

The following abbreviations are used in this manuscript:

2D	Two-dimensional
GO	Graphene oxide
SI	Special Issue
VMSF	Vertically ordered mesoporous silica films
DFT	Density functional theory

List of Contributions

1. Stankus, V.; Vasiliauskas, A.; Guobienė, A.; Andrulevičius, M.; Meškinis, Š. Synthesis and Characterization of Boron Nitride Thin Films Deposited by High-Power Impulse Reactive Magnetron Sputtering. *Molecules* **2024**, *29*, 5247. https://doi.org/10.3390/molecules29225247.
2. Xing, J.; Wang, H.; Yan, F. Carbon Nitride Nanosheets as an Adhesive Layer for Stable Growth of Vertically-Ordered Mesoporous Silica Film on a Glassy Carbon Electrode and Their Application for CA15-3 Immunosensor. *Molecules* **2024**, *29*, 4334. https://doi.org/10.3390/molecules29184334.
3. Aramayo, M.A.F.; Ferreira Fernandes, R.; Santos Dias, M.; Bozzo, S.; Steinberg, D.; Rocha Diniz da Silva, M.; Maroneze, C.M.; de Carvalho Castro Silva, C. Eco-Friendly Waterborne Polyurethane Coating Modified with Ethylenediamine-Functionalized Graphene Oxide for Enhanced Anticorrosion Performance. *Molecules* **2024**, *29*, 4163. https://doi.org/10.3390/molecules29174163.
4. Liu, W.; Wu, F.; Yi, Z.; Tang, Y.; Yi, Y.; Wu, P.; Zeng, Q. Broadband Solar Absorber and Thermal Emitter Based on Single-Layer Molybdenum Disulfide. *Molecules* **2024**, *29*, 4515. https://doi.org/10.3390/molecules29184515.

5. Wang, G.; Xie, W.; Guo, S.; Chang, J.; Chen, Y.; Long, X.; Zhou, L.; Ang, Y.S.; Yuan, H. Two-Dimensional GeC/MXY (M = Zr, Hf; X, Y = S, Se) Heterojunctions Used as Highly Efficient Overall Water-Splitting Photocatalysts. *Molecules* **2024**, *29*, 2793. https://doi.org/10.3390/molecules29122793.

6. Biehler, E.; Quach, Q.; Abdel-Fattah, T.M. Gold Nanoparticle Mesoporous Carbon Composite as Catalyst for Hydrogen Evolution Reaction. *Molecules* **2024**, *29*, 3707. https://doi.org/10.3390/molecules29153707.

7. Yaldagard, M.; Arkas, M. Enhanced Mass Activity and Durability of Bimetallic Pt-Pd Nanoparticles on Sulfated-Zirconia-Doped Graphene Nanoplates for Oxygen Reduction Reaction in Proton Exchange Membrane Fuel Cell Applications. *Molecules* **2024**, *29*, 2129. https://doi.org/10.3390/molecules29092129.

8. Ivanova, D.; Kolev, H.; Stefanov, B.I.; Kaneva, N. Enhanced Tribodegradation of a Tetracycline Antibiotic by Rare-Earth-Modified Zinc Oxide. *Molecules* **2024**, *29*, 3913. https://doi.org/10.3390/molecules29163913.

9. Fernández-Martínez, R.; Ortiz, I.; Gómez-Mancebo, M.B.; Alcaraz, L.; Fernández, M.; López, F.A.; Rucandio, I.; Sánchez-Hervás, J.M. Transformation of Graphite Recovered from Batteries into Functionalized Graphene-Based Sorbents and Application to Gas Desulfurization. *Molecules* **2024**, *29*, 3577. https://doi.org/10.3390/molecules29153577.

10. Oliveira, A.C.d.J.; Rodrigues, C.A.P.; de Almeida, M.C.; Társico, E.T.; Scalize, P.S.; de Oliveira, T.F.; Solar, V.A.; Valdés, H. Ethylene Elimination Using Activated Carbons Obtained from Baru (*Dipteryx alata* vog.) Waste and Impregnated with Copper Oxide. *Molecules* **2024**, *29*, 2717. https://doi.org/10.3390/molecules29122717.

11. Politano, G.G. Optical Properties of Graphene Nanoplatelets on Amorphous Germanium Substrates. *Molecules* **2024**, *29*, 4089. https://doi.org/10.3390/molecules29174089.

12. Alauwaji, R.M.; Dakhlaoui, H.; Algraphy, E.; Ungan, F.; Wong, B.M. Binding Energies and Optical Properties of Power-Exponential and Modified Gaussian Quantum Dots. *Molecules* **2024**, *29*, 3052. https://doi.org/10.3390/molecules29133052.

13. Huang, H.; He, F.; Liu, Q.; Yu, Y.; Zhang, M. Magnetic Exchange Mechanism and Quantized Anomalous Hall Effect in Bi_2Se_3 Film with a $CrWI_6$ Monolayer. *Molecules* **2024**, *29*, 4101. https://doi.org/10.3390/molecules29174101.

14. Uetake, Y.; Takemura, H. Complex Formation of Ag^+ and Li^+ with Host Molecules Modeled on Intercalation of Graphite. *Molecules* **2024**, *29*, 3987. https://doi.org/10.3390/molecules29173987.

15. Mukhametov, A.; Samikov, I.; Korznikova, E.A.; Kistanov, A.A. Density Functional Theory-Based Indicators to Estimate the Corrosion Potentials of Zinc Alloys in Chlorine-, Oxidizing-, and Sulfur-Harsh Environments. *Molecules* **2024**, *29*, 3790. https://doi.org/10.3390/molecules29163790.

16. Wang, J.; Wang, A.; Liu, J.; Niu, Q.; Zhang, Y.; Liu, P.; Liu, C.; Wang, H.; Zeng, X.; Zeng, G. Polyethyleneimine Modified Two-Dimensional GO/MXene Composite Membranes with Enhanced Mg^{2+}/Li^+ Separation Performance for Salt Lake Brine. *Molecules* **2024**, *29*, 4326. https://doi.org/10.3390/molecules29184326.

17. Yan, X.; Wang, J.; Chen, C.; Zheng, K.; Zhang, P.; Shen, C. Remote Sulfonylation of Anilines with Sodium Sulfinates Using Biomass-Derived Copper Catalyst. *Molecules* **2024**, *29*, 4815. https://doi.org/10.3390/molecules29204815.

References

1. Avouris, P.; Heinz, T.F.; Low, T. *2D Materials: Properties and Devices*; Cambridge University Press: Cambridge, UK, 2017. [CrossRef]
2. Tian, H.; Ren, C.; Wang, S. Valleytronics in two-dimensional materials with line defect. *Nanotechnology* **2022**, *33*, 212001. [CrossRef] [PubMed]
3. Yang, E.H. (Ed.) *Synthesis, Modelling and Characterization of 2D Materials and Their Heterostructures*; Elsevier: Amsterdam, The Netherlands, 2020.

4. Enoki, T.; Ando, T. *Physics and Chemistry of Graphene: Graphene to Nanographene*, 2nd ed.; Jenny Stanford Publishing: Singapore, 2020.
5. Anasori, B.; Gogotsi, Y. (Eds.) *2D Metal Carbides and Nitrides (MXenes): Structure, Properties and Applications*; Springer: Cham, Switzerland, 2019.
6. Molecules | Special Issue: Two-Dimensional Materials: From Synthesis to Applications, 2nd Edition. Available online: https://www.mdpi.com/journal/molecules/special_issues/VY337364WE (accessed on 25 January 2025).

Disclaimer/Publisher's Note: The statements, opinions and data contained in all publications are solely those of the individual author(s) and contributor(s) and not of MDPI and/or the editor(s). MDPI and/or the editor(s) disclaim responsibility for any injury to people or property resulting from any ideas, methods, instructions or products referred to in the content.

Synthesis and Characterization of Boron Nitride Thin Films Deposited by High-Power Impulse Reactive Magnetron Sputtering

Vytautas Stankus *, Andrius Vasiliauskas, Asta Guobienė, Mindaugas Andrulevičius and Šarūnas Meškinis *

Institute of Materials Science, Kaunas University of Technology, K. Baršausko St. 59, LT-51423 Kaunas, Lithuania; andrius.vasiliauskas@ktu.lt (A.V.); asta.guobiene@ktu.lt (A.G.); mindaugas.andrulevicius@ktu.lt (M.A.)
* Correspondence: vytautas.stankus@ktu.lt (V.S.); sarunas.meskinis@ktu.lt (Š.M.); Tel.: +370-61033946 (V.S.); +370-61554257 (Š.M.)

Abstract: In the present research, hexagonal boron nitride (h-BN) films were deposited by reactive high-power impulse magnetron sputtering (HiPIMS) of the pure boron target. Nitrogen was used as both a sputtering gas and a reactive gas. It was shown that, using only nitrogen gas, hexagonal-boron-phase thin films were synthesized successfully. The deposition temperature, time, and nitrogen gas flow effects were studied. It was found that an increase in deposition temperature resulted in hydrogen desorption, less intensive hydrogen-bond-related luminescence features in the Raman spectra of the films, and increased h-BN crystallite size. Increases in deposition time affect crystallites, which form larger conglomerates, with size decreases. The conglomerates' size and surface roughness increase with increases in both time and temperature. An increase in the nitrogen flow was beneficial for a significant reduction in the carbon amount in the h-BN films and the appearance of the h-BN-related features in the lateral force microscopy images.

Keywords: hexagonal boron nitride; reactive high-power magnetron sputtering; Raman; X-ray photoelectron spectroscopy; AFM

1. Introduction

Two-dimensional (2D) nanomaterials, such as graphene, boron nitride (BN), and molybdenum disulfide (MoS_2) nanosheets, have many unique properties that can be useful for various applications, such as composites, nanoelectromechanical systems, and sensing, optoelectronic, and electronic applications. BN can form several different allotropes with either sp^2 or sp^3 bonding. The sp^2-bonded BN crystallizes in a hexagonal (h-BN) or rhombohedral (r-BN) phase, and sp^3 BN crystallizes in a cubic (c-BN) or wurtzite (w-BN) phase [1]. Hexagonal boron nitride (h-BN) is a layered 2D nanomaterial that is structurally analogous to graphene [2]. It has excellent physical properties, such as an ultra-wide bandgap (~5.96 eV) [3], a high breakdown field (11.8 MV cm^{-1}) [4], high thermal conductivity (1000 W $m^{-1}K^{-1}$) [5], good thermal and chemical stability [6], and piezoelectricity [7]. BN nanostructures also present excellent mechanical properties [8]. The atomically thin layer can be assembled with various other 2D layers to create tunneling-based devices, vertical or in-plane heterostructures, and bistable memory devices [9–11]. In particular, graphene and h-BN share very similar hexagonal crystal lattice parameters, enabling epitaxial growth of low-defect-density graphene on boron nitride [12]. Therefore, graphene/h-BN heterojunctions and multilayer-based microelectronic and photonic devices are intensively studied [9]. In addition, hBN itself is a promising material for such applications as ultraviolet-light emitters [13], single-photon emitters [14], gas barrier films [14], and tunnel magnetic resistance devices [15]. Boron nitride thin films are synthesized using various deposition techniques. Chemical vapor deposition is the most widely used method for the large-scale production of h-BN layers at a low cost [16–18]. However, this technique uses transition metals (Cu, Ni, Fe, Pt, and Ir) as substrates, and a transfer process from metal substrates to a suitable surface (generally a dielectric substrate) is required for most device applications, which probably induces impurities and mechanical damage, thereby degrading the performance of h-BN-based devices. The h-BN films deposited by metalorganic vapor-phase epitaxy [19–21] and molecular beam epitaxy [22,23] are usually grown on sapphire, which requires high substrate temperatures (>1000 °C) to compensate for this substrate's poor catalytic activity. Furthermore, h-BN has been synthesized at lower temperatures by applying physical vapor deposition methods such as radio frequency (RF) magnetron sputtering [24]. However, RF magnetron sputtering is known for low power efficiency, high cost, and constraints in terms of scaling to large surface areas [25]. Thus, the development of other physical-vapor-deposition-based hexagonal boron nitride deposition methods is necessary.

Notably, high-power impulse magnetron sputtering (HIPIMS) is already successfully used for large-area industrial-scale coating deposition [26]. Compared to RF magnetron sputtering, high-power impulse magnetron sputtering ensures higher thin-film density [27,28], much better control of the structure and stoichiometry [29], enhanced adhesion [28], and higher stability [30]. However, there are few studies on h-BN film deposition by HIPIMS [31–36]. In addition, in [31,32,34], LaB_6 targets were used for h-BN growth, because the pure boron target is insulating and, usually, it should be heated to a temperature of 500 °C or higher to ignite the unipolar sputtering discharge [31,32,34,37]. In the case of hexagonal boron nitride deposition by boron target reactive HIPIMS, only Ar and N_2 gas flow ratio effects [36] and the influence of the boron isotope used as a sputtering target material [33] were studied.

The aim and objectives of this research work are to investigate the synthesis of h-BN thin films directly on noncatalytic Si(100) substrates by applying the reactive unipolar high-power impulse magnetron sputtering (HiPIMS) technique, using a pure boron cathode and nitrogen gas, and to investigate the influence of the deposition temperature and time and the flow rate of nitrogen gas on the structure and composition of thin films. It was revealed that there is no need for high-temperature heating of the boron target to ignite the sputtering discharge, and the target temperature of 100 °C is enough for that purpose. Taking into account the finding in [36] that h-BN can be grown using nitrogen gas alone instead of the Ar/N_2 gas mixture and the fact that the use of the Ar/N_2 gas mixture provides no benefits, in the present research, boron nitride films were deposited using N_2 as both reactive and sputtering gas. However, it was revealed that the nitrogen gas flow must be maximized to minimize the concentrations of unwanted impurities, such as carbon and oxygen, in the film. It was found that the h-BN films' deposition temperature and time also influenced the h-BN structure and composition. Control of the h-BN nanocrystallite size and decreased intensity of the samples' Raman spectra luminescence hump and background were achieved.

2. Results and Discussion

In the present research, the effects of the deposition temperature and time and the nitrogen gas flow on the structure of h-BN films were investigated by Raman scattering spectroscopy. Figure 1a shows the original Raman spectra of the deposited h-BN films grown at different substrate temperatures, with a 60 min deposition time and a 152 sccm nitrogen gas flow.

The main peak at ~1370 cm^{-1} is observed at all deposition temperatures. It can be assigned as an h-BN E2g peak related to in-plane, Raman-active vibrations [38,39]. No h-BN-related peaks were observed in the spectra of the samples grown at temperatures below 480 °C. In spectra of the films grown at lower temperatures, a strong luminescence hump and luminescence background are seen, which decreases and almost disappears at higher substrate temperatures. Figure 1b shows a view of the hexagonal boron nitride-related peaks with a removed background and deconvoluted curves. The peak's central position is down-shifted with the deposition temperature. Fitted values of full width at half maximum (FWHM) and the peak's central position are shown in Figure 1c. FWHM decreases from 38.01 ± 0.96 to 28.13 ± 0.96 cm^{-1} wavenumbers with increasing temperature, while central position, as was mentioned above, shifts from 1371.3 ± 0.3 to 1368.8 ± 0.3 cm^{-1} wavenumber. Figure 1d shows crystal sizes calculated using Equations (1) and (2). The size of the crystallites calculated using FWHM increases from 4.50 ± 0.46 to 6.50 ± 0.46 nm with increasing deposition temperature from 480 to 1070 °C. A similar effect of crystallite size depending on substrate temperature was reported for h-BN films deposited by RF sputtering [40]. Crystallite sizes calculated using h-BN peak position rise with synthesis temperature from 4.50 ± 1.21 to ~10.00 ± 1.21 nm. We suppose that peak position shifting is affected by two factors—change in the crystallite sizes [41–43] and strains induced in the film [44,45]. So, the larger size of the crystallites calculated using the h-BN peak position can be explained by strain appearance in the films grown at higher temperatures, which results in additional peak shifts. There are theoretical studies about the grain size and stress relationship in BN which have established that strength, toughness, Young's Modulus, and energy release rate all have a declining trend along with a decrease in grain size. At the same time, the ultimate strain increases as grain sizes decrease. These properties stem from the heterogeneity of BN, and the effect of this heterogeneity on the behavior of grain boundaries [46] and tensile strength and strain decreased after introducing vacancy defects in the hBNNR structure [47]. In our case, the grain boundaries play the same role as vacancies—the larger the crystallites, the less grain boundaries between them.

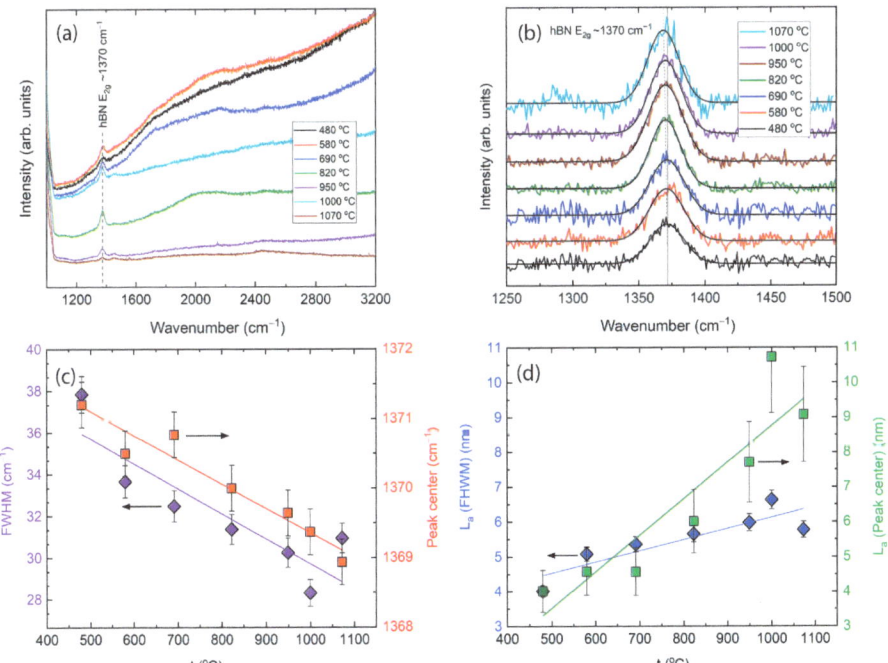

Figure 1. (**a**) Raman spectra of the h-BN films grown at different substrate temperatures and 60 min deposition time. (**b**) Raman spectra of the main peak with removed background and deconvoluted curves. (**c**) FWHM and peak center position dependence on deposition temperature. Crystallite sizes calculated using FWHM and central peak position shifting (**d**).

Figure 1 shows that larger crystallite values were calculated using the peak position compared to those obtained using the peak FWHM for samples deposited at temperatures higher than 900 °C. Boron nitride tensile strain results in E2g Raman peak downshift [38] and compressive strain—in upshift [48]. Thus, the presence of the tensile strain in boron nitride samples grown at temperatures higher than 900 °C can be supposed. That can be explained by the thermal stress appearance during the cooling due to the different thermal expansion coefficients of boron nitride and silicon [49,50]. Therefore, to avoid possible adverse effects of excessive thermal stress while maximizing h-BN crystallite size, the boron nitride growth temperature was set at 820 °C in subsequent experiments. Deposition time and nitrogen gas flow effects were investigated.

Figure 2a shows the original Raman spectra of the deposited h-BN films, grown at different times (from 30 to 180 min), 820 °C deposition temperature, and 152 sccm nitrogen gas flow. We see the main peak of hBN, attributed to ~1370 cm^{-1} of wavenumber. We can see the increase in luminescence hump with increasing deposition time. Figure 2b shows the high resolution of main peaks with removed background and deconvoluted curves. The broadened and slightly shifted peaks can be seen; we also see an increase in intensity with increasing deposition time. Fitted values of FWHM and the peak's central position are shown in Figure 2c, which shows that FWHM increases from 28.80 ± 0.96 to 34.50 ± 0.96 cm^{-1} of wavenumber with increasing deposition time, while central position, as was mentioned above, shifts from 1371.2 ± 0.3 to 1368.9 ± 0.3 cm^{-1} wavenumber. Figure 2d shows calculated crystal sizes. Using the calculation from FWHM, the crystallites' size decreases from 6.5 ± 0.46 to 4.8 ± 0.46 nm with increasing deposition time from 30 to 180 min. Using calculations from peak center shifting, we see that crystallite sizes change from 5.5 ± 0.96 to 2.3 ± 0.96 nm, and the inclination of dependence is different. As was described above, the different inclinations can be explained by the strain effect appearing in crystallites.

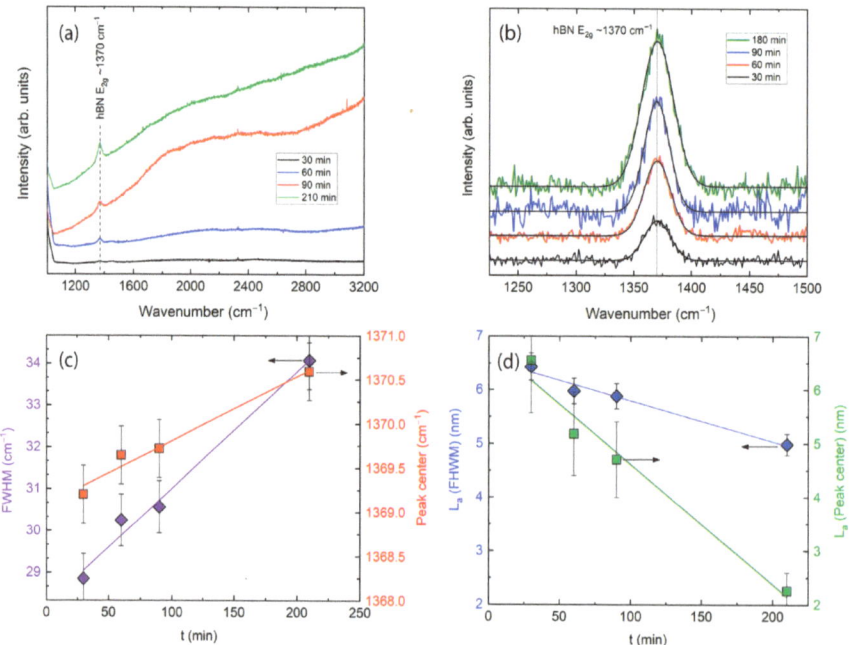

Figure 2. (**a**) Raman spectra of the h-BN films grown at different deposition times at constant substrate temperature 820 °C. (**b**) Raman spectra of the main peak with removed background and deconvoluted curves. (**c**) The dependence of the FWHM and peak center position on deposition time. Crystallite sizes (calculated using FWHM and central peak position shifting) (**d**).

Figure 3a shows the original Raman spectra of the deposited h-BN films, grown using different nitrogen gas flows at a constant 820 °C deposition temperature and deposition time of 60 min. We see the main peak of hBN, attributed to ~1370 cm^{-1} of wavenumber. A luminescence hump is clearly visible in the spectra of the sample deposited with 152 sccm of nitrogen gas flow. Increasing flow to 197 sccm gives the disappearance of the luminescence hump and background. Figure 3b shows the high resolution of main peaks with removed background and deconvoluted curves. An increase in intensity with increasing nitrogen gas flow is observed. Fitted values of FWHM and the peak's central position are presented in Figure 3c, which shows that FWHM decreases from 32.50 ± 0.96 to 31.40 ± 0.96 cm^{-1} with increasing gas flow. The h-BN peak position is slightly downshifting. Figure 3d shows calculated crystal sizes. With increasing nitrogen gas flow, the crystallite size calculated using FWHM values increases from 5.36 ± 0.46 to 5.65 ± 0.46 nm. The crystallite sizes estimated using h-BN peak position raised from 4.57 ± 0.93 to 6.02 ± 0.93 nm. As was described above, the different crystallite sizes calculated using FWHM and peak position can be explained by the strain effect appearing in crystallites.

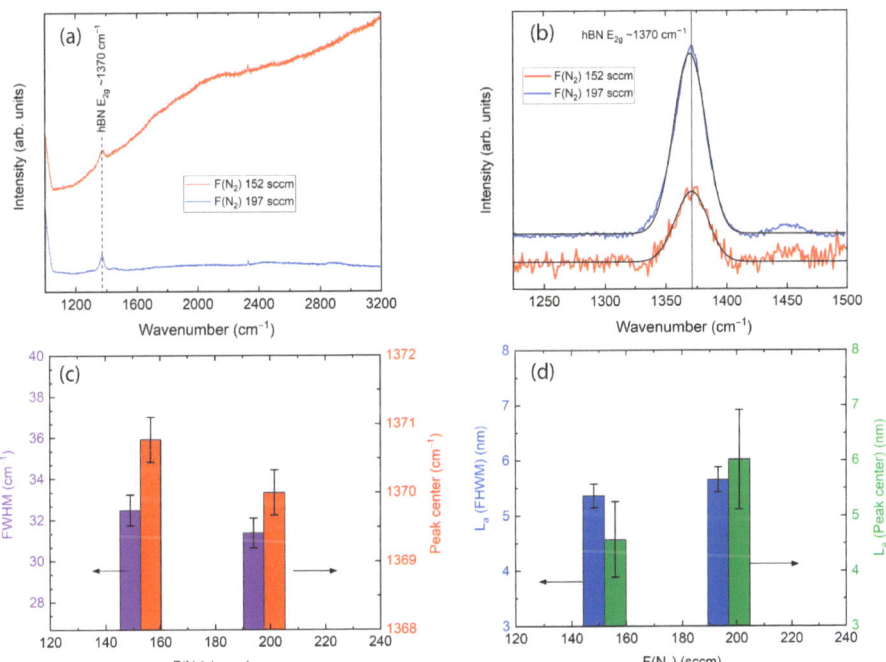

Figure 3. (a) Raman spectra of the h-BN films grown at different nitrogen gas flows at constant substrate temperature 820 °C and deposition time 60 min. (b) Raman spectra of the main peaks with removed background and deconvoluted curves. (c) FWHM and peak center position dependence on gas flow. Crystallite sizes calculated using FWHM and central peak position shifting (d).

Figure 4 shows AFM pictures, where (a–d) are from samples deposited at different temperatures (at a constant deposition time of 60 min) and (e–h) are from samples deposited at different times (from 30 to 180) when deposition temperature was constant at 820 °C. Meanwhile, Figure 4i,j shows surface roughness Rq dependent on deposition temperature and time. In (a–d), we see that deposition temperature influences the size of grains. So, at 820 °C, we see a fine-grained structure. It was determined by analyzing AFM images that there are 1–2 nm high and 20–40 nm wide elements and their derivatives (Figure 4a). At

950 °C temperature, pits and grains are seen. The pits' depth is 0.5 nm and the width is 25–30 nm. The height of the grains is 0.5–1 nm and the width is 25–30 nm (Figure 4b). A deposition temperature increase to 1000 °C results in 2–3.5 nm height and 130 nm width ribbons consisting of 200 nm long segments (Figure 4c). At 1070 °C temperature, we can observe interlaced grains of 50 nm width, 150 nm length, and 4–6 nm height (Figure 4d). In Figure 4e–h, we see that deposition time influences the size of grains. A fine-grained structure was grown after 30 min of deposition (at a constant 820 °C temperature). Grain height is up to 2 nm and width is 15–20 nm (Figure 4e). After 60 min deposition, 0.5 nm depth and 25–30 nm width pits and 0.5–1 nm height and 25–30 nm width grains are seen (Figure 4f). After 90 min, growth elements and their derivatives of 1–2 nm height and 20–40 nm width dominate (Figure 4g). The AFM image drastically changed after 180 min deposition—structural elements of 15–20 nm height and 320–360 nm width are seen (Figure 4h). Although the increase in grain size during the increase in time contradicts measurements of crystallites using Raman spectroscopy, it can be explained that grains are conglomerates that consist of nanocrystallites. That can be seen in Figure 4h, where grains consist of smaller objects corresponding to sizes determined by Raman spectroscopy. Figure 4i shows surface roughness Rq dependent on deposition temperature (at a constant deposition time of 60 min). Rq increases (from 0.5 to 1.25 nm) with increasing temperature. The effect of deposition time is similar—apart from fluctuations at 30–90 min, we see a strong roughness increase from ~0.5 to 4.7 nm (Figure 4j).

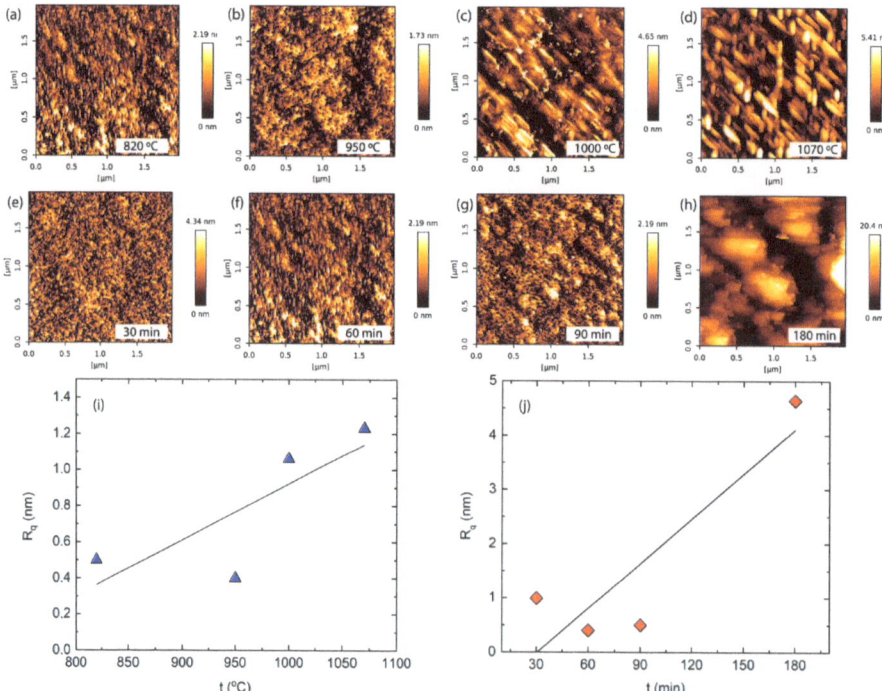

Figure 4. AFM images of samples deposited at (**a**) 820 °C, (**b**) 950 °C, (**c**) 1000 °C, and (**d**) 1070 °C temperatures and at a constant deposition time of 60 min. AFM images of samples deposited at (**e**) 30 min, (**f**) 60 min, (**g**) 90 min, and (**h**) 180 min times and at a constant deposition temperature of 820 °C. The surface roughness dependence on deposition temperature and time (**i**,**j**).

In Figure 5, AFM (a) and lateral force microscopy (LFM) (b) images of the sample deposited at 820 °C temperature, 60 min growth time, and 197 sccm nitrogen gas flow are shown. We see structural elements of 10–15 nm height, 180–200 nm width, and 200–240 nm length. Comparing the image of a sample deposited at the same conditions (Figure 4a) but with a different gas flow (152 sccm), we see that an increase in nitrogen gas flow strongly influences the grain size. The image is similar to Figure 4h; only lateral force microscopy shows (at the right) that the surface has visible hexagonal structures (marked areas). That is typical of a pure boron nitride surface [20].

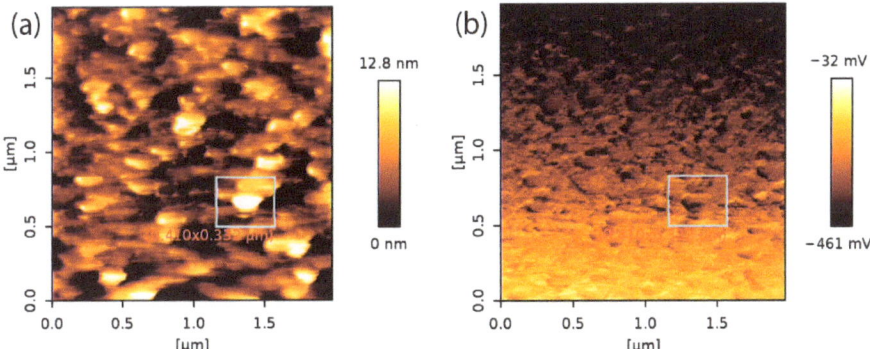

Figure 5. AFM (**a**) and LFM (**b**) images of sample deposited at 820 °C temperature, 60 min time, and 197 sccm nitrogen gas flow.

For the surface chemical composition evaluation, the samples were analyzed using XPS. The survey spectra for all samples were collected and compared. In Figure 6, spectra for several samples are depicted. The spectra showed very similar patterns for all samples; only the intensity of the main peaks for nitrogen and boron was different, according to the calculated surface atomic concentrations (Table 1).

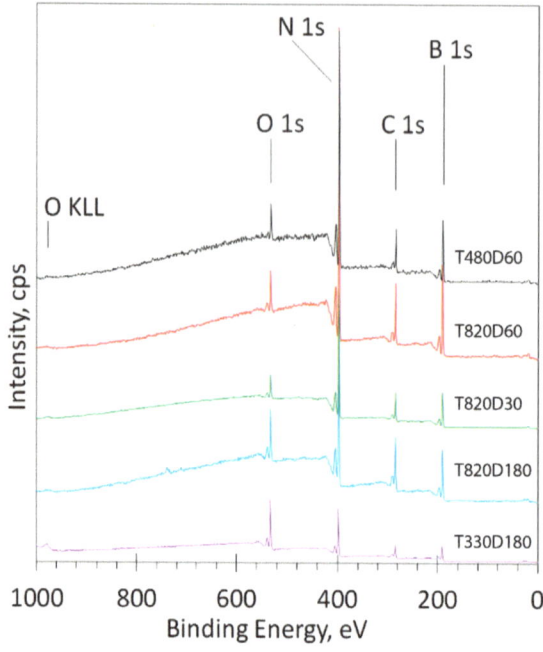

Figure 6. Comparison of the XPS survey spectra for several (as examples) samples; the elements and sample numbers are indicated for each curve.

Table 1. Calculated surface atomic concentrations. Samples are named T (temperature °C), D (deposition time min.), and N—increased (197 sccm) nitrogen flow.

Sample	O 1s	N 1s	C 1s	B 1s
T331D180	18.7	26.4	16.9	38.0
T820D180	7.9	32.1	21.7	38.3
T820D30	6.1	39.0	15.9	38.9
T820D60	4.3	37.1	15.3	43.4
T480D60	5.4	40.9	15.1	38.6
T820D60N	5.85	43.65	3.29	47.2

High-resolution XPS spectra in the N 1s and B 1s regions were scanned and deconvoluted for chemical bond detection (Figure 7). Figure 7a shows the spectra of the sample T820D60, deposited at 820 °C temperature, 60 min time, and 152 sccm nitrogen gas flow. Figure 7b shows the spectra of the sample T820D60N, deposited at the same conditions but with a larger nitrogen gas flow (197 sccm). In Figure 7a,b, the main peak at 397.8 eV indicates that most of the nitrogen is bonded to boron, as described in the literature [51–53]. The low-intensity peak at 398.5 eV was attributed to N-C bonds [51–53].

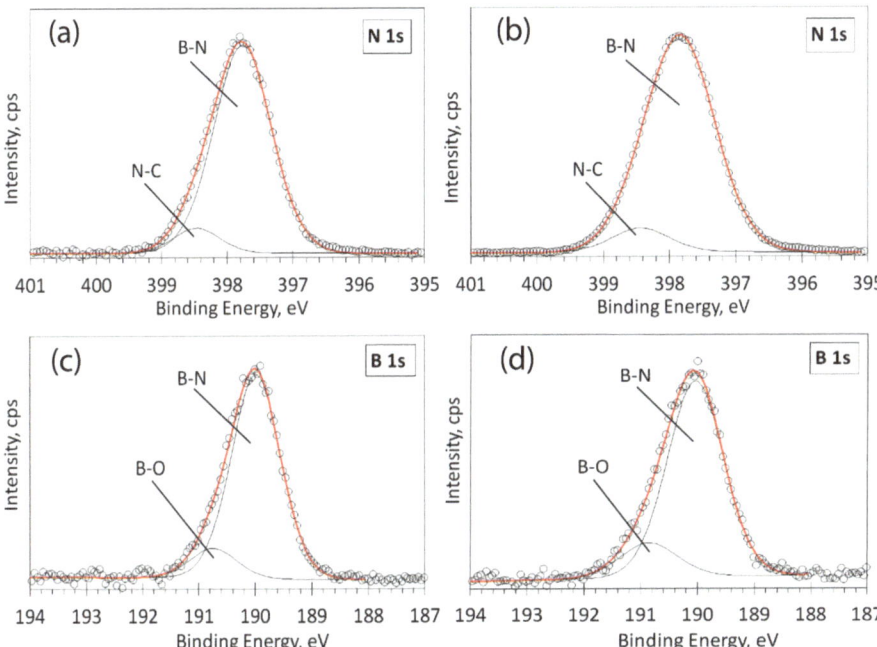

Figure 7. Deconvolution of high-resolution XPS spectra in the N 1s and B 1s regions: for the sample T820D60, deposited at 820 °C temperature, 60 min time, and 152 sccm nitrogen gas flow (**a**,**c**); for the sample T820D60N, deposited at 820 °C temperature, 60 min time, and 197 sccm nitrogen gas flow (**b**,**d**). Circles—acquired spectra; red line—envelope; thin black lines—fitted peaks.

Figure 7c,d presents the deconvolution of high-resolution XPS spectra in the B 1s region for the same T820D60 and T820D60N samples. The main peak at 190 eV indicates that most of the boron is bonded to nitrogen, in agreement with nitrogen bonds in the N 1s region. The position of this peak corresponds to known values of B-N bonds reported in the literature [51–53]. The low-intensity peak at 190.8 eV could be attributed to B-O bonds [51,52] due to adsorbed atmospheric oxygen.

3. Materials and Methods

The boron nitride thin films were synthesized by the reactive high-power impulse magnetron sputtering (Hippies) method. The initial vacuum pressure was 8×10^{-6} mBar. After reaching the initial vacuum, nitrogen (or nitrogen mixture with argon) gas was injected into the vacuum chamber. A too low or too high working pressure does not allow ignition of plasma. The working pressure was 9×10^{-3} and 1.8×10^{-2} mBar. It should be mentioned that, under normal conditions (room temperature), igniting the plasma was impossible due to the high resistivity of the boron cathode. Therefore, typically, the boron cathode is sputtered using RF magnetron sputtering systems [54–56] or impurities-added boron LaB_6 [32] and B_4C [57] cathodes are used. In our case, for ignition of plasma and carrying out the sputtering process, the boron cathode was heated with a heat lamp (at an angle of 45° and a distance of 20 cm from the cathode) in a vacuum before starting the process. The boron cathode was isolated from the cooling of the magnetron using thin (0.5 mm) quartz plates. After heating the boron cathode to a temperature of 100 °C, the resistivity of boron decreases up to 30 times (as was measured before). As a result, the plasma ignites and the plasma discharge maintains the elevated temperature of the cathode. An unbalanced magnetron (Milko Angelov Consulting Co., Plovdiv, Bulgaria) with a high-purity (99.99%) boron target (Kurt J. Lesker Company GmbH, Dresden, Germany) was used. The pulse DC power controller SPIK2000A (Melec GmbH, Baden-Baden, Germany) was applied to generate high-power pulses. Prime-grade double-sided polished n-type monocrystalline Si (100) wafers (Sil'tronix Silicon Technologies, Archamps, France) were used as a substrate. The substrate was placed parallel to the plane of the cathode at a distance of 15 cm. Impulse parameters were chosen: t_{On} = 17 µs, t_{Of} = 150 µs, impulse current I = 1.2 A. The average current was constant during all processes, about ~0.12 A, and the average voltage was ~930 V. The pulse parameters were chosen as such because obtaining the boron nitride phase was impossible when the t_{off}-to-t_{on} ratio was too low. The deposition time was chosen from 30 to 180 min (the shorter time gives too thin a film for Raman measurements, and growth time over 180 min results in no apparent changes in structure). The h-BN thin films' synthesis requires an appropriate temperature, and it is in a relatively wide range (500~1000 °C) [58–61], depending on the method and other parameters. Our purpose was to investigate the broadest possible range of temperatures for the case of our method. Samples were deposited on substrates at different temperatures (200 to 1050 °C) using different deposition times. Detailed deposition conditions are listed in Table 2. The film thickness was determined using a laser ellipsometer Gaertner L-115 operating with a He–Ne laser (λ = 632.8 nm). Raman scattering measurements were performed using a Raman microscope inVia (Renishaw Wotton-under-Edge, UK). The excitation beam from a diode laser of 532 nm wavelength was focused on the sample using a 50 × objective (NA = 0.75, Leica, Solms, Germany). Laser power at the sample surface was 1.75 mW, integration time was 10 s or 100 s, and the signal was accumulated once. The Raman Stokes signal was dispersed with a diffraction grating (2400 grooves/mm), and data were recorded using a Peltier—cooled charge-coupled device (CCD) detector (1024 × 256 pixels). The Raman setup in both Raman wavenumber and spectral intensity was calibrated using silicon. We used the Levenberg–Marquardt method to calculate the best-fit parameters that minimize the weighted mean square error between the observations in Y and the best nonlinear fit. The two main parameters of the Gauss function, full width at half maximum (FWHM) and peak center, were calculated using this method.

From the fitted Raman spectra, using two parameters (FHWM and center position (shifting of central position from largest values of crystallites—Δ)), the crystallite size L_a of the hBN films can be estimated by extending the Nemanich model for hBN microcrystallites to hBN films [41]. This method also was reported in a few other studies [62–64]. According to the Nemanich model for hBN microcrystallites [41] from the FWHM and Δ values:

$$L_a = \frac{1417}{\text{FWHM} - 8.70} \tag{1}$$

$$L_a = \frac{380 \cdot 10^{-8}}{\Delta + 0.29} \tag{2}$$

Raman scattering measurements were performed at least 3 times in different sample places, and the average values were calculated. The luminescence hump and luminescence background of the Raman scattering spectra of different h-BN films were estimated by calculating the ratio of the luminescence hump maximum intensity and h-BN Raman peak intensity, as well as the ratio of the luminescence background line slope and h-BN Raman peak intensity. Atomic force microscopy (AFM) experiments were carried out at room temperature using a NanoWizardIII atomic force microscope (JPK Instruments, Bruker Nano GmbH, Berlin, Germany). At the same time, the data were analyzed using JPKSPM Data Processing software (Version spm-4.3.13, JPK Instruments, Bruker Nano GmbH). The AFM images were collected using an ACTA (Applied NanoStructures, Inc., Mountain View, CA, USA) probe (silicon cantilever shape—pyramidal, the radius of curvature (ROC) < 10.0 nm and cone angle 20°; reflex side coating—Al with a thickness of 50 ± 5 nm, force constant ~40 N m^{-1}, and resonance frequency in the range of 300 kHz). Height, amplitude, and lateral imaging were recorded using steps with scan sizes of 2 µm and scan speeds of 1 Hz. The integral gain was set as 2, while the proportional gain was set as 5.

Pixels for samples and lines were 516 × 516, operating in contact mode. The film's surface composition was analyzed using the X-ray photoelectron spectroscopy (XPS) method. An X-ray photoelectron spectrometer XSAM800 (Kratos, Manchester, UK) equipped with a nonmonochromatic Al Kα radiation (1486.6 eV) excitation source was used for surface atomic calculations and survey spectra. A hemispherical electron energy analyzer was set to fixed analyzer transition (FAT) mode and 20 eV pass energy. A 0.5 eV increment of binding energy was used to acquire the survey. The energy scale of the system was calibrated using the peak positions of Au 4f7/2, Ag 3d5/2, and Cu 2p3/2. The base pressure in the analytical chamber was less than 5.8×10^{-8} Pa. Thermo Scientific ESCALAB 250Xi spectrometer with monochromatic Al Kα radiation (hν = 1486.6 eV) excitation was used for high-resolution spectra measurements and curve-fitting procedure. The hemispherical electron energy analyzer pass energy value of 20 eV was used. The energy scale of the system was calibrated with respect to Au 4f7/2, Ag 3d5/2, and Cu 2p3/2 peak positions. ESCALAB 250Xi Avantage software V5 was used for the peak deconvolution. All spectra fitting procedures were performed using symmetrical peaks and a 70:30 Gauss–Lorentz function ratio, except for the graphitic carbon peak, which was fitted using an asymmetrical peak shape and a Lorentzian–Gaussian function at a 70:30 ratio.

Table 2. Deposition conditions. Samples named T (temperature °C), D (deposition time min.), and N—increased (197 sccm) nitrogen flow.

Sample	N$_2$ Gas Flow, sccm	Working Pressure, mmBar	Deposition Temperature, °C	Deposition Time, min	Thickness, nm
T330D180	152	9.3×10^{-3}	330	180	255 ± 15
T820D180	152	9.3×10^{-3}	820	180	210 ± 40
T820D30	152	9.4×10^{-3}	820	30	60 ± 10
T820D90	152	9.4×10^{-3}	820	90	190 ± 20
T1000D60	152	9.3×10^{-3}	1000	60	75 ± 5
T1070D60	152	9.4×10^{-3}	1070	60	80 ± 20
T950D60	152	9.3×10^{-3}	950	60	77 ± 7
T820D60	152	9.3×10^{-3}	820	60	117 ± 3
T690D60	152	9.2×10^{-3}	690	60	150 ± 10
T580D60	152	9.2×10^{-3}	580	60	165 ± 15
T480D60	152	9.4×10^{-3}	480	60	152 ± 12
T820D60N	197	1.8×10^{-2}	820	60	107 ± 7

4. Discussion and Conclusions

As was mentioned above, a significant luminescence hump and luminescence background was seen in the Raman spectra of most h-BN films studied in this research. In this case, h-BN films contained at least 15 at.% of carbon. It should be mentioned that, in the case of the h-BN films deposited by reactive magnetron sputtering, a significant amount of carbon or oxygen impurities were found in numerous studies. Notably, the carbon content in h-BN films grown by RF reactive magnetron sputtering was as high as 20.83 at.%, and it decreased to about 8.98 at.% after the surface cleaning by argon ion [65]. In [66], the total amount of oxygen and carbon in the magnetron-sputtering-deposited h-BN films was much higher than in our study, in the 31–69 at.% range. A significant amount of carbon impurity in h-BN films was reported in [67]. In [68], the B-O component in the B1s peak was stronger than in our case, and the C-N fitting component area of the N1s peak was similar to that observed in our study. Chng, S. S. et al. found a significant amount of oxygen in most magnetron-sputtering-deposited h-BN films investigated in their study [36]. Carbon atomic concentration in h-BN films was decreased below 5 at.% only after selecting the appropriate additional hydrogen gas flow [69]. Thus, in the case of magnetron-sputter-deposited h-BN films, deposition conditions must be optimized to avoid film contamination by carbon or oxygen. In our case, the carbon amount in the films was minimized after the increase in the nitrogen gas flow and the related significant increase in the work pressure. Thus, the effects of the residual gas, along with the possible presence of the carbon-containing adsorbates, can be supposed.

Regarding the peculiarities of the Raman spectra of the h-BN films deposited in our study, it should be mentioned that the Raman spectra, very similar to those of the h-BN films grown in the present study at lower temperatures, were reported in [70]. Notably, Raman scattering spectra of h-BN films produced by vacuum annealing of the borazine amine polymer at 1600 °C temperature contained both h-BN-related sharp peak and a very broad luminescence hump with a maximum at ~2400 cm^{-1} [70]. It should be mentioned that a very broad Raman peak without any characteristic bands was observed for BC_xN (0 < x < 2) films deposited by plasma-enhanced chemical vapor deposition in the 1000–3000 cm^{-1} range [61]. It was attributed to the fluorescence from the h-BN defects without indicating the nature of those defects. A luminescence hump or luminescence background was reported for the h-BN flake implanted by high-energy Ga ions and annealed at 820 °C [71]. It was explained by defect migration due to the annealing at 850 °C and the resulting transformation of the boron vacancies to the anti-site nitrogen vacancy complex (N_BV_N) defects [71]. However, in our case, the luminescence hump and background are more pronounced for films deposited at a temperature below 850 °C, contradicting the [71] hypothesis. A broad Raman peak with a maximum in the 1150–1400 cm^{-1} range was reported for amorphous BN films containing up to 15 at.% of carbon [72]. It should be noted that the B-H stretching modes can be found at 2291 cm^{-1} and 2382 cm^{-1}, respectively [73]. Meanwhile, positions of the N-H bond vibration-related bands seem to be beyond the luminescence hump range reported in the present study (3176 cm^{-1}, 3251 cm^{-1}, and 3312 cm^{-1} wavenumbers) [73]. Thus, the luminescence hump can be partially related to the presence of the B-H bonds. On the other hand, in the present study, a luminescence hump was found for samples containing >15 at.% of carbon. At the same time, it was absent for samples containing less than 5 at.% of carbon. It is in good accordance with the studies mentioned above, in which a luminescence hump was reported for boron nitride films containing a significant amount of carbon [61,72]. The position of the C-C-bond-related Raman peaks is usually below 1700 cm^{-1} in amorphous carbon films [74]. However, C-H bond vibrations related to Raman peaks can be observed in the 2000–2200 cm^{-1} range [75]. The luminescence background can also be associated with the C-H bonds. Particularly, the ratio of the slope of the Raman spectra luminescence background line and G peak intensity is proportional to the bonded hydrogen amount in the diamond-like carbon films [74,76], and even a significant luminescence background slope with no characteristic Raman peaks was reported for hydrogenated amorphous carbon films containing >40 at.% of hydrogen [74,77]. The bonded hydrogen amount in the film should decrease with the increase in deposition temperature due to the desorption of hydrogen atoms caused by the B-H and C-H bond breakage [78–82]. That is in accordance with the present study, as seen in Figure 8. Thus, the observed luminescence hump can be explained by the formation of B-H and C-H bonds, and the presence of C-H bonds can explain the luminescence background. In this case, the increase in deposition temperature results in faster hydrogen desorption and a subsequent decrease in the h-BN luminescence-related features of the Raman film spectra. However, the increase in nitrogen gas flow was the most effective measure resulting in a significant decrease in the carbon amount in the film and a disappearance of the luminescence hump and luminescence background. The main factor can be supposed to be the increase in work pressure, while the base pressure remained the same, causing the residual gas to have a decreased influence on the growing film composition. Therefore, much fewer carbon and hydrogen atoms were incorporated into the films, and the number of C-H as well as B-H bonds was significantly decreased.

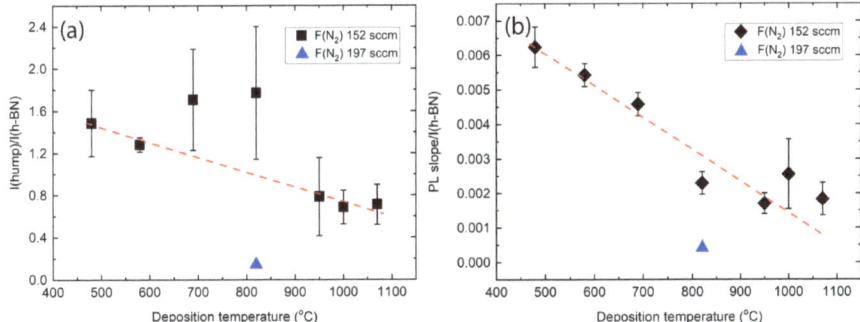

Figure 8. Luminescence hump maximum intensity and h-BN intensity ratio (**a**) and photoluminescence background line slope ratio with the h-BN peak intensity (**b**). Blue point refers to the h-BN film deposited using 197 sccm nitrogen gas flow.

In conclusion, hexagonal boron nitride films were deposited by high-power impulse reactive magnetron sputtering. Too low a nitrogen gas flow resulted in the formation of films containing a significant amount of carbon and the formation of C-H and B-H bonds. Increased synthesis temperature resulted in hydrogen desorption, less intensive hydrogen-bond-related luminescence features in Raman spectra of the films, and increased h-BN crystallite size. At the same time, in boron nitride samples grown at temperatures higher than 900 °C, tensile strain can be induced due to the thermal stress. The rise in nitrogen gas flow resulted in a significantly reduced carbon amount, the disappearance of the luminescence features in Raman scattering spectra of deposited films, and the appearance of h-BN-related features in the lateral force microscopy images of the boron nitride films. That was explained by decreased residual gas influence due to increased work pressure. Thus, h-BN film deposition temperature and nitrogen gas flow must be optimized to grow h-BN films containing fewer impurities and a more considerable amount of the h-BN phase.

Author Contributions: V.S.: Investigation, Writing—original draft, Writing—review and editing, Visualization. A.V.: Methodology, Investigation, Visualization. A.G.: Methodology, Investigation, Visualization. M.A.: Investigation, Writing—original draft, Visualization. Š.M.: Conceptualization, Methodology, Writing—original draft, Writing—review and editing, Visualization, Project administration, Funding acquisition. All authors have read and agreed to the published version of the manuscript.

Funding: This study was supported by the Research Council of Lithuania (Grant No. P-MIP-22-235).

Institutional Review Board Statement: Not applicable.

Informed Consent Statement: Not applicable.

Data Availability Statement: The original contributions presented in this study are included in the article; further inquiries can be directed to the corresponding authors.

Conflicts of Interest: The authors declare that they have no known competing financial interests or personal relationships that could have appeared to influence the work reported in this paper.

References

1. Taniguchi, T.; Sato, T.; Utsumi, W.; Kikegawa, T.; Shimomura, O. In-situ X-ray Observation of Phase Transformation of Rhombohedral Boron Nitride under Static High Pressure and High Temperature. *Diam. Relat. Mater.* **1997**, *6*, 1806–1815. [CrossRef]
2. Golberg, D.; Bando, Y.; Huang, Y.; Terao, T.; Mitome, M.; Tang, C.; Zhi, C. Boron Nitride Nanotubes and Nanosheets. *ACS Nano* **2010**, *4*, 2979–2993. [CrossRef] [PubMed]
3. Hong, S.; Lee, C.-S.; Lee, M.-H.; Lee, Y.; Ma, K.Y.; Kim, G.; Yoon, S.I.; Ihm, K.; Kim, K.-J.; Shin, T.J.; et al. Ultralow-Dielectric-Constant Amorphous Boron Nitride. *Nature* **2020**, *582*, 511–514. [CrossRef] [PubMed]

4. Cui, Z.; He, Y.; Tian, H.; Khanaki, A.; Xu, L.; Shi, W.; Liu, J. Study of Direct Tunneling and Dielectric Breakdown in Molecular Beam Epitaxial Hexagonal Boron Nitride Monolayers Using Metal–Insulator–Metal Devices. *ACS Appl. Electron. Mater.* **2020**, *2*, 747–755. [CrossRef]
5. Cai, Q.; Scullion, D.; Gan, W.; Falin, A.; Cizek, P.; Liu, S.; Edgar, J.H.; Liu, R.; Cowie, B.C.C.; Santos, E.J.G.; et al. Outstanding Thermal Conductivity of Single Atomic Layer Isotope-Modified Boron Nitride. *Phys. Rev. Lett.* **2020**, *125*, 085902. [CrossRef]
6. Kostoglou, N.; Polychronopoulou, K.; Rebholz, C. Thermal and Chemical Stability of Hexagonal Boron Nitride (h-BN) Nanoplatelets. *Vacuum* **2015**, *112*, 42–45. [CrossRef]
7. Ares, P.; Cea, T.; Holwill, M.; Wang, Y.B.; Roldán, R.; Guinea, F.; Andreeva, D.V.; Fumagalli, L.; Novoselov, K.S.; Woods, C.R. Piezoelectricity in Monolayer Hexagonal Boron Nitride. *Adv. Mater.* **2020**, *32*, 1905504. [CrossRef]
8. Falin, A.; Cai, Q.; Santos, E.J.G.; Scullion, D.; Qian, D.; Zhang, R.; Yang, Z.; Huang, S.; Watanabe, K.; Taniguchi, T.; et al. Mechanical Properties of Atomically Thin Boron Nitride and The Role of Interlayer Interactions. *Nat. Commun.* **2017**, *8*, 15815. [CrossRef]
9. Weng, Q.; Wang, X.; Wang, X.; Bando, Y.; Golberg, D. Functionalized hexagonal boron nitride nanomaterials: Emerging properties and applications. *Chem. Soc. Rev.* **2016**, *45*, 3989–4012. [CrossRef]
10. Britnell, L.; Gorbachev, R.V.; Jalil, R.; Belle, B.D.; Schedin, F.; Mishchenko, A.; Georgiou, T.; Katsnelson, M.I.; Eaves, L.; Morozov, S.V.; et al. Field-Effect Tunneling Transistor Based on Vertical Graphene Heterostructures. *Science* **2012**, *335*, 947–950. [CrossRef]
11. Gao, T.; Song, X.; Du, H.; Nie, Y.; Chen, Y.; Ji, Q.; Sun, J.; Yang, Y.; Zhang, Y.; Liu, Z. Temperature-Triggered Chemical Switching Growth of In-Plane and Vertically Stacked Graphene-Boron Nitride Heterostructures. *Nat. Commun.* **2015**, *6*, 6835. [CrossRef] [PubMed]
12. Han, Z.; Li, M.; Li, L.; Jiao, F.; Wei, Z.; Geng, D.; Hu, W. When Graphene Meets White Graphene—Recent Advances in the Construction of Graphene and h-BN Heterostructures. *Nanoscale* **2021**, *13*, 13174–13194. [CrossRef] [PubMed]
13. Watanabe, K.; Taniguchi, T.; Kanda, H. Direct-Bandgap Properties and Evidence for Ultraviolet Lasing of Hexagonal Boron Nitride Single Crystal. *Nat. Mater.* **2004**, *3*, 404–409. [CrossRef] [PubMed]
14. Li, L.H.; Cervenka, J.; Watanabe, K.; Taniguchi, T.; Chen, Y. Strong Oxidation Resistance of Atomically Thin Boron Nitride Nanosheets. *ACS Nano* **2014**, *8*, 1457–1462. [CrossRef]
15. Piquemal-Banci, M.; Galceran, R.; Godel, F.; Caneva, S.; Martin, M.-B.; Weatherup, R.S.; Kidambi, P.R.; Bouzehouane, K.; Xavier, S.; Anane, A.; et al. Insulator-to-Metallic Spin-Filtering in 2D-Magnetic Tunnel Junctions Based on Hexagonal Boron Nitride. *ACS Nano* **2018**, *12*, 4712–4718. [CrossRef]
16. Ma, K.Y.; Kim, M.; Shin, H.S. Large-Area Hexagonal Boron Nitride Layers by Chemical Vapor Deposition: Growth and Applications for Substrates, Encapsulation, and Membranes. *Acc. Mater. Res.* **2022**, *3*, 748–760. [CrossRef]
17. Liu, H.; You, C.Y.; Li, J.; Galligan, P.R.; You, J.; Liu, Z.; Cai, Y.; Luo, Z. Synthesis of Hexagonal Boron Nitrides by Chemical Vapor Deposition and Their Use as Single Photon Emitters. *Nano Mater. Sci.* **2021**, *3*, 291–312. [CrossRef]
18. Fukamachi, S.; Solís-Fernández, P.; Kawahara, K.; Tanaka, D.; Otake, T.; Lin, Y.-C.; Suenaga, K.; Ago, H. Large-Area Synthesis and Transfer of Multilayer Hexagonal Boron Nitride for Enhanced Graphene Device Arrays. *Nat. Electron.* **2023**, *6*, 126–136. [CrossRef]
19. Dąbrowska, A.K.; Binder, J.; Prozheev, I.; Tuomisto, F.; Iwański, J.; Tokarczyk, M.; Korona, K.P.; Kowalski, G.; Stępniewski, R.; Wysmołek, A. Defects in Layered Boron Nitride Grown by Metal Organic Vapor Phase Epitaxy: Luminescence and Positron Annihilation Studies. *J. Lumin.* **2024**, *269*, 120486. [CrossRef]
20. Yang, X.; Nitta, S.; Nagamatsu, K.; Bae, S.-Y.; Lee, H.-J.; Liu, Y.; Pristovsek, M.; Honda, Y.; Amano, H. Growth of Hexagonal Boron Nitride on Sapphire Substrate by Pulsed-Mode Metalorganic Vapor Phase Epitaxy. *J. Cryst. Growth* **2018**, *482*, 1–8. [CrossRef]
21. Li, X.; Sundaram, S.; El Gmili, Y.; Ayari, T.; Puybaret, R.; Patriarche, G.; Voss, P.L.; Salvestrini, J.P.; Ougazzaden, A. Large-Area Two-Dimensional Layered Hexagonal Boron Nitride Grown on Sapphire by Metalorganic Vapor Phase Epitaxy. *Cryst. Growth Des.* **2016**, *16*, 3409–3415. [CrossRef]
22. Cheng, T.S.; Summerfield, A.; Mellor, C.J.; Khlobystov, A.N.; Eaves, L.; Foxon, C.T.; Beton, P.H.; Novikov, S.V. High-Temperature Molecular Beam Epitaxy of Hexagonal Boron Nitride with High Active Nitrogen Fluxes. *Materials* **2018**, *11*, 1119. [CrossRef] [PubMed]
23. Vuong, T.Q.P.; Cassabois, G.; Valvin, P.; Rousseau, E.; Summerfield, A.; Mellor, C.J.; Cho, Y.; Cheng, T.S.; Albar, J.D.; Eaves, L.; et al. Deep Ultraviolet Emission in Hexagonal Boron Nitride Grown by High-Temperature Molecular Beam Epitaxy. *2D Mater.* **2017**, *4*, 021023. [CrossRef]
24. Rigato, V.; Spolaore, M.; Della Mea, G. Deposition of Boron Nitride Coatings by Reactive Rf Magnetron Sputtering: Correlation Between Boron and Nitrogen Contents and the Flux of Energetic Ar+ Ions at the Substrate. *MRS Proc.* **2011**, *396*, 557. [CrossRef]
25. Oks, E.; Anders, A.; Nikolaev, A.; Yushkov, Y. Sputtering of Pure Boron Using a Magnetron Without a Radio-Frequency Supply. *Rev. Sci. Instrum.* **2017**, *88*, 4. [CrossRef]
26. Vetter, J.; Shimizu, T.; Kurapov, D.; Sasaki, T.; Mueller, J.; Stangier, D.; Esselbach, M. Industrial Application Potential of High Power Impulse Magnetron Sputtering for Wear and Corrosion Protection Coatings. *J. Appl. Phys.* **2023**, *134*, 16. [CrossRef]
27. Olejníček, J.; Šmíd, J.; Perekrestov, R.; Kšírová, P.; Rathouský, J.; Kohout, M.; Dvořáková, M.; Kment, Š.; Jurek, K.; Čada, M.; et al. Co_3O_4 Thin Films Prepared by Hollow Cathode Discharge. *Surf. Coat. Technol.* **2019**, *366*, 303–310. [CrossRef]
28. Kipkirui, N.G.; Lin, T.-T.; Kiplangat, R.S.; Lee, J.-W.; Chen, S.-H. HiPIMS and RF magnetron sputtered $Al_{0.5}CCrFeNi_2Ti_{0.5}$ HEA Thin-Film Coatings: Synthesis and Characterization. *Surf. Coat. Technol.* **2022**, *449*, 128988. [CrossRef]

29. Hossain, M.D.; Borman, T.; McIlwaine, N.S.; Maria, J.-P. Bipolar High-Power Impulse Magnetron Sputtering Synthesis of High-entropy carbides. *J. Am. Ceram. Soc.* **2022**, *105*, 3862–3873. [CrossRef]
30. Loquai, S.; Baloukas, B.; Klemberg-Sapieha, J.E.; Martinu, L. HiPIMS-Deposited Thermochromic VO_2 Films with High Environmental Stability. *Sol. Energy Mater. Sol. Cells* **2017**, *160*, 217–224. [CrossRef]
31. Whiteside, M.; Arulkumaran, S.; Chng, S.S.; Shakerzadeh, M.; Teo, H.T.E.; Ng, G.I. On the Recovery of 2DEG Properties in Vertically Ordered h-BN Deposited AlGaN/GaN Heterostructures on Si Substrate. *Appl. Phys. Express* **2020**, *13*, 065508. [CrossRef]
32. Cometto, O.; Sun, B.; Tsang, S.H.; Huang, X.; Koh, Y.K.; Teo, E.H.T. Vertically Self-Ordered Orientation of Nanocrystalline Hexagonal Boron Nitride Thin Films for Enhanced Thermal Characteristics. *Nanoscale* **2015**, *7*, 18984–18991. [CrossRef] [PubMed]
33. Chng, S.S.; Zhu, M.; Du, Z.; Wang, X.; Whiteside, M.; Ng, Z.K.; Shakerzadeh, M.; Tsang, S.H.; Teo, E.H.T. Dielectric Dispersion and Superior Thermal Characteristics in Isotope-Enriched Hexagonal Boron Nitride Thin Films: Evaluation as Thermally Self-Dissipating Dielectrics for GaN Transistors. *J. Mater. Chem. C* **2020**, *8*, 9558–9568. [CrossRef]
34. Whiteside, M.; Arulkumaran, S.; Ng, G.I. Demonstration of Vertically-Ordered h-BN/AlGaN/GaN Metal-Insulator-Semiconductor High-Electron-Mobility Transistors on Si Substrate. *Mater. Sci. Eng. B* **2021**, *270*, 115224. [CrossRef]
35. Zhang, H.; Ju, X.; Jiang, H.; Yang, D.; Wei, R.; Hu, W.; Lu, X.; Zhu, M. Implementation of High Thermal Conductivity and Synaptic Metaplasticity in Vertically-Aligned Hexagonal Boron Nitride-Based Memristor. *Sci. China Mater.* **2024**, *67*, 1907–1914. [CrossRef]
36. Chng, S.S.; Zhu, M.; Wu, J.; Wang, X.; Ng, Z.K.; Zhang, K.; Liu, C.; Shakerzadeh, M.; Tsang, S.; Teo, E.H.T. Nitrogen-Mediated Aligned Growth of Hexagonal BN Films for Reliable High-Performance InSe Transistors. *J. Mater. Chem. C* **2020**, *8*, 4421–4431. [CrossRef]
37. Hahn, J.; Friedrich, M.; Pintaske, R.; Schaller, M.; Kahl, N.; Zahn, D.R.T.; Richter, F. Cubic Boron Nitride Films by d.c. and r.f. Magnetron Sputtering: Layer Characterization and Process Diagnostics. *Diam. Relat. Mater.* **1996**, *5*, 1103–1112. [CrossRef]
38. Androulidakis, C.; Koukaras, E.N.; Poss, M.; Papagelis, K.; Galiotis, C.; Tawfick, S. Strained Hexagonal Boron Nitride: Phonon shift and Gr\"uneisen parameter. *Phys. Rev. B* **2018**, *97*, 241414. [CrossRef]
39. Li, L.H.; Chen, Y. Atomically Thin Boron Nitride: Unique Properties and Applications. *Adv. Funct. Mater.* **2016**, *26*, 2594–2608. [CrossRef]
40. Chen, X.; Luan, K.; Zhang, W.; Liu, X.; Zhao, J.; Hou, L.; Gao, Y.; Song, J.; Chen, Z. Effect of Employing Chromium as a Buffer Layer on the Crystallinity of Hexagonal Boron Nitride Films Grown by LPCVD. *J. Mater. Sci. Mater. Electron.* **2021**, *32*, 13961–13971. [CrossRef]
41. Nemanich, R.J.; Solin, S.A.; Martin, R.M. Light Scattering Study of Boron Nitride Microcrystals. *Phys. Rev. B* **1981**, *23*, 6348–6356. [CrossRef]
42. Chen, X.; Sun, H.; Zhang, W.; Tan, C.; Liu, X.; Zhao, J.; Hou, L.; Gao, Y.; Song, J.; Chen, Z. The effects of Post-Annealing Technology on Crystalline Quality and Properties of Hexagonal Boron Nitride Films Deposited on Sapphire Substrates. *Vacuum* **2022**, *199*, 110935. [CrossRef]
43. Chen, X.; Tan, C.; Liu, X.; Luan, K.; Guan, Y.; Liu, X.; Zhao, J.; Hou, L.; Gao, Y.; Chen, Z. Growth of Hexagonal Boron Nitride Films on Silicon Substrates by Low-Pressure Chemical Vapor Deposition. *J. Mater. Sci. Mater. Electron.* **2021**, *32*, 3713–3719. [CrossRef]
44. Zhou, H.; Zhu, J.; Liu, Z.; Yan, Z.; Fan, X.; Lin, J.; Wang, G.; Yan, Q.; Yu, T.; Ajayan, P.; et al. High Thermal Conductivity of Suspended Few-Layer Hexagonal Boron nitride Sheets. *Nano Res.* **2014**, *7*, 1–9. [CrossRef]
45. Wang, W.; Li, Z.; Marsden, A.J.; Bissett, M.A.; Young, R.J. Interlayer and Interfacial Stress Transfer in hBN Nanosheets. *2D Mater.* **2021**, *8*, 035058. [CrossRef]
46. Becton, M.; Wang, X. Grain-Size Dependence of Mechanical Properties in Polycrystalline Boron-Nitride: A Computational Study. *Phys. Chem. Chem. Phys.* **2015**, *17*, 21894–21901. [CrossRef]
47. Paul, R.; Tasnim, T.; Dhar, R.; Mojumder, S.; Saha, S.; Motalab, M.A. Study of Uniaxial Tensile Properties of Hexagonal Boron Nitride Nanoribbons. In Proceedings of the TENCON 2017–2017 IEEE Region 10 Conference, Penang, Malaysia, 5–8 November 2017; pp. 2783–2788.
48. Bera, K.; Chugh, D.; Patra, A.; Tan, H.H.; Jagadish, C.; Roy, A. Strain Distribution in Wrinkled hBN Films. *Solid State Commun.* **2020**, *310*, 113847. [CrossRef]
49. Duan, X.; Yang, Z.; Chen, L.; Tian, Z.; Cai, D.; Wang, Y.; Jia, D.; Zhou, Y. Review on the Properties of Hexagonal Boron Nitride Matrix Composite Ceramics. *J. Eur. Ceram. Soc.* **2016**, *36*, 3725–3737. [CrossRef]
50. Zhang, X.; Yue, J.; Chen, G.; Yan, H. Study on Stress and Strain of Cubic Boron Nitride Thin Films. *Thin Solid Films* **1998**, *315*, 202–206. [CrossRef]
51. Chen, M.; Zhang, Q.; Fang, C.; Shen, Z.; Lu, Y.; Liu, T.; Tan, S.; Zhang, J. Influence of Sapphire Substrate with Miscut Angles on Hexagonal Boron Nitride Films Grown by Halide Vapor Phase Epitaxy. *CrystEngComm* **2023**, *25*, 4604–4610. [CrossRef]
52. Sharma, K.P.; Sharma, S.; Khaniya Sharma, A.; Paudel Jaisi, B.; Kalita, G.; Tanemura, M. Edge Controlled Growth of Hexagonal Boron Nitride Crystals on Copper Foil by Atmospheric Pressure Chemical Vapor Deposition. *CrystEngComm* **2018**, *20*, 550–555. [CrossRef]
53. Naumkin, A.V.; Kraut-Vass, A.; Powell, C.J.; Gaarenstroom, S.W.; National Institute of Standards and Technology. *NIST X-ray Photoelectron Spectroscopy Database, Version 4.1.*; Measurement Services Division of the National Institute of Standards and Technology (NIST) Technology Services: Gaithersburg, MD, USA, 2012.
54. Deng, J.; Wang, B.; Tan, L.; Yan, H.; Chen, G. The Growth of Cubic Boron Nitride Films by RF Reactive Sputter. *Thin Solid Films* **2000**, *368*, 312–314. [CrossRef]

55. Singh, M.; Vasudev, H.; Kumar, R. Microstructural Characterization of BN Thin Films Using RF Magnetron Sputtering Method. *Mater. Today Proc.* **2020**, *26*, 2277–2282. [CrossRef]
56. Mieno, M.; Yoshida, T. Preparation of Cubic Boron Nitride Films by RF Sputtering. *Jpn. J. Appl. Phys.* **1990**, *29*, L1175. [CrossRef]
57. Schütze, A.; Bewilogua, K.; Lüthje, H.; Kouptsidis, S.; Gaertner, M. Improvement of the Adhesion of Sputtered Cubic Boron Nitride Films. *Surf. Coat. Technol.* **1997**, *97*, 33–38. [CrossRef]
58. Liu, D.; Chen, X.; Yan, Y.; Zhang, Z.; Jin, Z.; Yi, K.; Zhang, C.; Zheng, Y.; Wang, Y.; Yang, J.; et al. Conformal Hexagonal-Boron Nitride Dielectric Interface for Tungsten Diselenide Devices with Improved Mobility and Thermal Dissipation. *Nat. Commun.* **2019**, *10*, 1188. [CrossRef]
59. Wei, D.; Peng, L.; Li, M.; Mao, H.; Niu, T.; Han, C.; Chen, W.; Wee, A.T.S. Low Temperature Critical Growth of High Quality Nitrogen Doped Graphene on Dielectrics by Plasma-Enhanced Chemical Vapor Deposition. *ACS Nano* **2015**, *9*, 164–171. [CrossRef]
60. Wei, D.; Lu, Y.; Han, C.; Niu, T.; Chen, W.; Wee, A.T.S. Critical Crystal Growth of Graphene on Dielectric Substrates at Low Temperature for Electronic Devices. *Angew. Chem.* **2013**, *125*, 14371. [CrossRef]
61. Yi, K.; Jin, Z.; Bu, S.; Wang, D.; Liu, D.; Huang, Y.; Dong, Y.; Yuan, Q.; Liu, Y.; Wee, A.T.S.; et al. Catalyst-Free Growth of Two-Dimensional BCxN Materials on Dielectrics by Temperature-Dependent Plasma-Enhanced Chemical Vapor Deposition. *ACS Appl. Mater. Interfaces* **2020**, *12*, 33113–33120. [CrossRef]
62. Snure, M.; Paduano, Q.; Hamilton, M.; Shoaf, J.; Mann, J.M. Optical Characterization of Nanocrystalline Boron Nitride Thin Films Grown by Atomic Layer Deposition. *Thin Solid Films* **2014**, *571*, 51–55. [CrossRef]
63. Ahmed, K.; Dahal, R.; Weltz, A.; Lu, J.-Q.; Danon, Y.; Bhat, I.B. Growth of Hexagonal Boron Nitride on (111) Si for Deep UV Photonics and Thermal Neutron Detection. *Appl. Phys. Lett.* **2016**, *109*, 113501. [CrossRef]
64. Singhal, R.; Echeverria, E.; McIlroy, D.N.; Singh, R.N. Synthesis of Hexagonal Boron Nitride Films on Silicon and Sapphire Substrates by Low-Pressure Chemical Vapor Deposition. *Thin Solid Films* **2021**, *733*, 138812. [CrossRef]
65. Quan, H.; Wang, X.; Zhang, L.; Liu, N.; Feng, S.; Chen, Z.; Hou, L.; Wang, Q.; Liu, X.; Zhao, J.; et al. Stability to Moisture of Hexagonal Boron Nitride Films Deposited on Silicon by RF Magnetron Sputtering. *Thin Solid Films* **2017**, *642*, 90–95. [CrossRef]
66. Hirata, Y.; Yoshii, K.; Yoshizato, M.; Akasaka, H.; Ohtake, N. Developing a Synthesis Process for Large-Scale h-BN Nanosheets Using Magnetron Sputtering and Heat Annealing. *Adv. Eng. Mater.* **2023**, *25*, 2300933. [CrossRef]
67. Kang, Y.; Chen, L.; Liu, C.; Tang, X.; Zhu, X.; Gao, W.; Yin, H. Enhancement of n-type Conductivity of Hexagonal Boron Nitride Films by In-Situ Co-Doping of Silicon and Oxygen. *J. Phys. Condens. Matter* **2022**, *34*, 384002. [CrossRef]
68. Chen, R.; Li, Q.; Zhang, Q.; Wang, M.; Fang, W.; Zhang, Z.; Yun, F.; Wang, T.; Hao, Y. Electronic Properties of Vertically Stacked h-BN/B1−xAlxN Heterojunction on Si(100). *ACS Appl. Mater. Interfaces* **2023**, *15*, 16211–16220. [CrossRef]
69. BenMoussa, B.; D'Haen, J.; Borschel, C.; Barjon, J.; Soltani, A.; Mortet, V.; Ronning, C.; D'Olieslaeger, M.; Boyen, H.G.; Haenen, K. Hexagonal Boron Nitride Nanowalls: Physical Vapour Deposition, 2D/3D Morphology and Spectroscopic Analysis. *J. Phys. D Appl. Phys.* **2012**, *45*, 135302. [CrossRef]
70. Rye, R.R.; Tallant, D.R.; Borek, T.T.; Lindquist, D.A.; Paine, R.T. Mechanistic Studies of the Conversion of Borazine Polymers to Boron Nitride. *Chem. Mater.* **1991**, *3*, 286–293. [CrossRef]
71. Venturi, G.; Chiodini, S.; Melchioni, N.; Janzen, E.; Edgar, J.H.; Ronning, C.; Ambrosio, A. Selective Generation of Luminescent Defects in Hexagonal Boron Nitride. *Laser Photonics Rev.* **2024**, *18*, 2300973. [CrossRef]
72. Gago, R.; Jiménez, I.; Agulló-Rueda, F.; Albella, J.M.; Czigány, Z.; Hultman, L. Transition from Amorphous Boron Carbide to Hexagonal Boron Carbon Nitride Thin Films Induced by Nitrogen Ion Assistance. *J. Appl. Phys.* **2002**, *92*, 5177–5182. [CrossRef]
73. Kupenko, I.; Dubrovinsky, L.; Dmitriev, V.; Dubrovinskaia, N. In Situ Raman Spectroscopic Study of the Pressure Induced Structural Changes in Ammonia Borane. *J. Chem. Phys.* **2012**, *137*, 074506. [CrossRef] [PubMed]
74. Ferrari, A.C.; Robertson, J.; Ferrari, A.C.; Robertson, J. Raman Spectroscopy of Amorphous, Nanostructured, Diamond–Like Carbon, and Nanodiamond. *Philos. Trans. R. Soc. Lond. Ser. A Math. Phys. Eng. Sci.* **2004**, *362*, 2477–2512. [CrossRef]
75. Tabata, H.; Fujii, M.; Hayashi, S.; Doi, T.; Wakabayashi, T. Raman and Surface-Enhanced Raman Scattering of a Series of Size-Separated Polyynes. *Carbon* **2006**, *44*, 3168–3176. [CrossRef]
76. Casiraghi, C.; Piazza, F.; Ferrari, A.C.; Grambole, D.; Robertson, J. Bonding in Hydrogenated Diamond-Like Carbon by Raman Spectroscopy. *Diam. Relat. Mater.* **2005**, *14*, 1098–1102. [CrossRef]
77. Casiraghi, C.; Ferrari, A.C.; Robertson, J. Raman Spectroscopy of Hydrogenated Amorphous Carbons. *Phys. Rev. B* **2005**, *72*, 085401. [CrossRef]
78. Hoang, D.-Q.; Pobedinskas, P.; Nicley, S.S.; Turner, S.; Janssens, S.D.; Van Bael, M.K.; D'Haen, J.; Haenen, K. Elucidation of the Growth Mechanism of Sputtered 2D Hexagonal Boron Nitride Nanowalls. *Cryst. Growth Des.* **2016**, *16*, 3699–3708. [CrossRef]
79. Akkerman, Z.L.; Kosinova, M.L.; Fainer, N.I.; Rumjantsev, Y.M.; Sysoeva, N.P. Chemical Stability of Hydrogen-Containing Boron Nitride Films Obtained by Plasma Enhanced Chemical Vapour Deposition. *Thin Solid Films* **1995**, *260*, 156–160. [CrossRef]
80. Bounouh, Y.; Thèye, M.L.; Dehbi-Alaoui, A.; Matthews, A.; Stoquert, J.P. Influence of Annealing on the Hydrogen Bonding and the Microstructure of Diamondlike and Polymerlike Hydrogenated Amorphous Carbon Films. *Phys. Rev. B* **1995**, *51*, 9597–9605. [CrossRef]
81. Bounouh, Y.; Zellama, K.; Zeinert, A.; Benlahsen, M.; Clin, M.; Thèye, M.L. Modes of Hydrogen Incorporation in Hydrogenated Amorphous Carbon (a–C:H), Modifications with Annealing Temperature. *J. Phys. III Fr.* **1997**, *7*, 2159–2164. [CrossRef]
82. Wang, W.J.; Wang, T.M.; Chen, B.L. Hydrogen Release from Diamondlike Carbon Films Due to Thermal Annealing in Vacuum. *Nucl. Instrum. Methods Phys. Res. Sect. B Beam Interact. Mater. At.* **1996**, *117*, 140–144. [CrossRef]

Disclaimer/Publisher's Note: The statements, opinions and data contained in all publications are solely those of the individual author(s) and contributor(s) and not of MDPI and/or the editor(s). MDPI and/or the editor(s) disclaim responsibility for any injury to people or property resulting from any ideas, methods, instructions or products referred to in the content.

Article

Carbon Nitride Nanosheets as an Adhesive Layer for Stable Growth of Vertically-Ordered Mesoporous Silica Film on a Glassy Carbon Electrode and Their Application for CA15-3 Immunosensor

Jun Xing [1], Hongxin Wang [2] and Fei Yan [2,*]

[1] Shanxi Bethune Hospital, Shanxi Academy of Medical Sciences, Tongji Shanxi Hospital, Third Hospital of Shanxi Medical University, Taiyuan 030032, China; xingjun2022@sxmu.edu.cn
[2] Department of Chemistry, School of Chemistry and Chemical Engineering, Zhejiang Sci-Tech University, Hangzhou 310018, China; 202230107404@mails.zstu.edu.cn
* Correspondence: yanfei@zstu.edu.cn

Citation: Xing, J.; Wang, H.; Yan, F. Carbon Nitride Nanosheets as an Adhesive Layer for Stable Growth of Vertically-Ordered Mesoporous Silica Film on a Glassy Carbon Electrode and Their Application for CA15-3 Immunosensor. *Molecules* **2024**, *29*, 4334. https://doi.org/10.3390/molecules29184334

Academic Editors: Sake Wang, Nguyen Tuan Hung and Minglei Sun

Received: 13 June 2024
Revised: 2 September 2024
Accepted: 6 September 2024
Published: 12 September 2024

Copyright: © 2024 by the authors. Licensee MDPI, Basel, Switzerland. This article is an open access article distributed under the terms and conditions of the Creative Commons Attribution (CC BY) license (https://creativecommons.org/licenses/by/4.0/).

Abstract: Vertically ordered mesoporous silica films (VMSF) are a class of porous materials composed of ultrasmall pores and ultrathin perpendicular nanochannels, which are attractive in the areas of electroanalytical sensors and molecular separation. However, VMSF easily falls off from the carbonaceous electrodes and thereby impacts their broad applications. Herein, carbon nitride nanosheets (CNNS) were served as an adhesive layer for stable growth of VMSF on the glassy carbon electrode (GCE). CNNS bearing plentiful oxygen-containing groups can covalently bind with silanol groups of VMSF, effectively promoting the stability of VMSF on the GCE surface. Benefiting from numerous open nanopores of VMSF, modification of VMSF's external surface with carbohydrate antigen 15-3 (CA15-3)-specific antibody allows the target-controlled transport of electrochemical probes through the internal silica nanochannels, yielding sensitive quantitative detection of CA15-3 with a broad detection range of 1 mU/mL to 1000 U/mL and a low limit of detection of 0.47 mU/mL. Furthermore, the proposed VMSF/CNNS/GCE immunosensor is capable of highly selective and accurate determination of CA15-3 in spiked serum samples, which offers a simple and effective electrochemical strategy for detection of various practical biomarkers in complicated biological specimens.

Keywords: carbon nitride nanosheets; vertically ordered mesoporous silica film; glassy carbon electrode; electrochemical immunosensor; carbohydrate antigen 15-3

1. Introduction

Vertically ordered mesoporous silica films (VMSF) have become increasingly prosperous electrode materials owing to their distinct selectivity at molecular scale and good anti-biofouling properties in complex biological samples [1–4]. Owing to the isolating nanoporous structure, VMSF-based electrochemical/electrochemiluminescence sensors involve the transport of targets or signal probes along the perpendicular nanochannels and/or biological recognition elements on the outer surface of VMSF [5–11]. According to the previous reports, indium tin oxide (ITO) coated glasses are very suitable for supporting VMSF and further functionalization of nanomaterials or recognition elements into VMSF's skeleton [12–16]. However, in comparison with commonly used carbonaceous electrodes (such as glassy carbon electrode (GCE) and three-dimensional graphene), ITO electrodes as the supporting substrate have shortcomings, including slow electron transport properties for small organic molecules and a narrow electrochemical window, which limit the analytical performances of VMSF-based sensors in practical analysis [17]. On account of the instability of VMSF on the carbonaceous electrodes, a group of nanomaterials have been employed to confer carbonaceous electrodes with adhesive properties for stable growth of VMSF, such as silane molecules [18], two-dimensional graphene nanosheets, and their hybrid materials [19–21]. In addition, pre-activation of carbonaceous electrodes by simple electrochemical procedure generates the oxygen-containing groups on the electrode [22,23], which is capable of stable fabrication of VMSF [24–27]. Therefore, developing diverse adhesive layers for stable fabrication of VMSF on the carbonaceous electrode surface is of great importance for extending the practical analytical applications of VMSF.

Ultrathin carbon nitride nanosheets (CNNS) possessing a two-dimensional (2D) graphene-like structure display unique characteristics, including large surface areas, excellent electron transport ability, and good catalysis properties, compared with their bulk counterparts [28]. Nowadays, CNNS has received tremendous research interest and shown significant promise in different applications, such as imaging, sensing, and biotherapy [29]. Moreover, CNNS, usually prepared by exfoliating bulk graphitic-phase carbon nitrides, can provide hydroxy groups for covalently binding with silanol groups of VMSF.

In this paper, CNNS is introduced on the GCE surface via π-π interaction using a simple drip-coating method, and its hydroxyl groups offer a suitable microenvironment for further stable fabrication of VMSF through covalent binding between silanol groups of VMSF and hydroxyl groups of CNNS. Such obtained VMSF/CNNS composite on the GCE has good stability and solves the problem of falling off of VMSF from GCE. Such VMSF/CNNS/GCE is fit for the construction of electrochemical biosensors. As a proof of concept, carbohydrate antigen 15-3 (CA15-3) is usually used as the indicator for lung cancer and breast cancer, which is selected as the analyte to examine the potential application capacity of VMSF/CNNS/GCE. In general, the normal CA15-3 amount in human serum is less than 30 U/mL, and an increased amount indicates an increased risk of developing cancer [30]. Quantitative detection of CA15-3 is realized by modification of CA15-3-specific antibody on the external surface of VMSF. The current signal variation of electrochemical probes (potassium ferricyanide/potassium ferrocyanide, $[Fe(CN)_6]^{3-/4-}$) after incubation of different concentrations of CA15-3 at the immunosensing interface is related to the CA15-3 concentration. Compared with traditional methods, our proposed electrochemical immunosensor has some advantages of low cost, rapid detection time, and convenient operation. Moreover, the VMSF/CNNS/GCE-based immunosensor has the same anti-biofouling characteristic as the VMSF-modified electrode, which offers a convenient sensing approach for monitoring the amount of CA15-3 and early screening of malignant tumors.

2. Results and Discussion

2.1. Fabrication of VMSF/CNNS/GCE-Based Immunosensor for Electrochemical Detection of CA15-3

VMSF consists of two regions, namely external surface and inner nanochannels, which can be used for functionalization of active groups and for mass transport of signal probes, respectively. Scheme 1 illustrates the specific construction procedures of CA15-3 immunosensors based on the VMSF/CNNS/GCE. As seen, CNNS is first dripped-coated on the commercial GCE surface through π-π interaction, and the EASA method is then conducted for the growth of VMSF using CNNS/GCE as the working electrode. In the growth process of VMSF, CNNS bearing negative charges and oxygen-containing groups is suitable for electrostatic adsorption of cationic surfactant micelles (SM) and formation of O-Si-O chemical bonds between CNNS and VMSF, exhibiting a good potential adhesive layer for stable growth of VMSF on GCE. When reaction ceases, templated SM are retained inside the nanospaces formed by VMSF's nanochannels, termed as SM@VMSF/CNNS/GCE. GPTMS carrying both epoxy groups and silane groups is introduced onto the external surface of SM@VMSF/CNNS/GCE, and SM is then extracted from the nanochannels, effectively guaranteeing the epoxy groups distributed on the external surface of VMSF/CNNS/GCE (O-VMSF/CNNS/GCE). Anti-CA15-3 antibody (Ab) enabling specifically recognition of CA15-3 is anchored on the surface of O-VMSF/CNNS/GCE through the chemical reaction between epoxy groups and amino groups. Electrochemical immunosensing interface, namely BSA/Ab/O-VMSF/CNNS/GCE, is obtained after the blocking of non-specific adsorption by BSA. The quantitative mechanism for CA15-3 relies on the current variation of electrochemical probes ($[Fe(CN)_6]^{3-/4-}$) produced by target CA15-3 binding on the

BSA/Ab/O-VMSF/CNNS/GCE. Target CA15-3 and its corresponding antibody can form immunocomposite on the VMSF surface, influencing the access of $[Fe(CN)_6]^{3-/4-}$ through the silica nanochannels to reach the underlying CNNS/GCE and finally resulting in the declined electrochemical current signals.

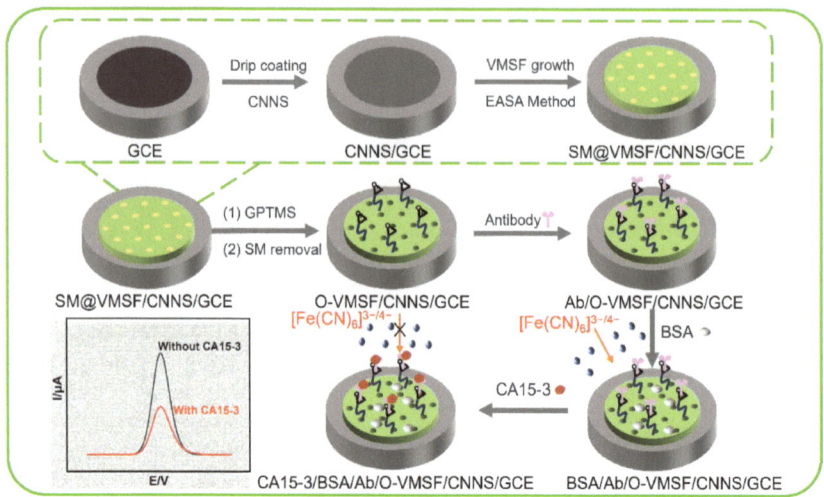

Scheme 1. Schematic illustration for the preparation of the VMSF/CNNS/GCE-based immunosensor and the electrochemical determination of CA15-3 with the help of electrochemical probes ($[Fe(CN)_6]^{3-/4-}$) in solution.

2.2. Characterization of CNNS

TEM were first used to show the morphology of the prepared CNNS. As shown in Figure 1a, CNNS prepared by H_2SO_4 exfoliation of bulk g-C_3N_4 has a 2D lamellar structure. Figure 1b compares the FT-IR spectra of g-C_3N_4 and CNNS. g-C_3N_4 displays several characteristic peaks of heterocycle skeleton. The peaks at around 1637 cm^{-1} and 1546 cm^{-1} in correspond to the stretching vibrations of C=N. The characteristic peaks at 1461 cm^{-1}, 1406 cm^{-1}, 1322 cm^{-1}, 1243 cm^{-1}, and 1205 cm^{-1} are assigned to the stretching vibrations of C−N. The peak at 810 cm^{-1} is attributed to the typical characteristic peak of the triazine ring. After H_2SO_4 treatment, all above characteristic peaks of CNNS have no obvious change, suggesting the remained skeleton of g-C_3N_4 heterocycle. The new peak at 1080 cm^{-1} resulted from the S-O stretching in SO_4^{2-}, showing the acidified effect of H_2SO_4 for g-C_3N_4. Moreover, compared to g-C_3N_4, the adsorption peak of CNNS at ~1600 cm^{-1} shifts slightly to the low wavenumber, suggesting the existence of carboxylate groups. The broad band at 3170 cm^{-1} belongs to the stretching vibrations of N-H and O-H. The above data indicate that CNNS has a 2D planar structure with oxygen-containing groups, which is suitable for improved stability of VMSF on the GCE surface.

2.3. Characterization of VMSF

TEM images show that the fabricated VMSF on CNNS/GCE surface is characteristic of numerous regularly aligned nanopores (Figure 2a) and nanochannels parallel to each other (Figure 2b). By measurement, the diameter and thickness of VMSF are 2~3 nm and 110 nm, respectively, which is similar to those of VMSF prepared on the other substrates (e.g., ITO and gold electrodes) [31–33]. Moreover, electrochemical technique (cyclic voltammetry (CV)) was used to provide information about intactness and mass transport ability for electrochemical probes of the fabricated VMSF/CNNS/GCE. As presented in Figure 2c,d, enhanced redox peak currents for both $Ru(NH_3)_6^{3+}$ and $Fe(CN)_6^{3-}$ are found at the CNNS/GCE, compared with bare GCE, indicating the good conductivity of synthesized CNNS. No Faradic currents of these two electrochemical probes can be seen at the SM@VMSF/CNNS/GCE, which is due to the impeded effect of SM confined into the silica nanochannels and further suggests the integrity of as-synthesized VMSF on the CNNS/GCE surface. By comparing the electrochemical current variation of two charged electrochemical probes before and after exclusion of SM from silica nanochannels, VMSF on the CNNS/GCE surface also exhibits its inherent charge permselectivity, namely amplifying the signal of positively charged $Ru(NH_3)_6^{3+}$ and suppressing the signal of negatively charged $Fe(CN)_6^{3-}$.

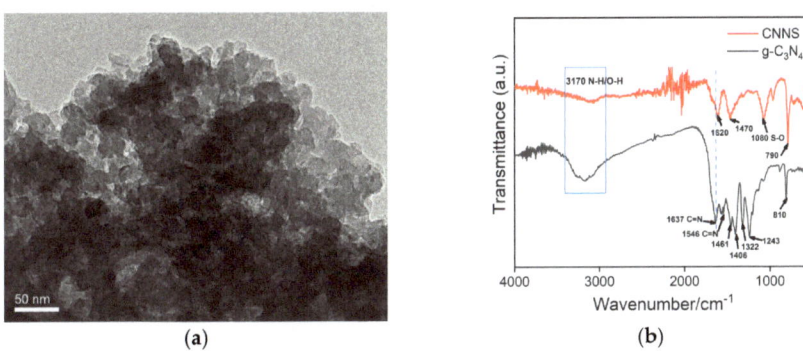

Figure 1. (a) TEM images of CNNS. (b) FT−IR spectra of g−C_3N_4 and CNNS.

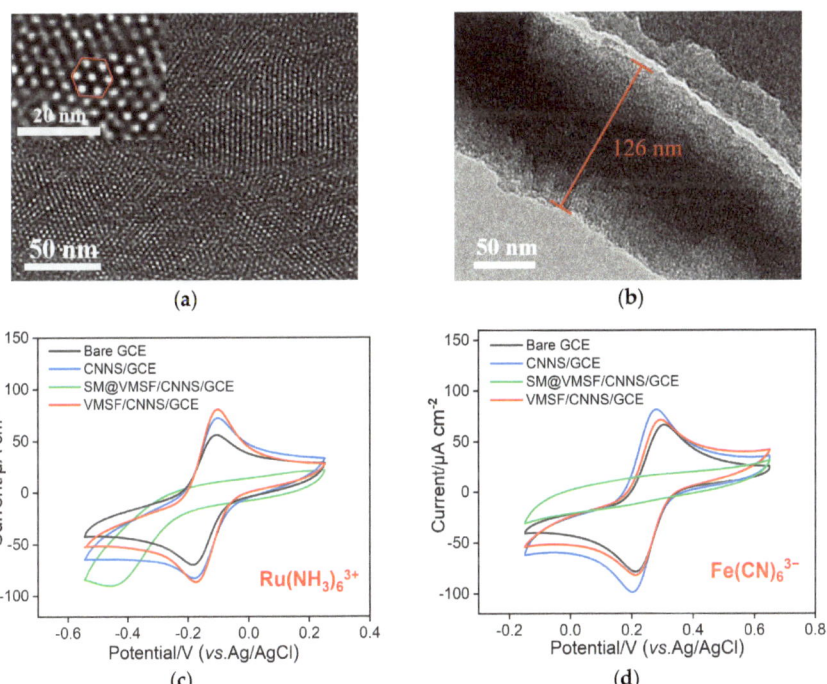

Figure 2. TEM images of VMSF: (a) top view and (b) cross−sectional view. CV curves of bare GCE, CNNS/GCE, SM@VMSF/CNNS/GCE and VMSF/CNNS/GCE in 0.05 M KHP containing 0.5 mM $Ru(NH_3)_6^{3+}$ (c) and 0.5 mM $Fe(CN)_6^{3-}$ (d).

2.4. Feasibility of BSA/Ab/O-VMSF/CNNS/GCE Immunosensor for Detection of CA15-3

CV and EIS are two kinds of commonly used techniques for characterization of electrode construction. As shown in Figure 3a, dropwise modification of GPTMS, Ab and BSA on the VMSF/CNNS/GCE can lead to the sequentially diminished electrochemical current signals of $[Fe(CN)_6]^{3-/4-}$, which arise from the hindered transport of $[Fe(CN)_6]^{3-/4-}$ through the silica nanochannels to the underlying GCE. Thanks to the immunocomposite consisting of CA15-3 and Ab formed at the sensing interface, the fabricated BSA/Ab/O-VMSF/CNNS/GCE immunosensor is used to detect 10 U/mL CA15-3, giving rise to the decreased electrochemical current signals of $[Fe(CN)_6]^{3-/4-}$ and indicating the successful fabrication of electrochemical immunosensor. Figure 3b displays the corresponding EIS plots, revealing the interfacial properties during the fabrication procedure of immunosensor. With the dropwise incubation of GPTMS, Ab, BSA, and CA15-3 on VMSF/CNNS/GCE, the electrode surface is covered with non-conductive substances, leading to a gradual increase in electron transfer resistance (R_{ct}) at the electrode/electrolyte interface, as evidenced by the increasing diameter of semicircle diameter in the high-frequency region (Figure 3b). EIS curves are fitted by equivalent circuit model shown in the inset of Figure 3b and the obtained R_{ct} values at the VMSF/CNNS/GCE, Ab/O-VMSF/CNNS/GCE, BSA/Ab/O-VMSF/CNNS/GCE and CA15-3/BSA/Ab/O-VMSF/CNNS/GCE are 614 Ω, 747 Ω, 843 Ω, 1055 Ω and 1543 Ω, respectively. The change trend in electron transfer resistance presented in Figure 3b is in accordance with the current responses shown in Figure 3a, further proving the feasibility of the designed BSA/Ab/O-VMSF/CNNS/GCE immunosensor for CA15-3 determination.

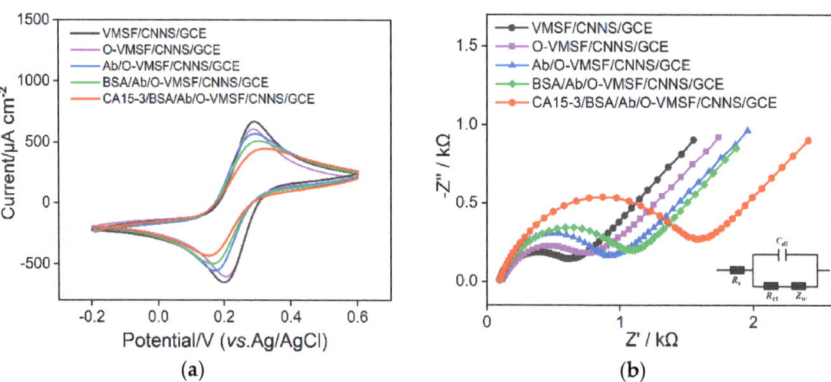

Figure 3. CV (**a**) and EIS curves (**b**) obtained during the fabrication procedure of the BSA/Ab/O−VMSF/CNNS/GCE immunosensor in 0.1 M KCl containing 2.5 mM $[Fe(CN)_6]^{3-/4-}$. The concentration of CA15−3 is 10 U/mL. Inset in (**b**) is the equivalent circuit diagram.

2.5. Optimization of Experimental Conditions for CA15-3 Detection Using BSA/Ab/O-VMSF/CNNS/GCE Immunosensor

The amount of CA15-3 antibody immobilized at the electrode surface can affect the analytical performance of CA15-3 determination. Therefore, incubation times for CA15-3 antibody and CA15-3 at the O-VMSF/CNNS/GCE and BSA/Ab/O-VMSF/CNNS/GCE, respectively, were optimized, and the results are shown in Figure 4. As seen, electrochemical current signals of $[Fe(CN)_6]^{3-/4-}$ obviously decrease and reach the plateau after a period of time. Therefore, the optimal incubation times for CA15-3 antibody and CA15-3 are 90 min and 60 min, respectively.

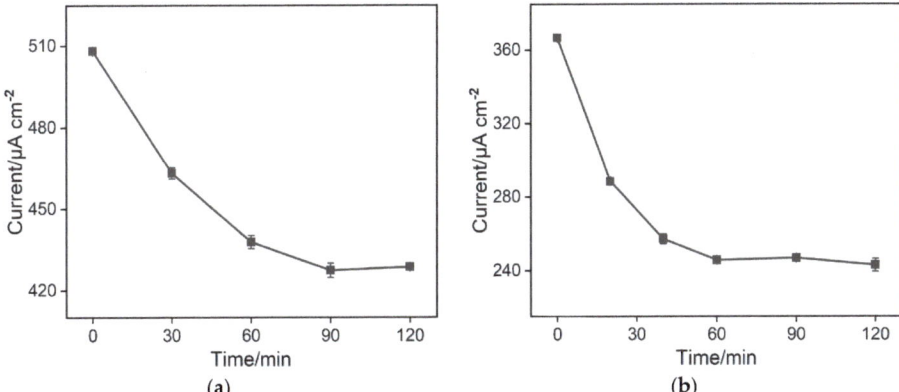

Figure 4. (a) Electrochemical current signals of 2.5 mM $[Fe(CN)_6]^{3-/4-}$ at the O−VMSF/CNNS/GCE after incubation with CA15−3 antibody at various incubation times. (b) Electrochemical current signals of 2.5 mM $[Fe(CN)_6]^{3-/4-}$ at the BSA/Ab/O−VMSF/CNNS/GCE immunosensor after incubation with 1 U/mL CA15−3 at various incubation times. The supporting electrolyte is 0.1 M KCl solution, and the error bars represent the relative standard deviation (RSD) of three measurements.

2.6. Electrochemical Detection of CA15-3 Using the Immunosensor

BSA/Ab/O-VMSF/CNNS/GCE was used to incubate with various concentrations of CA15-3 and the obtained CA15-3/BSA/Ab/O-VMSF/CNNS/GCE were tested in 0.1 M KCl containing 2.5 mM $[Fe(CN)_6]^{3-/4-}$. As depicted in Figure 5, the tested anodic peak current is proportional to the CA15-3 concentration in the range from 1 mU/mL to 1000 U/mL, giving rise to a good linear fitting equation of I (μA cm^{-2}) = −35.8 logC_{CA15-3} (U/mL) + 240 (R^2 = 0.995). The same trend is shown in CV and EIS data (Figure S1), but lower variation is obtained, suggesting the sensitive DPV technique. The limit of detection (LOD) for CA15-3 was calculated to be 0.47 mU/mL, which was lower than most of the electrochemical immunosensors reported previously (Table 1). Moreover, the fabrication of CA15-3/BSA/Ab/O-VMSF/CNNS/GCE has advantages of convenient operation and economic cost.

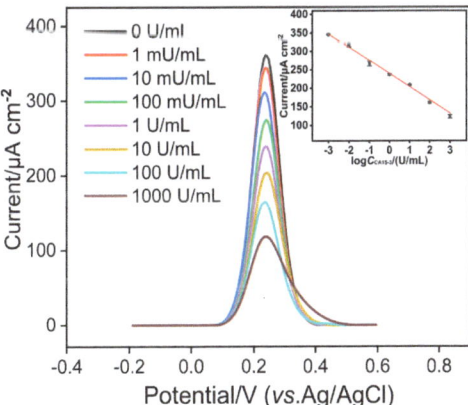

Figure 5. DPV responses of the prepared BSA/Ab/O−VMSF/CNNS/GCE to various concentrations of CA15−3 in 0.1 M KCl solution containing 2.5 mM $[Fe(CN)_6]^{3-/4-}$. Inset is the corresponding calibration curve for the detection of CA15−3. The error bars represent the RSD of three measurements.

Table 1. Analytical performances of various electrochemical sensors for detection of CA15-3.

Materials	Method	Liner Range (U/mL)	LOD (mU/mL)	Refs.
BSA/Ab/Ag/TiO$_2$/rGO/GCE	CA	0.1–300	70	[34]
BSA/Ab/DAP-AuNPs/P3ABA/2D-MoSe$_2$/GO/SPCE	DPV	0.14–500	140	[35]
EA/Ab/CuS-RGO/SPE	DPV	1–150	300	[36]
MCH/Ab/CoS$_2$-GR-AuNPs/SPE	DPV	0.1–150	30	[37]
BSA/Ab/Au@Pt NCs/Fc-g-CS/GCE	DPV	0.5–200	170	[38]
BSA/Ab/O-VMSF/CNNS/GCE	DPV	0.001–1000	0.47	this work

rGO: reduced graphene oxide; CA: chronoamperometry; DAP: deposited redox dye; P3ABA: poly(3-aminobenzylamine); GO: graphene oxide; SPCE: screen-printed carbon electrode; DPV: differential pulse voltammetry; EA: ethanolamine; SPE: screen printed electrode; MCH: 6-Mercapto-1-hexanol; GR: graphene; Au@Pt NCs: dendritic Au@Pt core–shell nanocrystals; Fc-g-CS: ferrocene grafted with chitosan.

To evaluate the analytical performance of the fabricated BSA/Ab/O-VMSF/CNNS/GCE immunosensor for CA15-3 detection, several important indicators, including selectivity, reproducibility, and stability, were studied. As shown in Figure 6a, various biomarkers (PSA, CA19-9, CA125, AFP, and CEA) and inorganic cations (Na$^+$ and K$^+$) were determined by BSA/Ab/O-VMSF/CNNS/GCE, and they could not produce obvious electrochemical responses. Only when incubated with target CA15-3 or a mixture consisting of CA15-3 and above interfering species are remarkably decreased electrochemical responses found at the BSA/Ab/O-VMSF/CNNS/GCE, indicating the good anti-interference and selectivity for CA15-3 detection. To verify the reproducibility of the designed BSA/Ab/O-VMSF/CNNS/GCE, seven BSA/Ab/O-VMSF/CNNS/GCE were prepared in different batches under the same procedures and incubated with 10 U/mL CA15-3 (Figure 6b), giving rise to comparable electrochemical responses with an RSD value of 1.3%. Figure 6c shows the storage stability of BSA/Ab/O-VMSF/CNNS/GCE for CA15-3 detection within a week, yielding ignorable variation on the electrochemical responses with an RSD of 0.8%. Excellent reproducibility and stability of BSA/Ab/O-VMSF/CNNS/GCE are revealed in Figure 6b,c, implying the great potential of the proposed BSA/Ab/O-VMSF/CNNS/GCE for real sample analysis.

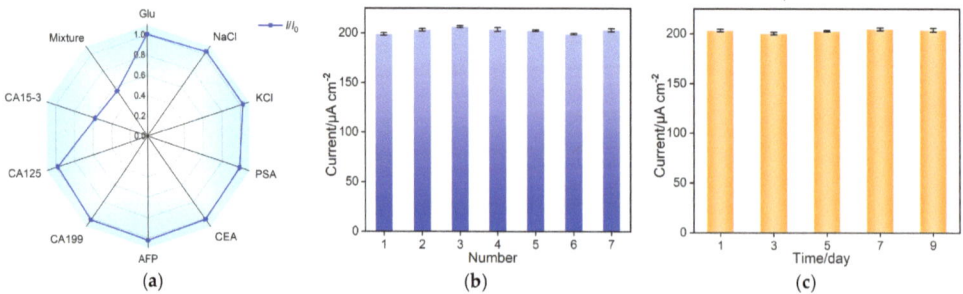

Figure 6. (a) Selectivity of the fabricated BSA/Ab/O−VMSF/CNNS/GCE immunosensor: electrochemical responses of BSA/Ab/O−VMSF/CNNS/GCE to interference substances (100 ng/mL PSA, 10 U/mL CA19−9, 10 mU/mL CA125, 100 ng/mL AFP, 0.1 ng/mL CEA, 100 μM NaCl, 100 μM KCl, and 100 μM Glu), CA15-3 (10 U/mL) and their mixture. I_0 and I refer to the electrochemical signals obtained at the BSA/Ab/O-VMSF/CNNS/GCE immunosensor before and after incubation with interfering substances, target CA15−3 or their mixture, respectively. (b) Reproducibility of the seven BSA/Ab/O−VMSF/CNNS/GCE immunosensors prepared in different batches after incubation with 10 U/mL of CA15−3. (c) Stability of the developed BSA/Ab/O−VMSF/CNNS/GCE immunosensor to 10 U/mL CA15−3 after storage for different days. The detection solution in (**a**–**c**) is 0.1 M KCl solution containing 2.5 mM [Fe(CN)$_6$]$^{3-/4-}$ and error bars represent the standard deviations of three measurements.

2.7. Real Sample Analysis

The standard addition method was employed to assess the practical potential of the BSA/Ab/O-VMSF/CNNS/GCE immunosensor in fetal bovine serum. Fetal bovine serum is used as a model for "real sample" and first subjected to simple dilution using 0.01 M PBS (pH 7.4) and then spiked with several known concentrations of CA15-3, followed by quantitative determination by our fabricated BSA/Ab/O-VMSF/CNNS/GCE. Recovery and RSD are two important indexes to examine the potential analytical performance of the proposed electrochemical immunosensor in real samples. Recovery is defined as the ratio of the detected concentration by our immunosensor to the spiked known concentration. RSD indicates the degree of dispersion or consistency of three tested results under the same experimental conditions. Low RSD confirms the high precision of the sensors. The results in Table 2 suggest that RSD values obtained from these samples are below 1.9% and recoveries are in the range of 100~106%. Therefore, our proposed BSA/Ab/O-VMSF/CNNS/GCE shows good reliability, which is suitable for analysis of CA15-3 in real samples.

Table 2. Determination of CA15-3 in fetal bovine serum.

Sample	Added [b] (ng/mL)	Found (ng/mL)	Recovery (%)	RSD (%, n = 3)
serum [a]	0.0100	0.0100	100	1.9
	1.00	1.06	106	1.1
	100	103	103	0.66

[a] Fetal bovine serum sample detected in this study is diluted by a factor of 50 using PBS (0.01 M, pH 7.4). [b] The concentration of CA15-3 indicated in this Table is obtained after the dilution.

3. Materials and Methods

3.1. Chemicals and Materials

Carbohydrate antigen 15-3 (CA15-3), anti-CA15-3 antibody, carcinoembryonic antigen (CEA), alpha-fetoprotein (AFP), carbohydrate antigen 125 (CA125), carbohydrate antigen 19-9 (CA19-9), and fetal bovine serum were purchased from Beijing KeyGen Biotech Co., Ltd. (Beijing, China). Prostate-specific antigen (PSA) was procured from Beijing Biodragon Immunotechnologies Co., Ltd. (Beijing, China). Tetraethyl orthosilicate (TEOS, 98%), 3-glycidoxypropyltrimethoxysilane (GPTMS, 97%), cetyltrimethylammonium bromide (CTAB, 99%), bovine serum albumin (BSA), sodium dihydrogen phosphate dih7ydrate ($NaH_2PO_4 \cdot 2H_2O$, 99%), disodium hydrogen phosphate dodecahydrate ($Na_2HPO_4 \cdot 12H_2O$, 99%), glucose (Glu, 100%), sodium hydroxide (NaOH, 97%), sodium chloride (NaCl, 99.5%), potassium chloride (KCl, 99.5%), potassium ferricyanide ($K_3Fe(CN)_6$, 99.5%), and potassium ferrocyanide ($K_4Fe(CN)_6$, 99.5%) were purchased from Shanghai Aladdin Bio-Chem Technology Co., Ltd. (Shanghai, China). Sulfuric acid (H_2SO_4, 98%), acetone (99.5%), anhydrous ethanol (99.8%), and concentrated hydrochloric acid (HCl, 36–38%) were obtained from Hangzhou Shuanglin Reagent Co., Ltd. (Hangzhou, China). Melamine (99%) was purchased from Jiangsu Yonghua Fine Chemicals Co., Ltd. (Suzhou, China) Phosphate-buffered saline (PBS, 0.01 M, pH = 7.4) was prepared using $NaH_2PO_4 \cdot 2H_2O$ and $Na_2HPO_4 \cdot 12H_2O$. All the aqueous solutions used here were prepared using ultrapure water (18.2 MΩ cm) from Milli-Q Systems (Millipore Inc., Burlington, MA, USA). All chemical reagents were of analytical grade.

3.2. Characterizations and Instrumentations

The morphological structures of g-C_3N_4, CNNS, and VMSF were characterized using transmission electron microscopy (TEM, model HT7700, Hitachi, Tokyo, Japan). To prepare TEM samples, the VMSF layer was carefully scraped off from the electrode using a scalpel and dispersed in anhydrous ethanol with subsequent ultrasonic dispersion. Subsequently, the resulting dispersion was dripped onto a copper grid. Before morphology characterization under 200 kV, the sample was allowed to air dry naturally. All electrochemical experiments, including cyclic voltammetry (CV), electrochemical impedance spectroscopy (EIS), and differential pulse voltammetry (DPV), were conducted on an Autolab electrochemical workstation (model PGSTAT302N, Metrohm Autolab, Herisau, Switzerland). A conventional three-electrode system was employed, with an Ag/AgCl as the reference electrode, a platinum wire as the counter electrode, and the modified electrode as the working electrode. The frequency range for EIS measurements was from 0.1 Hz to 100 kHz, with a perturbation amplitude of 5 mV. Fourier transform infrared spectroscopy (FT-IR) was measured using a Vertex 70 spectrometer (Bruker, Billerica, MA, USA) through the KBr tablet method.

3.3. Preparation of VMSF/CNNS/GCE

The $g-C_3N_4$ was prepared by thermal polycondensation of melamine [39]: 5 g of melamine was weighed and placed in a crucible, heated to 520 °C at the rate of 6 °C per minute in an air atmosphere, calcined at this temperature for 4 h, and ground into a yellow solid powder to obtain $g-C_3N_4$. CNNS was prepared by stripping $g-C_3N_4$ with H_2SO_4. Generally, 2 g of $g-C_3N_4$ powder is added to 40 mL of H_2SO_4 (98 wt%) and stirred at room temperature for 10 h. Then, the mixture was slowly poured into 100 mL of deionized water and put under ultrasound for 8 h. After pouring out the clear supernatant, the obtained precipitate was repeatedly washed with deionized water by centrifugation at 12,000 rpm, and a stable colloidal suspension of CNNS was obtained. Finally, CNNS powder was obtained by freeze-drying.

GCE was polished by alumina powder with specifications of 0.5 μm, 0.3 μm, and 0.05 μm, successively. Then, ultrasonic cleaning GCE with anhydrous ethanol (99.7%) and ultrapure water in turn. 10 μL of CNNS colloid was dropped on the polished GCE and dried at 60 °C to obtain CNNS/GCE, and then VMSF was grown by the electrochemically assisted self-assembly (EASA) method [40–42]. Preparation of precursor solution containing silica: addition of 3050 μL TEOS into the mixed solution of 20 mL ethanol, 20 mL $NaNO_3$ (0.1 M, pH = 2.6) and 1.585 g CTAB. Then, the mixed solution was vigorously stirred for 2.5 h. After stirring, a three-electrode system was used, with CNNS/GCE as the working electrode, and a constant voltage of −2.2 V was applied for electrodeposition for 5 s. After that, the electrode was quickly taken out and washed with a large amount of ultrapure water, dried by nitrogen gas, and aged at 80 °C overnight. At this time, the silica nanopores contained surfactant micelles (SM), named SM@VMSF/CNNS/GCE. Due to the hydrophobicity of SM, SM@VMSF/CNNS/GCE was immersed in an HCl-ethanol solution (0.1 M) and stirred for 5 min to remove micelles, yielding VMSF/CNNS/GCE with open channels.

3.4. Fabrication of the VMSF/CNNS/GCE-Based Immunosensor

To immobilize the anti-CA15-3 antibody on the outer surface of the VMSF/CNNS/GCE, a bifunctional reagent GPTMS containing both epoxy groups and silane groups was selected as the crosslinking agent. SM@VMSF/CNNS/GCE was immersed in GPTMS (2.26 mM in ethanol) for 30 min, and epoxy groups were introduced on the outer surface of VMSF instead of internal nanochannels. After washing with ultrapure water, immersing in a solution containing 0.1 M HCl/ethanol, and stirring for 5 min to remove SM, the O-VMSF/CNNS/GCE with open inner nanochannels and epoxy groups on the external surface was obtained.

Then, 10 μL anti-CA15-3 antibody solution (1 μg/mL in 0.01 M PBS, pH = 7.4) was dripped on the surface of the O-VMSF/CNNS/GCE and incubated at 4 °C for 90 min. Afterwards, the electrode was washed with PBS (0.01 M, pH = 7.4) to remove unbound anti-CA15-3 antibodies on the electrode surface, obtaining Ab/O-VMSF/CNNS/GCE. BSA (1 wt%) solution was utilized to incubate with Ab/O-VMSF/CNNS/GCE at 4 °C for 30 min to block nonspecific sites. After being washed with ultrapure water, the obtained immunosensing electrode was denoted as BSA/Ab/O-/VMSF/CNNS/GCE.

3.5. Electrochemical Detection of CA15-3

The BSA/Ab/O-VMSF/CNNS/GCE immunosensor was incubated with different concentrations of CA15-3 at 4 °C for 60 min, respectively. The detection solution was 2.5 mM $[Fe(CN)_6]^{3-/4-}$ in 0.1 M KCl solution. DPV was used to measure the electrochemical signal of the BSA/Ab/O-VMSF/CNNS/GCE immunosensor before and after CA15-3 binding. Without complicated pretreatment, the fetal bovine serum as an actual sample was used to simulate the medium of actual human serum and diluted 50 times with PBS (0.01 M, pH = 7.4) buffer solution and directly used for the recovery experiment using the standard addition method.

4. Conclusions

In summary, through the introduction of the adhesive GNNS, stability of VMSF on GCE surface was significantly promoted, and application of such VMSF/GNNS nanocomposite on GCE for construction of immunosensors had been studied. Due to the oxygen-containing groups of CNNS, Si-O-Si chemical bonds between VMSF and CNNS can be formed, which ensures the stable growth of VMSF on the GCE surface and greatly improves the accuracy and reproductivity of the electroanalytical sensor. Owing to the plentiful open nanopores of VMSF, VMSF possesses good permeability for transport of electrochemical probes ($[Fe(CN)_6]^{3-/4-}$). When the CA15-3 antibody is anchored on the external surface of VMSF and incubated with target CA15-3, access of $[Fe(CN)_6]^{3-/4-}$ to the underlying GCE through silica nanochannels is blocked, generating the declined electrochemical current and finally enabling the quantitative detection of CA15-3. A broad detection range of 1 mU/mL to 1000 U/mL and a low LOD of 0.47 mU/mL are achieved by our fabricated VMSF/CNNS/GCE-based immunosensor. Furthermore, determination of CA15-3 in spiked fetal bovine serum samples was used to evaluate the potential practical application of the fabricated immunosensor, displaying good selectivity and accuracy. This immunosensor extends the adhesive layer for growth of VMSF on GCE surface and also offers a simple and effective electrochemical strategy for detection of various practical biomarkers in complicated biological specimens. To better meet the demand of rapid determination in clinical applications, miniaturization and intellectualization of our fabricated electrochemical immunosensor require improvement in future research by combination with microelectronics technology and smart phones.

Supplementary Materials: The following supporting information can be downloaded at: https://www.mdpi.com/article/10.3390/molecules29184334/s1. Figure S1: CV (a) and EIS responses (b) of the prepared BSA/Ab/O-VMSF/CNNS/GCE to various concentrations of CA15-3 in 0.1 M KCl solution containing 2.5 mM $[Fe(CN)_6]^{3-/4-}$. Inset in (b) is the equivalent circuit diagram.

Author Contributions: Conceptualization, J.X. and H.W.; investigation, J.X.; data curation, J.X. and H.W.; writing—original draft preparation, J.X. and H.W.; writing—review and editing, F.Y.; supervision, F.Y. All authors have read and agreed to the published version of the manuscript.

Funding: This study was funded by Shanxi Province "136 Revitalization Medical Project Construction Funds" and the Zhejiang Provincial Natural Science Foundation of China (LY21B050003).

Institutional Review Board Statement: Not applicable.

Informed Consent Statement: Not applicable.

Data Availability Statement: The data presented in this study are available on request from the corresponding author.

Conflicts of Interest: The authors declare no conflicts of interest.

References

1. Zhou, P.; Yao, L.; Chen, K.; Su, B. Silica Nanochannel Membranes for Electrochemical Analysis and Molecular Sieving: A Comprehensive Review. *Crit. Rev. Anal. Chem.* **2020**, *50*, 424–444. [CrossRef]
2. Wu, Y.; Shi, Z.; Liu, J.; Luo, T.; Xi, F.; Zeng, Q. Simple fabrication of electrochemical sensor based on integration of dual signal amplification by the supporting electrode and modified nanochannel array for direct and sensitive detection of vitamin B(2). *Front. Nutr.* **2024**, *11*, 1352938. [CrossRef]
3. Zhu, X.; Xuan, L.; Gong, J.; Liu, J.; Wang, X.; Xi, F.; Chen, J. Three-dimensional macroscopic graphene supported vertically-ordered mesoporous silica-nanochannel film for direct and ultrasensitive detection of uric acid in serum. *Talanta* **2022**, *238*, 123027. [CrossRef]

4. Wang, K.; Yang, L.; Huang, H.; Lv, N.; Liu, J.; Liu, Y. Nanochannel array on electrochemically polarized screen printed carbon electrode for rapid and sensitive electrochemical determination of clozapine in human whole blood. *Molecules* **2022**, *27*, 2739. [CrossRef]
5. Huang, J.; Fan, X.; Yan, F.; Liu, J. Vertical silica nanochannels and o-phenanthroline chelator for the detection of trace Fe(II). *ACS Appl. Nano Mater.* **2024**, *7*, 7743–7752. [CrossRef]
6. Zhou, X.; Gu, X.; Zhang, S.; Zou, Y.; Yan, F. Magnetic graphene oxide and vertically-ordered mesoporous silica film for universal and sensitive homogeneous electrochemiluminescence aptasensor platform. *Microchem. J.* **2024**, *200*, 110315. [CrossRef]
7. Yu, R.; Zhao, Y.; Liu, J. Solid electrochemiluminescence sensor by immobilization of emitter ruthenium(II)tris(bipyridine) in bipolar silica nanochannel film for sensitive detection of oxalate in serum and urine. *Nanomaterials* **2024**, *14*, 390. [CrossRef]
8. Huang, J.; Zhang, T.; Zheng, Y.; Liu, J. Dual-mode sensing platform for cancer antigen 15-3 determination based on a silica nanochannel array using electrochemiluminescence and electrochemistry. *Biosensors* **2023**, *13*, 317. [CrossRef]
9. Chen, D.; Luo, X.; Xi, F. Probe-integrated electrochemical immunosensor based on electrostatic nanocage array for reagentless and sensitive detection of tumor biomarker. *Front. Chem.* **2023**, *11*, 1121450. [CrossRef]
10. Luo, X.; Zhang, T.; Tang, H.; Liu, J. Novel electrochemical and electrochemiluminescence dual-modality sensing platform for sensitive determination of antimicrobial peptides based on probe encapsulated liposome and nanochannel array electrode. *Front. Nutr.* **2022**, *9*, 962736. [CrossRef]
11. Ma, N.; Luo, X.; Wu, W.; Liu, J. Fabrication of a disposable electrochemical immunosensor based on nanochannel array modified electrodes and gated electrochemical signals for sensitive determination of C-reactive protein. *Nanomaterials* **2022**, *12*, 3981. [CrossRef] [PubMed]
12. Li, Y.; Luo, Z.; Li, G.; Belwal, T.; Li, L.; Xu, Y.; Su, B.; Lin, X. Interference-free detection of caffeine in complex matrices using a nanochannel electrode modified with binary hydrophilic-hydrophobic PDMS. *ACS Sens.* **2021**, *6*, 1604–1612. [CrossRef]
13. Zhou, X.; Zou, Y.; Ru, H.; Yan, F.; Liu, J. Silica nanochannels as nanoreactors for confined synthesis of Ag NPs to boost electrochemical stripping chemiluminescence of luminol-O_2 system for sensitive aptasensor. *Anal. Chem.* **2024**, *96*, 10264–10273. [CrossRef]
14. Li, F.; Han, Q.; Xi, F. The fabrication of a probe-integrated electrochemiluminescence aptasensor based on double-layered nanochannel array with opposite charges for the sensitive determination of C-reactive protein. *Molecules* **2023**, *28*, 7867. [CrossRef]
15. Ma, X.; Zhang, Z.; Zheng, Y.; Liu, J. Solid-phase electrochemiluminescence enzyme electrodes based on nanocage arrays for highly sensitive detection of cholesterol. *Biosensors* **2024**, *14*, 403. [CrossRef] [PubMed]
16. Ma, K.; Zheng, Y.; An, L.; Liu, J. Ultrasensitive immunosensor for prostate-specific antigen based on enhanced electrochemiluminescence by vertically ordered mesoporous silica-nanochannel film. *Front. Chem.* **2022**, *10*, 851178. [CrossRef]
17. Xi, F.; Xuan, L.; Lu, L.; Huang, J.; Yan, F.; Liu, J.; Dong, X.; Chen, P. Improved adhesion and performance of vertically-aligned mesoporous silica-nanochannel film on reduced graphene oxide for direct electrochemical analysis of human serum. *Sens. Actuators B Chem.* **2019**, *288*, 133–140. [CrossRef]
18. Nasir, T.; Zhang, L.; Vila, N.; Herzog, G.; Walcarius, A. Electrografting of 3-aminopropyltriethoxysilane on a glassy carbon electrode for the improved adhesion of vertically oriented mesoporous silica thin films. *Langmuir* **2016**, *32*, 4323–4332. [CrossRef]
19. Zhou, H.; Ding, Y.; Su, R.; Lu, D.; Tang, H.; Xi, F. Silica nanochannel array film supported by β-cyclodextrin-functionalized graphene modified gold film electrode for sensitive and direct electroanalysis of acetaminophen. *Front. Chem.* **2022**, *9*, 812086. [CrossRef]
20. Lv, N.; Qiu, X.; Han, Q.; Xi, F.; Wang, Y.; Chen, J. Anti-biofouling electrochemical sensor based on the binary nanocomposite of silica nanochannel array and graphene for doxorubicin detection in human serum and urine samples. *Molecules* **2022**, *27*, 8640. [CrossRef]
21. Ma, K.; Yang, L.; Liu, J.; Liu, J. Electrochemical sensor nanoarchitectonics for sensitive detection of uric acid in human whole blood based on screen-printed carbon electrode equipped with vertically-ordered mesoporous silica-nanochannel film. *Nanomaterials* **2022**, *12*, 1157. [CrossRef] [PubMed]
22. Zhou, H.; Dong, G.; Sailjoi, A.; Liu, J. Facile pretreatment of three-dimensional graphene through electrochemical polarization for improved electrocatalytic performance and simultaneous electrochemical detection of catechol and hydroquinone. *Nanomaterials* **2022**, *12*, 65. [CrossRef]
23. Gong, J.; Tang, H.; Wang, M.; Lin, X.; Wang, K.; Liu, J. Novel three-dimensional graphene nanomesh prepared by facile electro-etching for improved electroanalytical performance for small biomolecules. *Mater. Des.* **2022**, *215*, 110506. [CrossRef]
24. Su, R.; Tang, H.; Xi, F. Sensitive electrochemical detection of p-nitrophenol by pre-activated glassy carbon electrode integrated with silica nanochannel array film. *Front. Chem.* **2022**, *10*, 954748. [CrossRef]
25. Deng, X.; Lin, X.; Zhou, H.; Liu, J.; Tang, H. Equipment of vertically-ordered mesoporous silica film on electrochemically pretreated three-dimensional graphene electrodes for sensitive detection of methidazine in urine. *Nanomaterials* **2023**, *13*, 239. [CrossRef] [PubMed]
26. Huang, L.; Su, R.; Xi, F. Sensitive detection of noradrenaline in human whole blood based on Au nanoparticles embedded vertically-ordered silica nanochannels modified pre-activated glassy carbon electrodes. *Front. Chem.* **2023**, *11*, 1126213. [CrossRef] [PubMed]
27. Zheng, W.; Su, R.; Lin, X.; Liu, J. Nanochannel array modified three-dimensional graphene electrode for sensitive electrochemical detection of 2,4,6-trichlorophenol and prochloraz. *Front. Chem.* **2022**, *10*, 954802. [CrossRef] [PubMed]

28. Wang, L.; Fan, Z.; Yue, F.; Zhang, S.; Qin, S.; Luo, C.; Pang, L.; Zhao, J.; Du, J.; Jin, B.; et al. Flower-like 3D MoS_2 microsphere/2D C_3N_4 nanosheet composite for highly sensitive electrochemical sensing of nitrite. *Food Chem.* **2024**, *430*, 137027. [CrossRef] [PubMed]
29. Dong, Y.; Wang, Q.; Wu, H.; Chen, Y.; Lu, C.H.; Chi, Y.; Yang, H.H. Graphitic carbon nitride materials: Sensing, imaging and therapy. *Small* **2016**, *12*, 5376–5393. [CrossRef]
30. Fejzic, H.; Mujagic, S.; Azabagic, S.; Burina, M. Tumor marker CA 15-3 in breast cancer patients. *Acta Med. Acad.* **2015**, *44*, 39–46. [CrossRef]
31. Ma, N.; Xu, S.; Wu, W.; Liu, J. Electrochemiluminescence aptasensor with dual signal amplification by silica nanochannel-based confinement effect on nanocatalyst and efficient emitter enrichment for highly sensitive detection of C-reactive protein. *Molecules* **2023**, *28*, 7664. [CrossRef] [PubMed]
32. Zhang, T.; Xu, S.; Lin, X.; Liu, J.; Wang, K. Label-free electrochemical aptasensor based on the vertically-aligned mesoporous silica films for determination of aflatoxin B1. *Biosensors* **2023**, *13*, 661. [CrossRef] [PubMed]
33. Yan, L.; Xu, S.; Xi, F. Disposal immunosensor for sensitive electrochemical fetection of prostate-specific antigen based on amino-rich nanochannels array-modified patterned indium tin oxide electrode. *Nanomaterials* **2022**, *12*, 3810. [CrossRef] [PubMed]
34. Shawky, A.M.; El-Tohamy, M. Signal amplification strategy of label-free ultrasenstive electrochemical immunosensor based ternary $Ag/TiO_2/rGO$ nanocomposites for detecting breast cancer biomarker CA 15-3. *Mater. Chem. Phys.* **2021**, *272*, 124983. [CrossRef]
35. Pothipor, C.; Bamrungsap, S.; Jakmunee, J.; Ounnunkad, K. A gold nanoparticle-dye/poly(3-aminobenzylamine)/two dimensional $MoSe_2$/graphene oxide electrode towards label-free electrochemical biosensor for simultaneous dual-mode detection of cancer antigen 15-3 and microRNA-21. *Colloids Surf. B Biointerfaces* **2022**, *210*, 112260. [CrossRef]
36. Amani, J.; Khoshroo, A.; Rahimi-Nasrabadi, M. Electrochemical immunosensor for the breast cancer marker CA 15–3 based on the catalytic activity of a CuS/reduced graphene oxide nanocomposite towards the electrooxidation of catechol. *Microchim. Acta* **2017**, *185*, 79. [CrossRef]
37. Khoshroo, A.; Mazloum-Ardakani, M.; Forat-Yazdi, M. Enhanced performance of label-free electrochemical immunosensor for carbohydrate antigen 15-3 based on catalytic activity of cobalt sulfide/graphene nanocomposite. *Sens. Actuators B Chem.* **2018**, *255*, 580–587. [CrossRef]
38. Wang, A.-J.; Zhu, X.-Y.; Chen, Y.; Luo, X.; Xue, Y.; Feng, J.-J. Ultrasensitive label-free electrochemical immunoassay of carbohydrate antigen 15-3 using dendritic Au@Pt nanocrystals/ferrocene-grafted-chitosan for efficient signal amplification. *Sens. Actuators B Chem.* **2019**, *292*, 164–170. [CrossRef]
39. Qian, J.; Yuan, A.; Yao, C.; Liu, J.; Li, B.; Xi, F.; Dong, X. Highly efficient photo-reduction of *p*-nitrophenol by protonated graphitic carbon nitride nanosheets. *ChemCatChem* **2018**, *10*, 4747–4754. [CrossRef]
40. Walcarius, A.; Sibottier, E.; Etienne, M.; Ghanbaja, J. Electrochemically assisted self-assembly of mesoporous silica thin films. *Nat. Mater.* **2007**, *6*, 602–608. [CrossRef]
41. Zhang, Y.; Zhang, S.; Liu, J.; Qin, D. Label-free homogeneous electrochemical aptasensor based on size exclusion/charge-selective permeability of nanochannel arrays and 2D nanorecognitive probe for sensitive detection of alpha-fetoprotein. *Molecules* **2023**, *28*, 6935. [CrossRef]
42. Yan, Z.; Zhang, S.; Liu, J.; Xing, J. Homogeneous electrochemical aptamer sensor based on two-dimensional nanocomposite probe and nanochannel modified electrode for sensitive detection of carcinoembryonic antigen. *Molecules* **2023**, *28*, 5186. [CrossRef]

Disclaimer/Publisher's Note: The statements, opinions and data contained in all publications are solely those of the individual author(s) and contributor(s) and not of MDPI and/or the editor(s). MDPI and/or the editor(s) disclaim responsibility for any injury to people or property resulting from any ideas, methods, instructions or products referred to in the content.

Article

Eco-Friendly Waterborne Polyurethane Coating Modified with Ethylenediamine-Functionalized Graphene Oxide for Enhanced Anticorrosion Performance

Mariel Amparo Fernandez Aramayo [1,2,*], Rafael Ferreira Fernandes [1,2], Matheus Santos Dias [1,2], Stella Bozzo [1,2], David Steinberg [1,2], Marcos Rocha Diniz da Silva [1,2], Camila Marchetti Maroneze [1,2] and Cecilia de Carvalho Castro Silva [1,2,*]

1. Mackenzie School of Engineering, Mackenzie Presbyterian University, Consolação Street 930, São Paulo 01302-907, Brazil; reomldm@gmail.com (R.F.F.); theeus.santos@gmail.com (M.S.D.); stellabozzo98@gmail.com (S.B.); david.steinberg@mackenzie.br (D.S.); marcosrochadiniz@hotmail.com (M.R.D.dS.); camila.maroneze@mackenzie.br (C.M.M.)
2. MackGraphe-Mackenzie Institute for Research in Graphene and Nanotechnologies, Mackenzie Presbyterian University, Consolação Street 930, São Paulo 01302-907, Brazil
* Correspondence: aramayomariel.m@gmail.com (M.A.F.A.); cecilia.silva@mackenzie.br (C.d.C.C.S.)

Citation: Aramayo, M.A.F.; Ferreira Fernandes, R.; Santos Dias, M.; Bozzo, S.; Steinberg, D.; Rocha Diniz da Silva, M.; Maroneze, C.M.; de Carvalho Castro Silva, C. Eco-Friendly Waterborne Polyurethane Coating Modified with Ethylenediamine- Functionalized Graphene Oxide for Enhanced Anticorrosion Performance. *Molecules* 2024, 29, 4163. https://doi.org/10.3390/molecules29174163

Academic Editors: Sake Wang, Minglei Sun and Nguyen Tuan Hung

Received: 24 July 2024
Revised: 19 August 2024
Accepted: 23 August 2024
Published: 3 September 2024

Copyright: © 2024 by the authors. Licensee MDPI, Basel, Switzerland. This article is an open access article distributed under the terms and conditions of the Creative Commons Attribution (CC BY) license (https://creativecommons.org/licenses/by/4.0/).

Abstract: This study explores the potential of graphene oxide (GO) as an additive in waterborne polyurethane (WPU) resins to create eco-friendly coatings with enhanced anticorrosive properties. Traditionally, WPU's hydrophilic nature has limited its use in corrosion-resistant coatings. We investigate the impact of incorporating various GO concentrations (0.01, 0.1, and 1.3 wt%) and functionalizing GO with ethylenediamine (EDA) on the development of anticorrosive coatings for carbon steel. It was observed, by potentiodynamic polarization analysis in a 3.5% NaCl solution, that the low GO content in the WPU matrix significantly improved anticorrosion properties, with the 0.01 wt% GO-EDA formulation showing exceptional performance, high E_{corr} (−117.82 mV), low i_{corr} (3.70×10^{-9} A cm^{-2}), and an inhibition corrosion efficiency (η) of 99.60%. Raman imaging mappings revealed that excessive GO content led to agglomeration, creating pathways for corrosive species. In UV/condensation tests, the 0.01 wt% GO-EDA coating exhibited the most promising results, with minimal corrosion products compared to pristine WPU. The large lateral dimensions of GO sheets and the cross-linking facilitated by EDA enhanced the interfacial properties and dispersion within the WPU matrix, resulting in superior barrier properties and anticorrosion performance. This advancement underscores the potential of GO-based coatings for environmentally friendly corrosion protection.

Keywords: eco-friendly; waterborne polyurethane; coating; graphene oxide; low additive content; functionalization; anticorrosion; UV resistance; water resistance

1. Introduction

Metal corrosion is a widespread issue that significantly affects human life in terms of economics, the environment, and health safety [1]. Anticorrosive coatings provide durability, cost-efficiency, and excellent protection for metal surfaces, mainly carbon steel, which represents the most used metal alloy in the world [2,3]. In this context, organic coatings are considered the most effective method for protecting carbon steel from corrosion, because they create a physical barrier against the aggressive environment [4–7]. The production of durable coatings with anticorrosive properties, while adhering to green chemistry principles (GCP), is currently an emerging challenge and a market demand [8–11]. Waterborne polyurethane (WPU) resins exemplify GCP, featuring low volatile organic compound content, which contributes to reducing emissions and air pollution [12,13]. WPU resins demonstrate excellent abrasion resistance. However, the hydrophilic segments in the polymer backbone, which contribute to the superior colloidal stability of WPU, also lead to some detrimental surface effects, including reduced water resistance, increased susceptibility to UV degradation, and vulnerability to corrosion. These factors collectively shorten the long-term application of WPU-based materials [14,15]. A great strategy to improve the long-term performance of WPU-based materials for anticorrosion applications is the development of nanocomposites [16,17]. The addition of nanoparticles in the polymer matrix really plays an important role in the improvement of the barrier properties and the water resistance [18,19]. In this scenario, graphene derivatives are appealing materials that have recently been applied to enhance the barrier properties of polymer-based materials [20–22].

Graphene and its derivatives, such as graphene oxide (GO) and reduced graphene oxide (rGO), have been widely utilized in the formulation of anticorrosive coatings due to their nano-barrier effect, which creates a labyrinth within the coatings, thereby prolonging the penetration path of corrosive species [13,23,24]. In particular, GO has attracted attention as an additive in coatings because of its impermeable properties, especially in water-based coatings [12,24,25]. This is due to its high dispersibility in water, which is attributed to its hydrophilic nature and the presence of oxygen-containing functional groups (hydroxyl, carboxyl, and epoxy groups) on its surface and edges [26]. However, achieving uniform GO dispersion within the polymeric matrix remains a common challenge highlighted in the literature [27]. This issue could potentially compromise the anticorrosive performance of the coatings since a poorly dispersed GO on the coating results in the creation of paths within the coating due to the agglomeration of the sheets [27–29]. Functionalization of GO is a promising strategy for promoting the uniform dispersibility of GO sheets in the coating polymer matrix [9,30]. Ning et al. functionalized GO with dodecylbenzenesulfonic acid, phosphoric acid, and polyaniline. This composite demonstrated improved dispersion and compatibility in WPU, thereby delaying the time for corrosion species to access the metal interface [31]. Wen et al. demonstrated the improvement in anticorrosion properties in WPU by incorporating 0.3 wt% GO covalently functionalized with isophorone diisocyanate [30]. Similarly, Cui et al. achieved comparable results by incorporating 0.2 wt% GO functionalized with polycarbodiimide into the WPU matrix [9]. Li et al. also described superior anticorrosion properties by incorporating 0.4 wt% rGO functionalized with titanate coupling agents into WPU [32]. A simple and cost-effective approach that has attracted attention involves using ethylenediamine (EDA) to functionalize graphene oxide (GO), because of its ability to manipulate the interlayer spacing between GO sheets [33]. Maslekar et al. determined the reactivity of EDA toward the main oxygen-containing functional groups of GO (epoxy, hydroxyl, and carboxylic acid) in lithium-ion batteries. Their findings highlighted a significant impact on properties, such as colloidal stability. For anticorrosive coatings, this enhanced colloidal stability is particularly valuable as it improves the uniform dispersion of GO flakes within the resin, thereby enhancing the effectiveness of the anticorrosive coating [33].

Most reports in the literature have achieved enhanced anticorrosion properties of WPU coatings by incorporating a high content (at least 0.2 wt%) of functionalized GO as an additive [8,9,16,18,30]. Song et al. reported that adding GO to WPU improved the water contact angle compared to pure WPU films, indicating enhanced hydrophobic properties. Nevertheless, they also noted that incorporating more than 0.5 wt% of GO decreased the tensile strength of the films [8]. Thus, it remains essential to balance effective functionalization to prevent agglomeration with determining the optimal concentration of GO within the polymeric matrix. However, a common limitation in many studies is the lack of long-term accelerated testing, which is crucial to fully evaluate the performance and durability of these coatings.

Aiming to improve GCP principles by reducing the amount of nanomaterial used, we present an investigation into the effect of incorporating different concentrations of GO (1.3, 0.1, and 0.01 wt%) in WPU resins and the impact of functionalizing a low content (0.01 wt%) of GO sheets with ethylenediamine (EDA) as an additive to develop anticorrosive coatings for carbon steel surfaces. The functionalization method using EDA, a commercially available single-component reagent, is not only straightforward and cost-effective but also highly efficient. It requires fewer materials and less time than the more complex methods reported in the literature [18,31,34–38]. Potentiodynamic polarization analysis in a 3.5% NaCl solution established a strong correlation between the low GO content in the WPU resin matrix and higher anticorrosion properties, especially for the 0.01 wt% GO-EDA. The large lateral size of the GO sheets, combined with the cross-linking between GO and WPU promoted by EDA, improved the interfacial properties between the GO and WPU polymer matrix. This resulted in better barrier properties and a less hydrophilic surface, consequently leading to a higher anticorrosion performance of the developed eco-friendly coating.

2. Results and Discussion

2.1. Characterization of GO

The successful oxidation of the graphite and the achievement of GO sheets were confirmed by the well-defined Raman spectrum of GO presented in Figure 1a and UV-Vis spectra of GO dispersion (Figure S1). In the Raman spectrum, Figure 1a, the D band at ~1350 cm^{-1} is related to the defects in GO structures (sp^3 carbons from the oxygen functional groups); the G band at ~1610 cm^{-1} is attributed to the sp^2 carbons (refers to the stretching of the C–C bonds), where its intensity and position indicate the degree of graphitization; and the absence of the 2D band (usually at ~2700 cm^{-1}) is related with the high structural disorder degree in the two-dimensional plane of GO [39,40]. Figure S1 shows the characteristic UV-Vis spectra for GO dispersion, exhibiting two bands at approximately 230 and 300 nm, respectively, attributed to $\pi \rightarrow \pi^*$ transitions, associated with C–C bonds in aromatic species, and n $\rightarrow \pi^*$ transitions, related to C–O bonds, resulting from oxidation processes [41–43].

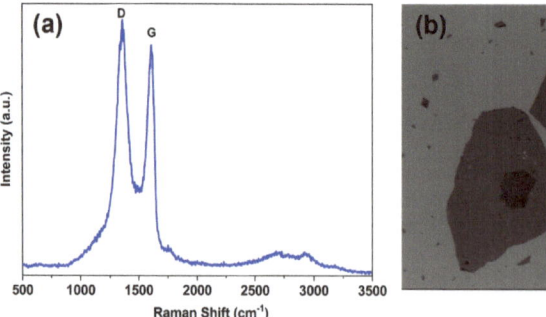

Figure 1. (**a**) Representative Raman spectrum of GO sheets used to prepare the nanocomposites with WPU, displaying the typical D and G bands. (**b**) SEM micrograph of GO sheets on a Si substrate.

The morphology and lateral size of the GO sheets were examined using scanning electron microscopy (SEM), as shown in Figures 1b and S2. This analysis underscores graphite's effective oxidation and exfoliation, resulting in GO sheets distinguished by their transparency and large lateral dimensions exceeding 25 µm. This transparency indicates the presence of only a few layers of GO [44,45].

2.2. Dispersion of GO in the WPU Matrix

To evaluate the stability of GO in the WPU resins, the dispersions were stored for three months. Figure 2 shows photographic images of the WPU resin and GO dispersions in WPU at various concentrations, both immediately after preparation and after three months of storage. After these months, a characteristic black color was observed in the dispersions, which can be attributed to the mild partial reduction of the GO to rGO, as also observed by Otsuka et al. for GO aqueous colloidal suspension after seven days [46]. All samples remained stable except for 0.01-GO and 0.1-GO, likely due to their low GO concentrations. However, mechanical agitation restored 0.1-GO stability. The functionalization with EDA was performed to promote the better dispersibility of GO sheets on the WPU matrix for the 0.01-GO sample. The EDA can interact with GO by hydrogen bonding and/or as a cross-linking agent between the GO sheets since the two terminal amino groups (-NH$_2$) can covalently bind to the carboxyl groups of GO sheets, promoting a higher dispersion of GO sheets into the WPU matrix [47]. This potentially improves the nanocomposite's barrier properties [47,48]. Additionally, since EDA acts as a hardener for the polyurethane resin, it may enhance the mechanical resistance of the material. However, it is crucial to use the appropriate amount (low content) of this chain extender to achieve optimal results, avoiding the self-polymerization process between the EDA and the WPU [49,50].

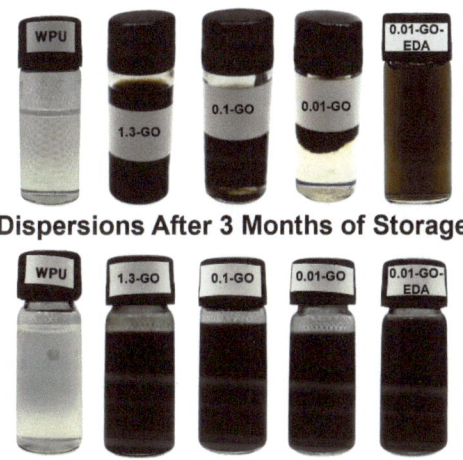

Figure 2. Photographic images of WPU resin and dispersions of GO in WPU resin at various concentrations after preparation and 3 months of storage.

Raman analysis and 2D imaging mappings of a section of the coating were performed to confirm the incorporation and dispersibility of GO in the WPU resin as a function of GO content. Figure 3(a-i) shows the representative Raman spectrum for the 0.01-GO sample. It is possible to identify the typical D (1356 cm^{-1}) and G (1607 cm^{-1}) band for GO sheets. The ID/IG intensity ratio measures the disorder degree and the average size of the sp^2 domains in graphite materials [39,40]. For the 0.010-GO sample, it was found to be 1.15. For the identification of the WPU, the presence of the Raman bands was detected with a maximum at 2932 cm^{-1} and 2863 cm^{-1}, which can be attributed to asymmetric and symmetric C-H stretching vibration of CH$_2$ groups, and the Raman band at 1732 cm^{-1} can be related to the C=O stretching vibration modes of the ester group of the polyurethanes [51]. Figure 3(b-ii) shows the respective Raman imaging mapping for the 0.01-GO sample, which was constructed based on the intensity of the G band. It is possible to observe some regions of high intensity in the G band (white color), which may be related to the non-uniform dispersibility of the GO sheets in the WPU in this concentration, as also observed in Figure 2. Supplementary Video S1, associated with the real-time Raman mapping of the 0.01-GO sample, makes it possible to observe some dark areas in the mapping, where is not possible to observe the D and G band of the GO Raman spectrum with enough intensity, corroborating with the low uniform distribution of the GO sheets in the WPU matrix. Figure 3(a-iii,b-iv) shows the effect of functionalizing the 0.01-GO sample with EDA. The Raman spectrum of the 0.01-GO-EDA sample exhibits the same Raman features as the 0.01-GO sample. However, after functionalization with EDA, the ID/IG intensity ratio increased from 1.15 to 1.34. This indicates a decrease in the size of the in-plane sp^2 crystalline domains, which may be related to the chemical functionalization and partial reduction in GO through its reaction with EDA [52]. The carboxyl and epoxy groups of GO can react with the amine groups of the EDA molecule, partially restoring the sp^2 crystalline domains but with smaller sizes [47]. This increases the intensity of the D band due to border defects (edges). The improvement in the homogeneous dispersibility of GO sheets after functionalization with EDA is further evidenced by the Raman imaging mapping (Figure 3(b-iv)) and Supplementary Video S2. A significant decrease in the area of GO agglomeration (white color) can be observed. Additionally, the presence of GO sheets is visible throughout the entire extent of the samples. For the samples with higher GO content in the WPU matrix, 0.1-GO and 1.3-GO, a similar Raman spectrum was observed (Figure 3(a-v,vii)). The D and G bands are present; however, the Raman feature related to the C=O stretching vibration modes (1732 cm^{-1}) of the ester group in the polyurethanes was not observable. Besides that, the intensity of the asymmetric and symmetric C-H stretching vibrations of CH$_2$ groups significantly decreased, likely due to the high content of GO sheets in the polymer matrix, which dominate the Raman signature signal. An ID/IG intensity ratio of 1.35 and 1.09 was found for the 0.1-GO and 1.3-GO samples, respectively. The Raman imaging mapping (Figure 3(b-vi,viii)) and Supplementary Videos S3 and S4 clearly show the agglomeration process that occurs with the GO sheets in the WPU matrix as the GO content increases. In both cases, regions with a high agglomeration of GO sheets (white areas in the Raman imaging mapping—Figure 3(b-vi,viii)) are evident. In the 0.1-GO samples, several areas without GO sheets (black areas in the Raman imaging mapping—Supplementary Video S3) were also observed.

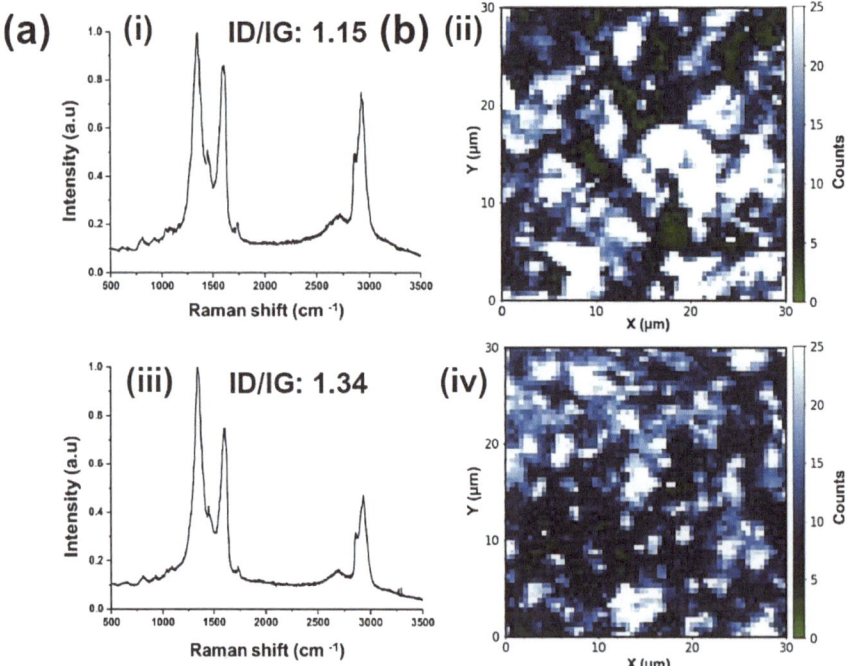

Figure 3. *Cont.*

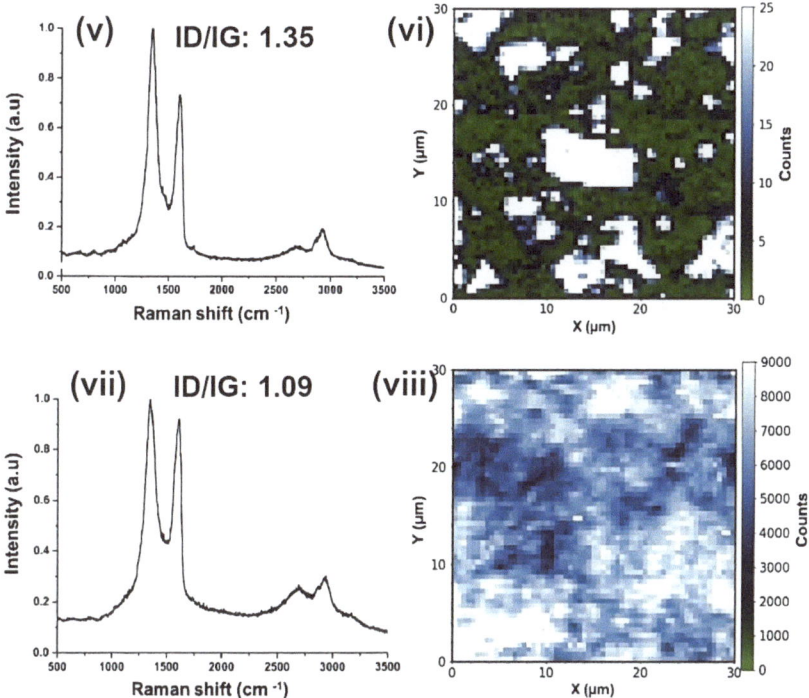

Figure 3. (a) Representative Raman spectrum, ID/IG intensity ratio, and corresponding (b) Raman imaging mapping, performed using the G band intensity as the reference for the samples, for (**i,ii**) 0.01-GO; (**iii,iv**) 0.01-GO-EDA; (**v,vi**) 0.1-GO; and (**vii,viii**) 1.3-GO.

2.3. Evaluation of Chemical Interaction of GO with WPU Resin

Figure 4 shows the FTIR spectra of GO, WPU, and the coated steel samples of 1.3-GO, 0.1-GO, 0.01-GO, and 0.01-GO-EDA. The FTIR spectrum of the GO sample exhibits the characteristic features of graphene oxide. Bands in 1720, 1620, and 1411 cm^{-1} correspond to the vibrational modes of carbonyl groups (C=O), alkene groups (C=C), and hydroxyl groups (–OH), respectively [53–56]. Additionally, the bands in 1045 and 979 cm^{-1} are attributed to the epoxide groups [53–56].

The spectrum of WPU (green curve) displays the functional groups attributed to polyurethane resin. The band at 3300 cm^{-1} is attributed to the N-H stretching vibration of the urethane bond of PU. The two bands at 2930 cm^{-1} and 2850 cm^{-1} correspond to the C-H stretching vibrations of urethane bonds. The absorption band at 1740 cm^{-1} is attributed to the C=O stretching vibration of the urethane group [8,9,14,15,57]. The FTIR spectra of WPU and WPU containing the different concentrations of GO are similar because the content of GO incorporated into the WPU matrix is minimal in relation to the presence of the WPU, causing the GO bands to overlap with the more intense bands of the WPU resin. The same behavior is observed for the 0.01 GO-WPU sample functionalized with EDA. It was not possible to verify the presence of the C–N stretching (related to the covalent bond between the oxygen functional groups of GO, such as epoxy and carboxyl, and the amino groups of EDA) [47] through FTIR analysis, as this feature was already obscured by the C–N stretching of WPU (~1453 cm^{-1}) [58]. Similarly, the –NH bands of the primary (3300 cm^{-1}) and secondary amines (1570 cm^{-1}) in EDA [47] were masked by the –NH stretching (3500–3300 cm^{-1}) and –NH bending (1583–1486 cm^{-1}) in the WPU [9].

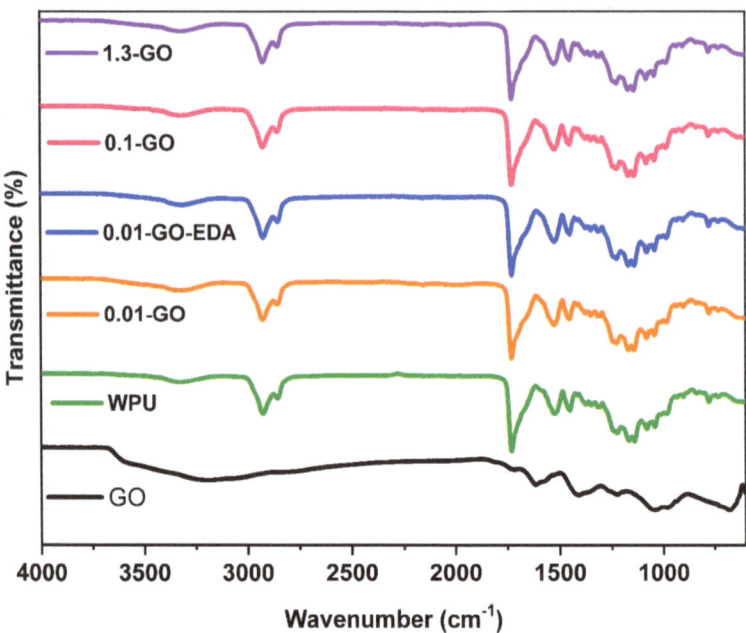

Figure 4. FTIR spectra of GO, WPU resin, and all concentrations of GO incorporated into the WPU resin.

2.4. Hydrophobicity Analysis

Figure 5 shows the results of the static contact angle measurements used to examine the hydrophobicity of GO-WPU coatings applied on carbon steel sheets. The waterborne polyurethane resin without GO already exhibits an intrinsic low hydrophobic character. Nevertheless, the incorporation of GO increased the contact angle, hence increasing the hydrophobicity of the coating. This improvement is primarily due to the graphitic structure of GO, which produces a higher hydrophobic surface, in comparison with the pristine WPU surface, that resists water spreading [24,59,60]. The increase in GO content from 0.01 to 1.3 wt%. increased the contact angle of the water on the WPU films from 67.35° to 71.82°, compared to 65.92° for the neat WPU. Song et al. found that the maximum water contact angle reached for 1.0 wt% GO was 55.38°, with flakes measuring 105.42 nm on average. In our study, the GO flakes possess an average size of around 25 μm, which may contribute to the enhanced contact angle due to improved hierarchical surface structures and roughness in the coating [21,24]. Additionally, Song et al. reported that an optimal GO content in WPU up to 0.5 wt% could prevent GO agglomeration in the resin [8]. In the present work, the coatings with the lowest concentration studied (0.01 wt% GO) and functionalized with EDA (sample 0.01-GO-EDA, in Figure 5) demonstrated the highest contact angle (71.92°). This can be attributed to the chemical functionalization and partial reduction of GO through its reaction with EDA. The carboxyl and epoxy groups of GO react with the amine groups of the EDA molecule, promoting the mild reduction in GO and improving the hydrophobicity of the surface [57]. This enhancement in hydrophobicity is a critical aspect of achieving high anticorrosion properties [8,22]. Reducing water permeation through WPU represents one of the biggest challenges in using this polymer to develop coatings with barrier properties [16]. In this work, we achieved promising results using a low content of GO (only 0.01 wt%) due to the combination of the sizeable lateral size of the sheets and the functionalization with EDA.

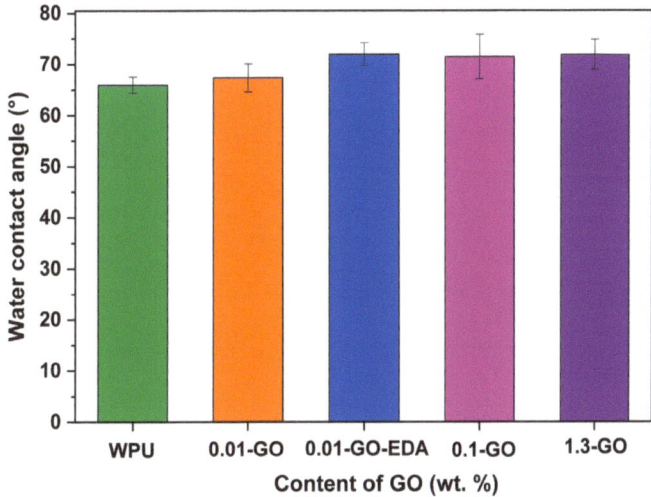

Figure 5. The water contact angle of WPU resin is a function of incorporating different GO and EDA-functionalized GO contents applied to the carbon steel substrates.

2.5. Evaluation of Anticorrosion Performance

Potentiodynamic polarization curves were measured for all studied conditions to evaluate the effect of GO incorporation on the anticorrosive performance of coated carbon steel in a 3.5% NaCl solution. The results are illustrated in Figure 6.

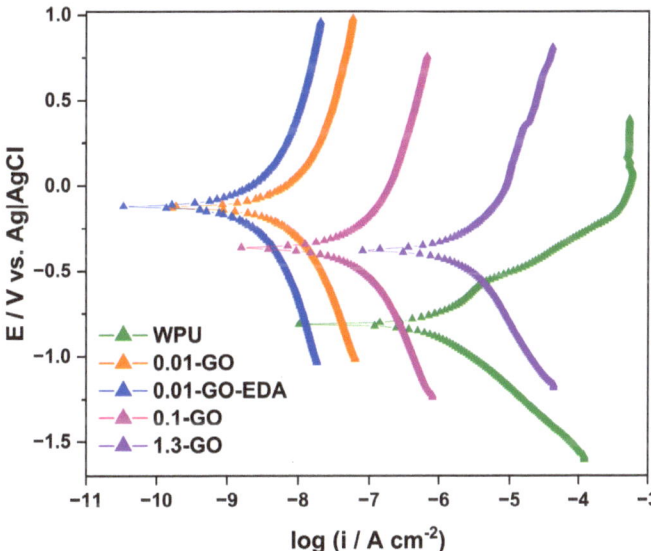

Figure 6. Potentiodynamic polarization curves of WPU resin and various concentrations of GO incorporated into the WPU resin in 3.5% NaCl aqueous solution.

From the corrosion parameters, such as the corrosion current density (i_{corr}) and the corrosion potential (E_{corr}) extracted from Figure 6, it was possible to calculate the corrosion rate and the coating inhibition efficiency (η), which is determined using Equation (1) [61,62]:

$$\eta = 1 - \frac{i^{GO}_{corr}}{i_{corr}} \quad (1)$$

where i^{GO}_{corr} is the corrosion current density for the WPU-coated samples containing graphene oxide, and i_{corr} is the corrosion current density for the coated samples with pristine WPU. All these corrosion parameters are summarized in Table 1.

Table 1. Electrochemical parameters obtained from potentiodynamic polarization curves for all studied conditions.

Sample	E_{corr}/(V) vs. Ag|AgCl	i_{corr}/(A cm^{-2})	Corrosion Rate (mm/Year)	η
WPU	−0.811	9.03×10^{-7}	1.01×10^{-2}	-
1.3-GO	−0.379	2.57×10^{-6}	2.88×10^{-2}	-
0.1-GO	−0.361	1.02×10^{-7}	1.15×10^{-3}	88.70%
0.01-GO	−0.126	9.34×10^{-9}	1.05×10^{-4}	99.00%
0.01-GO-EDA	−0.118	3.70×10^{-9}	4.15×10^{-5}	99.60%

The Figure 6 shows that the corrosion potential of the samples containing GO shifted to a more noble potential than that of the sample covered exclusively with WPU resin. This indicates a decrease in their susceptibility to corrosion (see Table 1) [63,64]. As observed from the contact angle results, the incorporation of GO increased the contact angle of the surface, improving the hydrophobicity of the WPU-modified coatings and, therefore, the water barrier permeation. However, it also altered the rate of oxidation–reduction reactions occurring on the coating surface, as indicated by the electrochemical test. When comparing the pristine WPU resin to WPU resin containing GO, the latter shows a two-order magnitude decrease in corrosion current density. This decrease can be attributed to the addition of GO and functionalized GO in the polymeric matrix, which improves the barrier properties of the coatings [12,25]. Among all samples studied, the 0.01-GO-EDA sample exhibited the best anticorrosion behavior, with the highest E_{corr} value (−117.82 mV) and the lowest i_{corr} value (3.70×10^{-9} A cm^{-2}). Generally, lower I_{corr} and higher E_{corr} indicate more protective properties against corrosion [63]. The corrosion rate and inhibition efficiency (η) data show that the 0.01-GO and 0.01-GO-EDA samples exhibit lower corrosion rates (1.05×10^{-4} mm/year and 4.15×10^{-5} mm/year, respectively) and higher inhibition efficiencies (99.00% and 99.60%), indicating enhanced corrosion protection compared to the 1.3-GO and 0.1-GO samples. The corrosion rate of pristine WPU is lower than that of the 1.3-GO sample, suggesting that a high GO content accelerates corrosion, thereby compromising the protection offered by the pristine WPU.

The functionalization of 0.01-GO-WPU with EDA enhances performance compared to the 0.01 wt% GO sample, which, although still effective, exhibits a slightly lower E_{corr} of −126.30 mV and a higher i_{corr} of 9.34×10^{-9} A cm^{-2}. This demonstrates that the functionalization of GO with EDA further improves the barrier properties of the coating. In contrast, the 0.1% GO sample shows a higher i_{corr} (1.02×10^{-7} A cm^{-2}) than the 0.01 wt% GO samples. This indicates that increased GO content does not necessarily improve anticorrosion behavior and may lead to more agglomeration than the 0.01 wt% GO samples. This suggests that increasing GO content does not necessarily improve anticorrosion performance and may lead to agglomeration issues of GO flakes. The 1.3-GO sample, which shows an even higher corrosion current density than both the 0.1-GO sample and pristine WPU resin, further confirms that excessive GO content can compromise the coating's barrier properties as observed in Figure 6 and Raman imaging mappings (Figure 3(b-vi,viii)) and videos (Supplementary Videos S3 and S4). Heiba et al. reported that incorporating a low content of GO can enhance corrosion protection due to better uniform distribution of GO sheets in the polymer matrix [63]. Chen et al. reported that the use of high-content of nanoparticles tended to agglomerate when added to a polymeric matrix [65]. In our study, the GO flakes measured approximately 25 μm in size, which can improve coating protection by establishing a nano-barrier effect, which creates a labyrinth within the coatings, thereby prolonging the penetration path of corrosive species [4,22,27]. However, inadequate dispersion of GO sheets can lead to the formation of preferential pathways within the coating, allowing for aggressive species to easily reach the metal, thereby reducing its barrier protection, as illustrated in Figure 7.

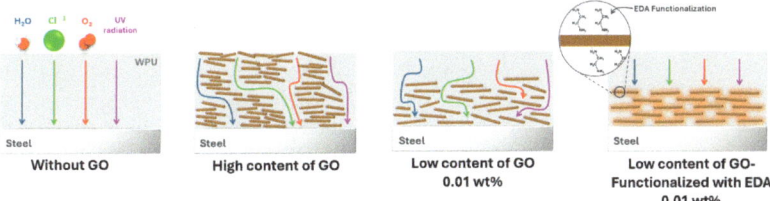

Figure 7. Schematic illustration depicting the effect of different contents of GO and EDA-functionalized GO on the barrier protection of the coating.

Our findings regarding the anticorrosion properties of the developed coating 0.01-GO-EDA based on WPU are much more promising than those of several reports in the literature, even in relation to those exploring the use of epoxy resins to develop the GO coatings. Table 2 summarizes recent works reported in the literature on developing coatings with anticorrosion properties based on the incorporation of graphene derivatives as an additive. The respective results regarding corrosion potential (E_{corr}) and corrosion current (i_{corr}) are presented.

Table 2. Recent works in the literature based on incorporating GO as an additive for developing coatings with anticorrosion properties. Evaluated performance based on corrosion potential (E_{corr}) and corrosion current (i_{corr}) extrapolated from potentiodynamic polarization curves.

Ref.	Coating	Graphene Derivative	Concentration (wt%)	Application Method	E_{corr} (V)	i_{corr} (A cm^{-2})
[34]	Waterborne hydroxyl acrylic	WHAR MGO	0.50	Bar coater	−0.27	0.90 × 10^{-6}
[35]	Epoxy	GO-PANI-PDA	-	Wire bar coater	−0.59	3.83 × 10^{-8}
[36]	Waterborne epoxy	CMCS-rGO	0.05	Bar coater	−0.63	3.05 × 10^{-10}
[18]	WPU	GO-PNNG	0.05	-	−0.06	4.98 × 10^{-10}
[37]	Polyvinyl alcohol	GO-PVA-SiC	10.00	Spray	−0.45	1.22 × 10^{-6}
[38]	Epoxy	PA-G-EP	1.00	Bar coater	−0.62	3.10 × 10^{-8}
This work	WPU	GO-EDA	0.01	Blade coater	−0.12	3.70 × 10^{-9}

WHAR MGO: dispersion of hydroxy acrylic resin with GO modified with 3-aminopropyltriethoxysilane; GO-PANI-PDA: GO modified with polyaniline and polydopamine; CMCS-rGO: rGO functionalized with carboxymethyl chitosan; GO-PNNG: GO modified with aminoethyl aminopropyl isobutyl polyhedral oligomeric silsesquioxane; GO-pva-SiC: GO modified with polyvinyl alcohol silicon carbide nanowires; PA-G-EP: nanocomposite based on epoxy, phytic acid, and graphene. WPU-GO-EDA: GO functionalized with ethylenediamine incorporated into waterborne polyurethane.

As demonstrated in Figure 6 and Table 1, the samples 0.01-GO and 0.01-GO-EDA exhibited the lowest corrosion current. Therefore, these concentrations and the pristine WPU resin were selected for the durability investigation test.

Figure 8 shows the surface morphology of the coated samples with artificial defects after 168 h and 515 h of exposure to a UV/condensation test. As observed in the images, prior to the accelerated aging test, the coated samples showed the characteristic brightness of WPU resin. Duong et al. reported that after 216 h of UV/condensation testing, polyurethane coatings (solvent-based) without GO degraded with loss in gloss, suggesting UV radiation-induced degradation. However, samples containing 0.01 wt% GO exhibited less gloss degradation. This can be attributed to the properties of GO in absorbing UV light, increasing the UV durability of the polyurethane coating [12]. In our study, after 515 h of exposure, we observed that the pristine WPU resin lost its brightness, while the 0.01-GO-EDA formulation exhibited less degradation, highlighting the beneficial role of the functionalized EDA-GO even at low concentrations. Furthermore, the involvement of EDA in creating a more compact cross-linking matrix and hydrophobic surface, contributes to a slower degradation of the film than using pristine WPU resin [49,50]. The evaluation of brightness in the 0.01-GO sample was complex because of the presence of corrosion products developed on its surface, likely caused by GO agglomeration that created preferential pathways through the coating to the metal, thereby triggering premature corrosion. In contrast, the 0.01-GO-EDA sample revealed a small quantity of formation of cracks in the resin after exposure, indicating that aggressive species did not easily penetrate the coating. This finding confirms that the functionalization of GO with EDA not only enhanced the formation of a dense three-dimensional network but also improved the dispersion of GO in the coating and the hydrophobicity of the coating surface, as illustrated in the schematic representation of Figure 7.

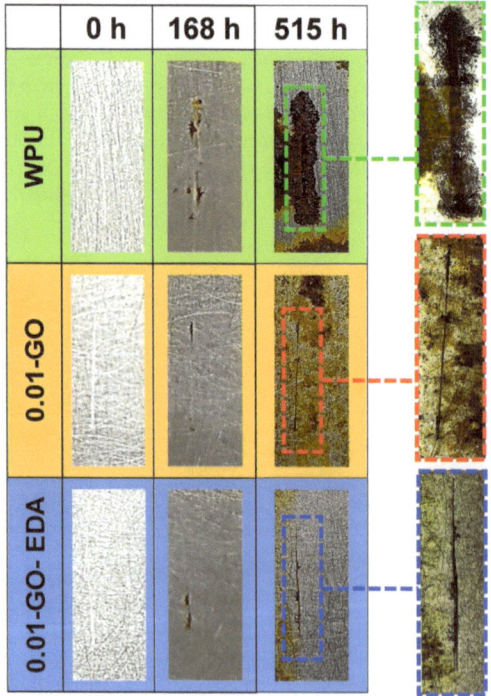

Figure 8. Photographic images of coated samples with artificial defects exposed to UV/condensation tests after 168 and 515 h, along with their respective optical microscopic images of the defect sites.

After 168 h of exposure, corrosion products were observed at the artificial defect sites in all samples, with a notable increase in quantity after 515 h. Although corrosion progressed in all studied conditions, the extent of propagation from the defect sites differed among samples. The pure resin showed the most severe corrosive attack, as evidenced by the most corrosion products at the defect site. The low anticorrosion performance of the pure resin coating after the aging test can be attributed to surface deterioration caused by

UV exposure, which compromised its barrier properties. In contrast, the 0.01-GO sample exhibited fewer corrosion products, and the 0.01-GO-EDA sample even less, indicating that the addition of GO and functionalized GO increased the corrosion resistance of the coating after UV/condensation aging. According to ISO 4628-8 standards, the corrosion around the scribe can be classified as grade 5 (severe) for the WPU-coated sample, and grade 2 (slight) for both the 0.01-GO and 0.01-GO-EDA samples [66]. This analysis indicates that the UV/condensation exposure induced physical changes in the coatings due to the combined effects of UV radiation, temperature, and condensation. Catastrophic macroscopic failures start from microscopic physical changes in the coating, often triggered by preceding chemical alterations [67].

3. Materials and Methods

3.1. Materials

Graphite crystals with lateral sizes of 9 mm to 12 mm from Nacional de Grafite (Itapecerica, Brazil) were used in the preparation of GO along with sulfuric acid (H_2SO_4) P.A. (98.0% purity), hydrochloric acid (HCl) P.A. (37.0% purity), potassium permanganate ($KMnO_4$) PA (99.0% of purity), and hydrogen peroxide (H_2O_2), 30 volumes, all of these chemicals from Synth (São Paulo, Brazil). The Ultra-pure water from the Milli-Q commercial system by Millipore Corporation (Burlington, MA, USA) with a resistivity of 18.2 MΩ.cm and TOC 2 ppb. The sodium nitrate ($NaNO_3$) PA (99.9% purity) from Merck (Darmstadt, Germany). The EDA (MW = 60.1 g mol^{-1}) (99% purity) was used to functionalize GO-WPU from Merck (Darmstadt, Germany).

AISI 1070 carbon steel sheets with dimensions of 10 mm $\times$ 30 mm $\times$ 2 mm were used as working electrodes. The type of carbon steel used was determined by energy-dispersive X-ray spectroscopy (EDS) analysis (field-emission scanning electron microscopy (SEM) model JSM-7800—Oxford Aztec, INCA) from JEOL Ltd. (Tokyo, Japan), presented in Supplementary Materials, Figure S3. Renner Sayerlack Company (Cajamar, Brazil) provided the coating, a single-component waterborne aliphatic polyurethane resin with a pH of 8.5, density of 1.05 g cm^{-3}, and viscosity of 150cP. The manufacturer specifies the volume solids percentage of the mixture to be 36 $\pm$ 2%.

The electrolyte utilized in the potentiodynamic polarization tests was 3.5% NaCl solution from Synth (São Paulo, Brazil).

3.2. Methods

3.2.1. Preparation of Graphene Oxide

Graphene oxide (GO) was synthesized according to the well-developed Hummer's modified method [68,69] with a total oxidation time of three days and purified according to the process described by Rocha [70]. In a 100 mL round-bottom flask, 0.5 g of graphite flakes and 16.9 mL of H_2SO_4 were added. This mixture was stirred for 30 min using a magnetic stirrer in an ice bath. After 30 min, 0.38 g of $NaNO_3$ was added. The solution was agitated for 5 min to ensure proper incorporation, and then 2.25 g of $KMnO_4$ was gradually added over 1 h while stirring and using an ice bath. After the $KMnO_4$ addition, the system was stirred for 24 h before resting for 72 h at room temperature to oxidize the material. After resting, 50 mL of H_2SO_4 at a concentration of 0.06 mol L^{-1} was gradually added over 1 h, with continued stirring and ice bath. Subsequently, 1.5 mL of 30% H_2O_2 was slowly added while maintaining stirring in an ice bath. The prepared GO dispersion was washed three times with a 10% hydrochloric acid aqueous solution to finish the process. The GO dispersion was purified using dialysis bags (porosity of 12 kDa, from Sigma Aldrich (San Luis, MO, USA)) in distilled water. The ultrapure water was changed until a pH of 5.5 was achieved. The GO dispersion was concentrated using a rotary evaporator (Q344M from QUIMIS Aparelhos Científicos, São Paulo, Brazil) to reach a concentration of 4.97 mg mL^{-1}.

3.2.2. Preparation of GO Waterborne Polyurethane Composite Coatings

The graphene oxide/waterborne polyurethane (GO/WPU) composite coating was prepared by uniformly dispersing GO in the resin. The procedure can be observed in Figure 9. After adding the GO to the WPU resin, it was subjected to magnetic stirring for 30 min, followed by sonication on an ultrasound bath for 20 min at 37 kHz.

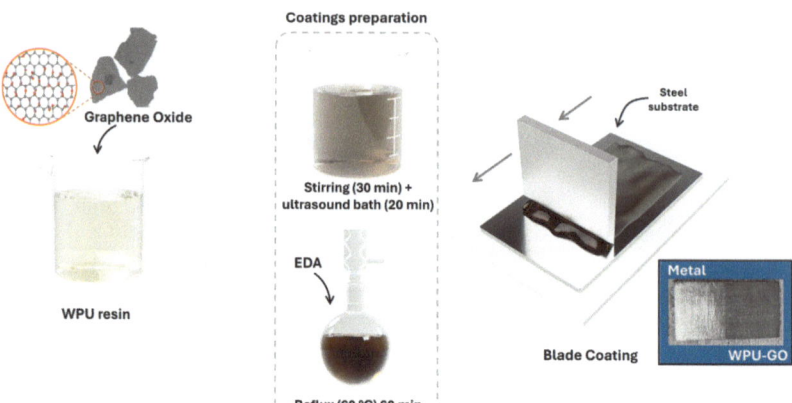

Figure 9. Schematic illustration of the preparation and application of GO-WPU and GO-EDA-WPU coatings on carbon steel sheets. Photograph image of a representative WPU-GO coating applied by blade coating on the carbon steel sheet showing the uniformity of the deposited film.

Based on the quantity of the solids in the pure resin, the GO was incorporated at concentrations of 1.3 wt%, 0.1 wt%, 0.01 wt%, and 0.01 wt% functionalized with EDA. These conditions are designated as 1.3-GO, 0.1-GO, 0.01-GO, and 0.01-GO-EDA, respectively. Control samples, referred to as WPU, were also evaluated, representing the waterborne polyurethane resin without GO.

The functionalization of 0.01 wt% GO with EDA was also evaluated. Inspired by the work of Jang et al. [47], who prepared a cross-linked GO membrane by functionalization with EDA, 30 µL (0.45 mmol) of EDA were added to 10 mL of WPU resin containing 0.01 wt% GO under constant stirring at 1300 rpm at room temperature for five minutes. Subsequently, the mixture was subjected to vigorous magnetic stirring and heating at 60 °C (silicone bath) in a reflux system for sixty minutes. After this period, the 0.01-GO-EDA sample was ready for use.

The carbon steel sheets were evenly sanded with 100-grit sandpaper. The coating film was then applied to the treated sheets using a Blade Coater, model BCC-02-V3 from Autocoat (Campinas, Brazil). This device uses a knife-shaped blade set at a 30-degree angle, with a height of 100 µm relative to the substrate and a deposition speed of 20 mm s^{-1} at an ambient temperature of 25 °C. The coatings were cured at room temperature for 24 h to obtain the final coated samples. After drying, the final single coating layers achieved an average thickness of 44.3 ± 3.8 µm, as measured by a hand-held electronic gauge, model 456C, from Elcometer (Manchester, UK).

3.2.3. Characterization

The morphology of the GO flakes was characterized by field-emission scanning electron microscopy (SEM) (Jeol, model JSM-7800), operated at 0.3 keV of accelerating voltage. The GO samples, with concentrations of 0.001 mg mL^{-1}, were prepared by drop casting on Si substrates. The type of carbon steel used was also evaluated by SEM images and determined by energy-dispersive X-ray spectroscopy (EDS) analysis (Jeol, model JSM-7800—Oxford Aztec, INCA).

The FTIR technique was employed to chemically characterize the GO-WPU composite coatings. All measurements were conducted in a Shimadzu (Tokyo, Japan) IRAffinity 1S FTIR spectrometer using attenuated total reflection (ATR). The spectra were obtained in the region from 600 to 4000 cm^{-1}.

The chemical and spatial characterization of GO in the WPU was achieved utilizing a Raman spectrometer from Witec, Oxford Instruments Group (High Wycombe, UK), Aplha 300R model, coupled to a confocal optical microscope, with a laser set at 532 nm, 0.5 mW of power, 600 gr mm^{-1} grating, and BLZ of 500 nm. The spectra were obtained with an integration time of 10 s and 10 accumulations, utilizing lenses with 10× and 50× magnification. The Raman mappings were performed to identify the effective dispersion of the GO in the resin. The microscopic image of the resin film containing the GO sheets was initially obtained. Then, using the conditions of 2 s of integration, 15 accumulations, and lenses with 10× and 50× magnification, areas of 30 × 30 μm were analyzed to identify the D and G bands, as well as the central band of the resin. The obtained spectra were analyzed using Python language by Google Colaboratory software (https://colab.research.google.com/, accessed on 11 May 2024) software, which attributed different colors to their composition and generated images of the analyzed region. The Raman spectra were obtained in the range of 500 to 3500 cm^{-1}. The successful achievement of GO dispersion was also evaluated by UV-Vis-NIR spectroscopy, which was performed on a Shimadzu spectrophotometer, model UV-3600, with a scanning range of 200 to 800 nm and an optical path length of 1.00 cm, controlled by UV-Probe software, version 2.62.

The wettability of coated carbon steel sheets was evaluated by static contact angle measurements utilizing the sessile drop method. An average of twenty measurements were taken to report the wettability. For each measurement, a 1 μL droplet of deionized water was deposited on the coated sample surface. The equipment employed was a Drop shape analyzer (DSA100) from KRUSS (Hamburg, Germany), operated with ADVANCE software, 2017 version.

The optical microscopy images of the carbon steel samples coated with the WPU, WPU-GO and WPU-GO-EDA after the weathering test were achieved by an optical microscope from Olympus, model BX51M (Tokyo, Japan) with a magnification of 5×.

3.2.4. Electrochemical Characterization and UV/Condensation Exposure Test

The electrochemical test workstation used to assess the protective properties of the coatings by potentiodynamic polarization (PDP) analysis was the Autolab SS101, manufactured by Methrom (Herisau, Switzerland). Figure 10 displays the three-electrode system utilized, showing the components of it. The coated samples served as the working electrode with an exposed area of 7.1 mm^2. A graphite rod electrode was utilized as the counter electrode, and an Ag/AgCl electrode as the reference electrode. The PDP curves were conducted in the potential range of 1 V on either side of Ecorr and at a scan rate of 0.01 V/s. The test corrosive solution was 3.5 wt% NaCl. All conditions were repeated three times under environmental conditions to confirm the reproducibility of the results.

The PDP analysis identified the two most effective concentrations of GO in WPU, which exhibited the lowest corrosion current, and these were compared with the pure WPU resin. These three conditions were then subjected to a weathering test that was conducted according to ASTM standard G51 [71]. First, an artificial defect, approximately 10 mm in length, was created on the surface of each coated carbon steel sample using a cutter, exposing the underlying steel. The samples were then placed in a UV/condensation chamber (QUV/spray with Solar Eye Irradiance Control from QLab) designed to simulate accelerated weathering conditions. During the test, the samples were cyclically exposed to UV-B radiation (310 nm) at 0.71 W/m^2 for 4 h at 60 °C, followed by 4 h of humidity condensation at 50 °C. The coated carbon steel panels (10 × 30 × 2 mm) were subjected to a total of 515 h of exposure. Samples were removed at intervals of 168 h and 515 h to assess their resistance to weathering. After the test, photographic and optical microscopic images of the coated samples were captured to evaluate the coating's performance.

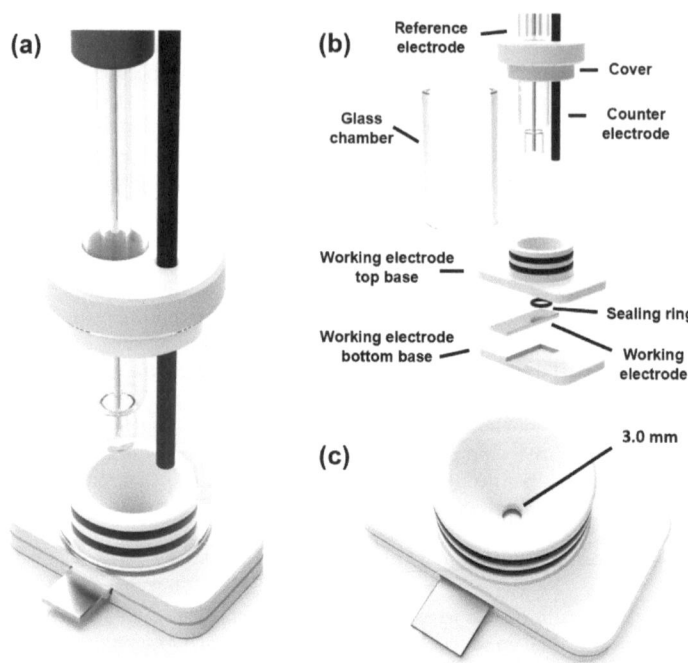

Figure 10. (a) Schematic diagram of a three-electrode cell used for electrochemical testing, (b) components of the cell, and (c) description of the cell base showing the test area for the samples.

4. Conclusions

In conclusion, we present an investigation into the effect of incorporating different concentrations of GO (1.3, 0.1, and 0.01 wt%) in WPU resins and the impact of functionalizing low-content (0.01 wt%) GO sheets with ethylenediamine (EDA) as an additive to develop anticorrosive coatings for carbon steel surfaces. Potentiodynamic polarization analysis in a 3.5% NaCl solution established a strong correlation between the low GO content in the WPU resin matrix and enhanced anticorrosion properties, particularly for the 0.01 wt% GO-EDA, which achieved high E_{corr} (−117.82 mV) and low i_{corr} (3.70×10^{-9} A cm^{-2}) values and an inhibition corrosion efficiency (η) of 99.60%. Raman imaging mapping analysis demonstrated that increasing the GO content as an additive in the WPU matrix leads to the agglomeration of GO sheets, creating pathways for corrosive species to permeate the WPU resin and reach the carbon steel surface. The sizeable lateral dimensions of the GO sheets, combined with the cross-linking between GO and WPU promoted by EDA, improved the interfacial properties between the GO and WPU polymer matrix. This resulted in better barrier properties, homogeneous dispersion, and a less hydrophilic surface, consequently leading to superior anticorrosion performance of the developed eco-friendly coating.

Supplementary Materials: The following supporting information can be downloaded at: https://www.mdpi.com/article/10.3390/molecules29174163/s1, Figure S1: UV-vis spectrum of graphene oxide (GO) dispersion. Figure S2: SEM micrographs of the GO sheets samples on Si substrates, obtained in different areas and magnification. (i,ii) 2700×, (iii) 3300×, (iv) 3700×, and (v,vi) 5000×, showing the sizeable lateral size of the GO sheets. Figure S3: SEM image (A), EDS mapping (chemical composition) (B), and EDS spectrum (C) of the corresponding area of the carbon steel sample used in this work. The sample is typical of AISI 1070 carbon steel, containing 7.4% carbon and 90.8% iron. Supplementary Video S1: Real-time Raman imaging mapping of the 0.01-GO sample; Supplementary Video S2: Real-time Raman imaging mapping of the 0.01-GO-EDA sample; Supplementary Video S3:

Real-time Raman imaging mapping of the 0.1-GO sample; Supplementary Video S4: Real-time Raman imaging mapping of the 1.3-GO sample.

Author Contributions: Conceptualization, C.d.C.C.S. and M.A.F.A.; methodology, M.A.F.A., R.F.F., M.S.D., S.B., D.S. and M.R.D.d.S.; software, M.A.F.A., R.F.F., M.S.D., S.B., D.S. and M.R.D.d.S.; validation, C.d.C.C.S. and M.A.F.A.; formal analysis, C.d.C.C.S., M.A.F.A., R.F.F., M.S.D., S.B., D.S., M.R.D.d.S. and C.M.M.; investigation, C.d.C.C.S., M.A.F.A., R.F.F., M.S.D., S.B., D.S., M.R.D.d.S. and C.M.M.; resources, C.d.C.C.S. and C.M.M.; data curation, C.d.C.C.S. and M.A.F.A.; writing—original draft preparation, C.d.C.C.S. and M.A.F.A.; writing—review and editing, C.d.C.C.S., M.A.F.A., R.F.F. and C.M.M.; visualization, C.d.C.C.S., M.A.F.A., R.F.F., M.S.D., S.B., D.S., M.R.D.d.S. and C.M.M.; resources, C.d.C.C.S. and C.M.M.; supervision, C.d.C.C.S.; project administration, C.d.C.C.S.; funding acquisition, C.d.C.C.S. and C.M.M. All authors have read and agreed to the published version of the manuscript.

Funding: This research was funded by Coordination for the Improvement of Higher Education Personnel (CAPES), Mackenzie Research Fund (MACKPESQUISA), grant number 2231019 and 231022; São Paulo Research Foundation (FAPESP), grant number 2023/12225-9; the National Council for Scientific and Technological Development (CNPq) (grant number 408248/2023-8, 313091/2022-6, and 384616/2023-2); INCT NanoVida (grant number 406079/2022-6); and Financiadora de Estudos e Projetos (Finep), grant number 1151/22.

Institutional Review Board Statement: Not applicable.

Informed Consent Statement: Not applicable.

Data Availability Statement: The data presented in this study are available upon request from the corresponding authors.

Acknowledgments: All the authors acknowledge Renner Sayerlack Company for providing the waterborne aliphatic polyurethane resins and Jan Vatavuk from Mackenzie School of Engineering, Mackenzie Presbyterian University for providing the carbon steel samples used in this work.

Conflicts of Interest: The authors declare no conflicts of interest.

References

1. Cui, M.; Wang, B.; Wang, Z. Nature-Inspired Strategy for Anticorrosion. *Adv. Eng. Mater.* **2019**, *21*, 1801379. [CrossRef]
2. Karimi, M. Review of Steel Material Engineering and Its Application in Industry. *J. Eng. Ind. Res.* **2023**, *4*, 61–67. [CrossRef]
3. Amegroud, H.; Boudalia, M.; Elhawary, M.; Garcia, A.J.; Bellaouchou, A.; Amin, H.M.A. Electropolymerized Conducting Polyaniline Coating on Nickel-Aluminum Bronze Alloy for Improved Corrosion Resistance in Marine Environment. *Colloids Surf. A Physicochem. Eng. Asp.* **2024**, *691*, 133909. [CrossRef]
4. Ollik, K.; Lieder, M. Review of the Application of Graphene-Based Coatings as Anticorrosion Layers. *Coatings* **2020**, *10*, 883. [CrossRef]
5. Farzi, G.; Davoodi, A.; Ahmadi, A.; Neisiany, R.E.; Anwer, M.K.; Aboudzadeh, M.A. Encapsulation of Cerium Nitrate within Poly(Urea-Formaldehyde) Microcapsules for the Development of Self-Healing Epoxy-Based Coating. *ACS Omega* **2021**, *6*, 31147–31153. [CrossRef]
6. Aramayo, M.A.F.; Aoki, I.V. Synthesis of Innovative Epoxy Resin and Polyamine Hardener Microcapsules and Their Age Monitoring by Confocal Raman Imaging. *J. Appl. Polym. Sci.* **2024**, *141*, e55342. [CrossRef]
7. Nazeer, A.A.; Madkour, M. Potential Use of Smart Coatings for Corrosion Protection of Metals and Alloys: A Review. *J. Mol. Liq.* **2018**, *253*, 11–22. [CrossRef]
8. Song, H.; Wang, M.; Wang, Y.; Zhang, Y.; Umar, A.; Guo, Z. Waterborne Polyurethane/Graphene Oxide Nanocomposites with Enhanced Properties. *Sci. Adv. Mater.* **2017**, *9*, 1895–1904. [CrossRef]
9. Cui, J.; Xu, J.; Li, J.; Qiu, H.; Zheng, S.; Yang, J. A Crosslinkable Graphene Oxide in Waterborne Polyurethane Anticorrosive Coatings: Experiments and Simulation. *Compos. B Eng.* **2020**, *188*, 107889. [CrossRef]
10. Boutoumit, A.; Elhawary, M.; Bellaouchou, A.; Boudalia, M.; Hammani, O.; José Garcia, A.; Amin, H.M.A. Electrochemical, Structural and Thermodynamic Investigations of Methanolic Parsley Extract as a Green Corrosion Inhibitor for C37 Steel in HCl. *Coatings* **2024**, *14*, 783. [CrossRef]
11. Eddahhaoui, F.-Z.; Najem, A.; Elhawary, M.; Boudalia, M.; Campos, O.S.; Tabyaoui, M.; José Garcia, A.; Bellaouchou, A.; Amin, H.M.A. Experimental and Computational Aspects of Green Corrosion Inhibition for Low Carbon Steel in HCl Environment Using Extract of Chamaerops Humilis Fruit Waste. *J. Alloys Compd.* **2024**, *977*, 173307. [CrossRef]
12. Duong, N.T.; An, T.B.; Thao, P.T.; Oanh, V.K.; Truc, T.A.; Vu, P.G.; Hang, T.T.X. Corrosion Protection of Carbon Steel by Polyurethane Coatings Containing Graphene Oxide. *Vietnam J. Chem.* **2020**, *58*, 108–112. [CrossRef]

13. Zhang, J.; Wang, J.; Wen, S.; Li, S.; Chen, Y.; Wang, J.; Wang, Y.; Wang, C.; Yu, X.; Mao, Y. Waterborne Polyurea Coatings Filled with Sulfonated Graphene Improved Anti-Corrosion Performance. *Coatings* **2021**, *11*, 251. [CrossRef]
14. Li, C.; Dong, Y.; Yuan, X.; Zhang, Y.; Gao, X.; Zhu, B.; Qiao, K. Waterborne Polyurethane Sizing Agent with Excellent Water Resistance and Thermal Stability for Improving the Interfacial Performance of Carbon Fibers/Epoxy Resin Composites. *Colloids Surf. A Physicochem. Eng. Asp.* **2024**, *681*, 132817. [CrossRef]
15. Trovati, G.; Sanches, E.A.; Neto, S.C.; Mascarenhas, Y.P.; Chierice, G.O. Characterization of Polyurethane Resins by FTIR, TGA, and XRD. *J. Appl. Polym. Sci.* **2010**, *115*, 263–268. [CrossRef]
16. Salzano de Luna, M. Recent Trends in Waterborne and Bio-Based Polyurethane Coatings for Corrosion Protection. *Adv. Mater. Interfaces* **2022**, *9*, 2101775. [CrossRef]
17. Medeiros, G.S.; Nisar, M.; Peter, J.; Andrade, R.J.E.; Fechine, G.J.M. Different Aspects of Polymer Films Based on Low-density Polyethylene Using Graphene as Filler. *J. Appl. Polym. Sci.* **2023**, *140*, 1–12. [CrossRef]
18. Zhang, F.; Liu, W.; Liang, L.; Wang, S.; Shi, H.; Xie, Y.; Yang, M.; Pi, K. The Effect of Functional Graphene Oxide Nanoparticles on Corrosion Resistance of Waterborne Polyurethane. *Colloids Surf. A Physicochem. Eng. Asp.* **2020**, *591*, 124565. [CrossRef]
19. Pinto, G.M.; Cremonezzi, J.M.O.; Ribeiro, H.; Andrade, R.J.E.; Demarquette, N.R.; Fechine, G.J.M. From two-dimensional Materials to Polymer Nanocomposites with Emerging Multifunctional Applications: A Critical Review. *Polym. Compos.* **2023**, *44*, 1438–1470. [CrossRef]
20. Cui, Y.; Kundalwal, S.I.; Kumar, S. Gas Barrier Performance of Graphene/Polymer Nanocomposites. *Carbon N. Y.* **2016**, *98*, 313–333. [CrossRef]
21. Nine, M.J.; Cole, M.A.; Johnson, L.; Tran, D.N.H.; Losic, D. Robust Superhydrophobic Graphene-Based Composite Coatings with Self-Cleaning and Corrosion Barrier Properties. *ACS Appl. Mater. Interfaces* **2015**, *7*, 28482–28493. [CrossRef]
22. Tan, B.; Thomas, N.L. A Review of the Water Barrier Properties of Polymer/Clay and Polymer/Graphene Nanocomposites. *J. Memb. Sci.* **2016**, *514*, 595–612. [CrossRef]
23. Li, X.; Li, D.; Chen, J.; Huo, D.; Gao, X.; Dong, J.; Yin, Y.; Liu, J.; Nan, D. Melamine-Modified Graphene Oxide as a Corrosion Resistance Enhancing Additive for Waterborne Epoxy Resin Coatings. *Coatings* **2024**, *14*, 488. [CrossRef]
24. Jena, G.; Philip, J. A Review on Recent Advances in Graphene Oxide-Based Composite Coatings for Anticorrosion Applications. *Prog. Org. Coat.* **2022**, *173*, 107208. [CrossRef]
25. Nayak, S.R.; Mohana, K.N.S. Corrosion Protection Performance of Functionalized Graphene Oxide Nanocomposite Coating on Mild Steel. *Surf. Interfaces* **2018**, *11*, 63–73. [CrossRef]
26. Liu, S.; Liu, H.; Shao, N.; Dong, Z. Modification of Electrochemical Exfoliation of Graphene Oxide with Dopamine and Tannic to Enhance Anticorrosion Performance of Epoxy Coatings. *Coatings* **2023**, *13*, 1809. [CrossRef]
27. Cui, G.; Bi, Z.; Zhang, R.; Liu, J.; Yu, X.; Li, Z. A Comprehensive Review on Graphene-Based Anti-Corrosive Coatings. *Chem. Eng. J.* **2019**, *373*, 104–121. [CrossRef]
28. Liang, A.; Jiang, X.; Hong, X.; Jiang, Y.; Shao, Z.; Zhu, D. Recent Developments Concerning the Dispersion Methods and Mechanisms of Graphene. *Coatings* **2018**, *8*, 33. [CrossRef]
29. Pu, N.W.; Wang, C.A.; Liu, Y.M.; Sung, Y.; Wang, D.S.; Ger, M. Der Dispersion of Graphene in Aqueous Solutions with Different Types of Surfactants and the Production of Graphene Films by Spray or Drop Coating. *J. Taiwan Inst. Chem. Eng.* **2012**, *43*, 140–146. [CrossRef]
30. Wen, J.G.; Geng, W.; Geng, H.Z.; Zhao, H.; Jing, L.C.; Yuan, X.T.; Tian, Y.; Wang, T.; Ning, Y.J.; Wu, L. Improvement of Corrosion Resistance of Waterborne Polyurethane Coatings by Covalent and Noncovalent Grafted Graphene Oxide Nanosheets. *ACS Omega* **2019**, *4*, 20265–20274. [CrossRef]
31. Ning, Y.J.; Zhu, Z.R.; Cao, W.W.; Wu, L.; Jing, L.C.; Wang, T.; Yuan, X.T.; Teng, L.H.; Bin, P.S.; Geng, H.Z. Anti-Corrosion Reinforcement of Waterborne Polyurethane Coating with Polymerized Graphene Oxide by the One-Pot Method. *J. Mater. Sci.* **2021**, *56*, 337–350. [CrossRef]
32. Li, Y.; Wang, Z.; Qiu, H.; Dai, Y.; Zheng, Q.; Li, J.; Yang, J. Self-Aligned Graphene as Anticorrosive Barrier in Waterborne Polyurethane Composite Coatings. *J. Mater. Chem. A Mater.* **2014**, *2*, 14139–14145. [CrossRef]
33. Maslekar, N.; Zetterlund, P.B.; Kumar, P.V.; Agarwal, V. Mechanistic Aspects of the Functionalization of Graphene Oxide with Ethylene Diamine: Implications for Energy Storage Applications. *ACS Appl. Nano Mater.* **2021**, *4*, 3232–3240. [CrossRef]
34. Fan, X.; Xia, Y.; Wu, S.; Zhang, D.; Oliver, S.; Chen, X.; Lei, L.; Shi, S. Covalently Immobilization of Modified Graphene Oxide with Waterborne Hydroxyl Acrylic Resin for Anticorrosive Reinforcement of Its Coatings. *Prog. Org. Coat.* **2022**, *163*, 106685. [CrossRef]
35. Huang, Y.; Zhang, B.; Wu, J.; Hong, R.; Xu, J. Preparation and Characterization of Graphene Oxide/Polyaniline/Polydopamine Nanocomposites towards Long-Term Anticorrosive Performance of Epoxy Coatings. *Polymers* **2022**, *14*, 3355. [CrossRef]
36. Shi, H.; Liu, W.; Xie, Y.; Yang, M.; Liu, C.; Zhang, F.; Wang, S.; Liang, L.; Pi, K. Synthesis of Carboxymethyl Chitosan-Functionalized Graphene Nanomaterial for Anticorrosive Reinforcement of Waterborne Epoxy Coating. *Carbohydr. Polym.* **2021**, *252*, 117249. [CrossRef]
37. Yu, S.; Yang, Y.; Ma, L.; Jia, W.; Zhou, Q.; Zhu, J.; Wang, J. SiC Nanowires Enhanced Graphene Composite Coatings with Excellent Tribological and Anticorrosive Properties. *Tribol. Int.* **2023**, *188*, 108894. [CrossRef]
38. Guo, X.; Xu, H.; Pu, J.; Yao, C.; Yang, J.; Liu, S. Corrosion Performance and Rust Conversion Mechanism of Graphene Modified Epoxy Surface Tolerant Coating. *Front. Mater.* **2021**, *8*, 767776. [CrossRef]

39. Dresselhaus, M.S.; Jorio, A.; Hofmann, M.; Dresselhaus, G.; Saito, R. Perspectives on Carbon Nanotubes and Graphene Raman Spectroscopy. *Nano Lett.* **2010**, *10*, 751–758. [CrossRef]
40. Ferrari, A.C. Raman Spectroscopy of Graphene and Graphite: Disorder, Electron–Phonon Coupling, Doping and Nonadiabatic Effects. *Solid State Commun.* **2007**, *143*, 47–57. [CrossRef]
41. Konios, D.; Stylianakis, M.M.; Stratakis, E.; Kymakis, E. Dispersion Behaviour of Graphene Oxide and Reduced Graphene Oxide. *J. Colloid Interface Sci.* **2014**, *430*, 108–112. [CrossRef]
42. Chen, D.; Feng, H.; Li, J. Graphene Oxide: Preparation, Functionalization, and Electrochemical Applications. *Chem. Rev.* **2012**, *112*, 6027–6053. [CrossRef] [PubMed]
43. Chen, J.; Yao, B.; Li, C.; Shi, G. An Improved Hummers Method for Eco-Friendly Synthesis of Graphene Oxide. *Carbon N. Y.* **2013**, *64*, 225–229. [CrossRef]
44. Parvin, N.; Kumar, V.; Joo, S.W.; Park, S.S.; Mandal, T.K. Recent Advances in the Characterized Identification of Mono-to-Multi-Layer Graphene and Its Biomedical Applications: A Review. *Electronics* **2022**, *11*, 3345. [CrossRef]
45. Zhang, Z.; Schniepp, H.C.; Adamson, D.H. Characterization of Graphene Oxide: Variations in Reported Approaches. *Carbon N. Y.* **2019**, *154*, 510–521. [CrossRef]
46. Otsuka, H.; Urita, K.; Honma, N.; Kimuro, T.; Amako, Y.; Kukobat, R.; Bandosz, T.J.; Ukai, J.; Moriguchi, I.; Kaneko, K. Transient Chemical and Structural Changes in Graphene Oxide during Ripening. *Nat. Commun.* **2024**, *15*, 1708. [CrossRef] [PubMed]
47. Jang, J.; Park, I.; Chee, S.-S.; Song, J.-H.; Kang, Y.; Lee, C.; Lee, W.; Ham, M.-H.; Kim, I.S. Graphene Oxide Nanocomposite Membrane Cooperatively Cross-Linked by Monomer and Polymer Overcoming the Trade-off between Flux and Rejection in Forward Osmosis. *J. Memb. Sci.* **2020**, *598*, 117684. [CrossRef]
48. Zhu, M.; Li, S.; Sun, Q.; Shi, B. Enhanced Mechanical Property, Chemical Resistance and Abrasion Durability of Waterborne Polyurethane Based Coating by Incorporating Highly Dispersed Polyacrylic Acid Modified Graphene Oxide. *Prog. Org. Coat.* **2022**, *170*, 106949. [CrossRef]
49. Rahman, M.M. Synthesis and Properties of Waterborne Polyurethane Adhesives: Effect of Chain Extender of Ethylene Diamine, Butanediol, and Fluoro-Butanediol. *J. Adhes. Sci. Technol.* **2013**, *27*, 2592–2602. [CrossRef]
50. Lei, L.; Zhong, L.; Lin, X.; Li, Y.; Xia, Z. Synthesis and Characterization of Waterborne Polyurethane Dispersions with Different Chain Extenders for Potential Application in Waterborne Ink. *Chem. Eng. J.* **2014**, *253*, 518–525. [CrossRef]
51. Bruckmoser, K.; Resch, K. Investigation of Ageing Mechanisms in Thermoplastic Polyurethanes by Means of IR and Raman Spectroscopy. In *Macromolecular Symposia*; Wiley-VCH: Weinheim, Germany, 2014; Volume 339, pp. 70–83.
52. Christopher, G.; Anbu Kulandainathan, M.; Harichandran, G. Comparative Study of Effect of Corrosion on Mild Steel with Waterborne Polyurethane Dispersion Containing Graphene Oxide versus Carbon Black Nanocomposites. *Prog. Org. Coat.* **2015**, *89*, 199–211. [CrossRef]
53. He, D.; Peng, Z.; Gong, W.; Luo, Y.; Zhao, P.; Kong, L. Mechanism of a Green Graphene Oxide Reduction with Reusable Potassium Carbonate. *RSC Adv.* **2015**, *5*, 11966–11972. [CrossRef]
54. Guo, W.; Chen, J.; Sun, S.; Zhou, Q. In Situ Monitoring the Molecular Diffusion Process in Graphene Oxide Membranes by ATR-FTIR Spectroscopy. *J. Phys. Chem. C* **2016**, *120*, 7451–7456. [CrossRef]
55. Hu, Y.; Cao, K.; Ci, L.; Mizaikoff, B. Selective Chemical Enhancement via Graphene Oxide in Infrared Attenuated Total Reflection Spectroscopy. *J. Phys. Chem. C* **2019**, *123*, 25286–25293. [CrossRef]
56. Surekha, G.; Krishnaiah, K.V.; Ravi, N.; Padma Suvarna, R. FTIR, Raman and XRD Analysis of Graphene Oxide Films Prepared by Modified Hummers Method. *J. Phys. Conf. Ser.* **2020**, *1495*, 012012. [CrossRef]
57. Khatoon, H.; Iqbal, S.; Ahmad, S. Covalently Functionalized Ethylene Diamine Modified Graphene Oxide Poly-Paraphenylene Diamine Dispersed Polyurethane Anticorrosive Nanocomposite Coatings. *Prog. Org. Coat.* **2021**, *150*, 105966. [CrossRef]
58. Bahadur, A.; Shoaib, M.; Saeed, A.; Iqbal, S. FT-IR Spectroscopic and Thermal Study of Waterborne Polyurethane-Acrylate Leather Coatings Using Tartaric Acid as an Ionomer. *e-Polymers* **2016**, *16*, 463–474. [CrossRef]
59. Suthar, V.; Asare, M.A.; de Souza, F.M.; Gupta, R.K. Effect of Graphene Oxide and Reduced Graphene Oxide on the Properties of Sunflower Oil-Based Polyurethane Films. *Polymers* **2022**, *14*, 4974. [CrossRef]
60. Cui, L.; Xiang, T.; Hu, B.; Lv, Y.; Rong, H.; Liu, D.; Zhang, S.; Guo, M.; Lv, Z.; Chen, D. Design of Monolithic Superhydrophobic Concrete with Excellent Anti-Corrosion and Self-Cleaning Properties. *Colloids Surf. A Physicochem. Eng. Asp.* **2024**, *685*, 133345. [CrossRef]
61. Xu, H.; Hu, H.; Wang, H.; Li, Y.; Li, Y. Corrosion Resistance of Graphene/Waterborne Epoxy Composite Coatings in CO_2-Saturated NaCl Solution. *R. Soc. Open Sci.* **2020**, *7*, 191943. [CrossRef]
62. Najem, A.; Campos, O.S.; Girst, G.; Raji, M.; Hunyadi, A.; García-Antón, J.; Bellaouchou, A.; Amin, H.M.A.; Boudalia, M. Experimental and DFT Atomistic Insights into the Mechanism of Corrosion Protection of Low-Carbon Steel in an Acidic Medium by Polymethoxyflavones from Citrus Peel Waste. *J. Electrochem. Soc.* **2023**, *170*, 093512. [CrossRef]
63. Heiba, A.R.; Taher, F.A.; Abou Shahba, R.M.; Abdel Ghany, N.A. Corrosion Mitigation of Carbon Steel in Acidic and Salty Solutions Using Electrophoretically Deposited Graphene Coatings. *J. Coat. Technol. Res.* **2021**, *18*, 501–510. [CrossRef]
64. Liu, Q.; Ma, R.; Du, A.; Zhang, X.; Yang, H.; Fan, Y.; Zhao, X.; Cao, X. Investigation of the Anticorrosion Properties of Graphene Oxide Doped Thin Organic Anticorrosion Films for Hot-Dip Galvanized Steel. *Appl. Surf. Sci.* **2019**, *480*, 646–654. [CrossRef]
65. Chen, L.; Song, R.G.; Li, X.W.; Guo, Y.Q.; Wang, C.; Jiang, Y. The Improvement of Corrosion Resistance of Fluoropolymer Coatings by SiO_2/Poly(Styrene-Co-Butyl Acrylate) Nanocomposite Particles. *Appl. Surf. Sci.* **2015**, *353*, 254–262. [CrossRef]

66. *ISO 4628-8*; Paints and Varnishes—Evaluation of Degradation of Coatings—Designation of Quantity and Size of Defects, and of Intensity of Uniform Changes in Appearance—Part 8: Assessment of Degree of Delamination and Corrosion around a Scribe or Other 2012. ISO: Geneva, Switzerland, 2012.
67. Wood, K.A. Optimizing the Exterior Durability of New Fluoropolymer Coatings. In *Progress in Organic Coatings*; Elsevier: Amsterdam, The Netherlands, 2001; Volume 43.
68. Hirata, M.; Gotou, T.; Horiuchi, S.; Fujiwara, M.; Ohba, M. Thin-Film Particles of Graphite Oxide 1: High-Yield Synthesis and Flexibility of the Particles. *Carbon N. Y.* **2004**, *42*, 2929–2937. [CrossRef]
69. Hummers, W.S.; Offeman, R.E. Preparation of Graphitic Oxide. *J. Am. Chem. Soc.* **1958**, *80*, 1339. [CrossRef]
70. Rocha, J.F.; Hostert, L.; Bejarano, M.L.M.; Cardoso, R.M.; Santos, M.D.; Maroneze, C.M.; Gongora-Rubio, M.R.; Silva, C.D.C.C. Graphene Oxide Fibers by Microfluidics Assembly: A Strategy for Structural and Dimensional Control. *Nanoscale* **2021**, *13*, 6752–6758. [CrossRef] [PubMed]
71. *ASTM G154*; Standard Practice for Operating Fluorescent Ultraviolet (UV) Lamp Apparatus for Exposure of Materials. ASTM: West Conshohocken, PA, USA, 2023.

Disclaimer/Publisher's Note: The statements, opinions and data contained in all publications are solely those of the individual author(s) and contributor(s) and not of MDPI and/or the editor(s). MDPI and/or the editor(s) disclaim responsibility for any injury to people or property resulting from any ideas, methods, instructions or products referred to in the content.

Article

Broadband Solar Absorber and Thermal Emitter Based on Single-Layer Molybdenum Disulfide

Wanhai Liu [1], Fuyan Wu [2], Zao Yi [2,*], Yongjian Tang [2], Yougen Yi [3], Pinghui Wu [4] and Qingdong Zeng [5]

1. School of Intelligent Manufacturing, Zhejiang Guangsha Vocational and Technical University of Construction, Jinhua 322100, China; wanh2006@126.com
2. Joint Laboratory for Extreme Conditions Matter Properties, Southwest University of Science and Technology, Mianyang 621010, China; 15284169398@163.com (F.W.); tangyongjian2000@sina.com (Y.T.)
3. College of Physics and Electronics, Central South University, Changsha 410083, China; yougenyi@csu.edu.cn
4. College of Physics & Information Engineering, Quanzhou Normal University, Quanzhou 362000, China; phwu@zju.edu.cn
5. School of Physics and Electronic-Information Engineering, Hubei Engineering University, Xiaogan 432000, China; zengqingdong2005@163.com
* Correspondence: yizaomy@swust.edu.cn

Abstract: In recent years, solar energy has become popular because of its clean and renewable properties. Meanwhile, two-dimensional materials have become a new favorite in scientific research due to their unique physicochemical properties. Among them, monolayer molybdenum disulfide (MoS_2), as an outstanding representative of transition metal sulfides, is a hot research topic after graphene. Therefore, we have conducted an in-depth theoretical study and design simulation using the finite-difference method in time domain (FDTD) for a solar absorber based on the two-dimensional material MoS_2. In this paper, a broadband solar absorber and thermal emitter based on a single layer of molybdenum disulfide is designed. It is shown that the broadband absorption of the absorber is mainly due to the propagating plasma resonance on the metal surface of the patterned layer and the localized surface plasma resonance excited in the adjacent patterned air cavity. The research results show that the designed structure boasts an exceptional broadband performance, achieving an ultrawide spectral range spanning 2040 nm, with an overall absorption efficiency exceeding 90%. Notably, it maintains an average absorption rate of 94.61% across its spectrum, and in a narrow bandwidth centered at 303 nm, it demonstrates a near-unity absorption rate, surpassing 99%, underscoring its remarkable absorptive capabilities. The weighted average absorption rate of the whole wavelength range (280 nm–2500 nm) at AM1.5 is above 95.03%, and even at the extreme temperature of up to 1500 K, its heat radiation efficiency is high. Furthermore, the solar absorber in question exhibits polarization insensitivity, ensuring its performance is not influenced by the orientation of incident light. These advantages can enable our absorber to be widely used in solar thermal photovoltaics and other fields and provide new ideas for broadband absorbers based on two-dimensional materials.

Keywords: ultra-wideband absorption; two-dimensional materials; molybdenum disulfide; thermal emission

Citation: Liu, W.; Wu, F.; Yi, Z.; Tang, Y.; Yi, Y.; Wu, P.; Zeng, Q. Broadband Solar Absorber and Thermal Emitter Based on Single-Layer Molybdenum Disulfide. *Molecules* **2024**, *29*, 4515. https://doi.org/10.3390/molecules29184515

Academic Editors: Sake Wang, Minglei Sun and Nguyen Tuan Hung

Received: 25 August 2024
Revised: 17 September 2024
Accepted: 20 September 2024
Published: 23 September 2024

Copyright: © 2024 by the authors. Licensee MDPI, Basel, Switzerland. This article is an open access article distributed under the terms and conditions of the Creative Commons Attribution (CC BY) license (https://creativecommons.org/licenses/by/4.0/).

1. Introduction

In contemporary society, with the sharp increase in energy demand, the supply of traditional fossil energy has been struggling to meet the needs of sustainable development, prompting people to focus on a wider range of renewable energy fields [1–3]. Among them, solar energy, as an emerging and clean form of renewable energy, has attracted much attention. In order to cope with the challenge of energy shortage, researchers have carried out in-depth and extensive research on various types of clean energy, including solar energy [4–6]. However, despite extensive research on solar absorbers, there are still many drawbacks. For example, the narrow width of the absorption band, the low absorption intensity, and the complex structure limit the application of absorbers in solar photovoltaic and other fields. Therefore, it is important to explore a broadband absorber with good oblique incidence characteristics and polarization angle independence as well as high thermal radiation efficiency [7–9].

In addition, ultra-wideband absorbers constructed from refractory materials show significant promise for applications where thermophotovoltaic devices are frequently subjected to high-temperature extremes. These absorbers not only operate stably at high temperatures, but also maintain excellent absorption stability, which lays a solid technical foundation for efficient energy conversion in high-temperature environments. Therefore, an in-depth exploration and optimization of the design strategy and preparation technology of such absorbers is of great significance to accelerate the innovation and development of solar energy and wider clean energy technologies. Titanium metal is known for its unique physical properties, including core advantages such as high strength, excellent heat and corrosion resistance, superior ductility, and relatively low density [10]. As a member of the refractory metals, titanium has a high melting point of 1668 °C, displays excellent thermal stability, and exhibits good antimagnetization properties in strong magnetic field environments [11]. In absorber applications, titanium stands out not only for its excellent stability, but also for its cost-effectiveness compared to precious metals such as gold (Au) and silver (Ag). Of particular interest is the ability of titanium as a resonant material to excite a broader bandwidth response in the infrared spectral region, a property that offers the possibility of realizing highly efficient absorption in an ultra-broad band, thus greatly broadening its potential for a wide range of practical applications.

Over the past few years, two-dimensional materials have garnered significant attention within the scientific community, primarily due to their unparalleled physical and chemical characteristics, such as high electron mobility, excellent heat resistance, and chemical stability. Graphene [12,13], as a leader in the field of two-dimensional materials [14–17], has attracted extensive research interest for its unique physical properties. Its ultra-thin characteristics are particularly remarkable, with a thickness of only 0.34 nm, which is almost equal to the diameter of a single carbon atom. As an allotrope composed of carbon elements, in its monolayer state, the carbon atoms of graphene are closely combined with three adjacent carbon atoms by SP hybridization, forming a unique planar hexagonal honeycomb structure [18]. This structure endows graphene with excellent mechanical, electrical, and thermal properties, which makes it show great potential in scientific research and industrial applications. Many studies have confirmed that these excellent properties of graphene provide a broad prospect for its application in many fields [19–22]. Graphene is unique in its zero-band gap property, which enables electrons to transition from valence band to conduction band at a very low energy state. However, it is the energy band structure of graphene that has zero bandgap [23], so it has no adjustable semiconductor conductivity at all, which limits the possibility of its further development [24].

Therefore, in contrast to pristine graphene, transition metal dichalcogenides (TMDCs) are generally regarded as superior absorbing materials. This is due to the fact that graphene possesses a zero bandgap, necessitating thermal excitation of electrons for conductivity, a characteristic that significantly constrains its practical applicability. TMDCs exhibit a tunable bandgap spanning from 1 to 2 eV, offering significant advantages in the fabrication of absorbers. Key attributes that render TMDCs particularly suitable for the production of solar absorbers include their remarkable stability, controllable thickness, and high absorption efficiency. Their ultrathin (monolayer) nature endows them with a direct bandgap within the visible spectrum, resulting in exceptional absorption capabilities. Among two-dimensional materials, MoS_2 stands out as a highly promising candidate for functional photonic devices due to its notable current cutoff ratio and tunable optoelectronic properties, as reported in [25,26]. Consequently, this study focuses on single-layer MoS_2 as a representative example for our investigation into two-dimensional materials.

Molybdenum disulfide (MoS_2), a transition metal sulfide, has received a lot of attention in recent years for its potential applications in optoelectronics. The unique properties of MoS$_2$, especially its broadband light absorption capability in the visible to near-infrared region, make it an ideal material for solar absorbers [27]. Compared with other two-dimensional materials, MoS$_2$ has a direct bandgap, which gives it a significant advantage in light absorption efficiency. In addition, the high electron mobility of MoS$_2$ facilitates the rapid transport of electrons under light excitation, which improves the response speed and efficiency of optoelectronic devices. In terms of chemical stability, MoS$_2$ is able to maintain its performance under a wide range of environmental conditions, which is essential for the fabrication of durable optoelectronic devices [28]. With the development of solution processing techniques, the preparation of MoS$_2$ thin films has become more economical and scalable. Spin-coating techniques have been used to prepare homogeneous MoS$_2$ films and characterize them by variable angle spectroscopic ellipsometry. In addition, by using the dimethylformamide/n-butylamine/2-aminoethanol solvent system, researchers have been able to synthesize wafer-scale MoS$_2$ thin films with controllable thickness using the solution method. These advances not only improve the quality of MoS$_2$ films, but also pave the way for their integration in solar cells, photodetectors, and other optoelectronic devices [29,30]. The bandgap of MoS$_2$ can be tuned by chemical doping or strain engineering, enabling precise control of the absorption spectrum. This tunability provides great flexibility in designing optoelectronic devices with specific spectral responses. With the further understanding of MoS$_2$ material properties and the continuous advancement of processing technologies, MoS$_2$-based optoelectronic devices are expected to play an important role in the future of sustainable energy and advanced electronics.

Currently, the state of absorbers in this domain is marred by numerous limitations: some absorbers are too complicated to design and too bulky for actual manufacturing [31]. The working range of the absorber is too narrow, the absorption capacity is limited, and numerous state-of-the-art absorbers utilize costly precious metals, particularly gold and silver, resulting in significant production expenses. Additionally, there is ample scope for enhancing their thermal endurance to ensure optimal performance under various conditions. Thus, in this paper, we propose a cost-effective solar absorber featuring straightforward manufacturing processes, a broad absorption spectrum, and superior absorption efficiency within the solar radiation spectrum. The solar energy absorber demonstrates remarkable absorption capabilities across a broad spectral domain, particularly within the wavelength interval spanning from 280 to 2320 nanometers (2040 nm in total), and its absorption efficiency keeps above 90%, with an average absorption efficiency as high as 94.61%, which fully proves its high-efficiency light energy absorption capacity. Therefore, our absorber will provide some references for similar structures of other solar absorbers and has potential for a wide range of applications such as solar thermal photovoltaic systems. Our work also opens up new avenues for the application of two-dimensional materials in the field of optoelectronics and energy conversion.

2. Results and Discussion

Figure 1 meticulously portrays the absorption characteristics of the structure through a comprehensive chart. Specifically, the red trace delineates the absorption efficiency, whereas the black trace signifies reflectivity, and the blue trace is indicative of transmittance. In view of the sufficient thickness of the substrate Ti, we observed that the transmittance was almost zero. Obviously, the solar energy absorber shows excellent absorption performance in a wide spectral range, particularly within the spectral range spanning from 280 to 2320 nanometers (a total of 2040 nm), within which the absorber maintains an absorption efficiency exceeding 90% with an average efficiency of 94.61%. Notably, it achieves a near-perfect absorption rate of 99% specifically at a wavelength of 303 nm, which fully proves its high-efficiency light energy absorption ability. Compared with other recent absorbents based on molybdenum disulfide, our absorbent has obvious advantages in bandwidth and average absorption rate, as shown in Table 1 [12,32–37].

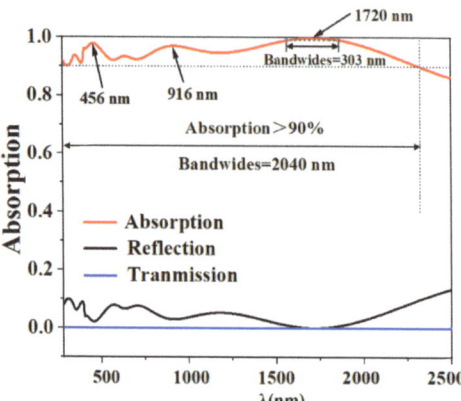

Figure 1. Depicts the spectral characteristics of absorption, reflectance, and transmittance, under plane incident light.

Table 1. Comparison with other literature.

Essay	Spectral Region Exhibiting an Absorption Rate Exceeding 90%	Average Absorption Efficiency	Maximum Absorption Rate
[32]	712 nm	97%	99.80%
[12]	<1000 nm	85%	97.00%
[38]	1110 nm	<90%	99.80%
[34]	475 nm	94%	97%
[35]	1547 nm	90%	98%
[36]	1200 nm	91%	\
[37]	1100 nm	99.6%	98.5%
This text	2040 nm	94.61%	99.87%

In order to probe into the rationale behind the ultra-broadband and near-perfect absorption achieved in this paper, we chose three representative bands for the follow-up study of electric field distribution. These three bands are $\lambda 1 = 456$ nm (in the visible region), $\lambda 2 = 916$ nm (in the vicinity of the infrared spectral region) and $\lambda 3 = 1720$ nm (also in the near-infrared region). Through this analysis, we expect to reveal the key factors and mechanisms to achieve efficient absorption.

A comprehensive comparison of the performance metrics of the absorber introduced in this study with those reported in prior research is presented in Table 1. The analysis shows that the perfect absorber designed by us has remarkable ultra-wide bandwidth characteristics, which obviously exceed other comparative absorbers, and its highest absorption rate is also better. In addition, although our solar energy absorber is not the highest in terms of average absorption efficiency, it still shows significant advantages in comprehensive performance of bandwidth and efficiency, which provides us with an ideal and practical choice.

2.1. The Influence of Different Structures on the Results

Initially, we delve into the consequences of integrating a single-layer molybdenum disulfide within the architecture. For illustrative purposes, we have modeled the structure with and without a single layer of molybdenum disulphide and with the transformed 2D material being graphene. For example, the absorption of light by monolayer graphene is inherently low, about 2.3 per cent, due to its energy band structure [39]. Graphene is a zero-bandgap material with an electronic structure similar to a Dirac cone, which makes it less responsive to light in the visible and near-infrared bands, as shown in Figure 2a. As evident from Figure 2a, the incorporation of this monolayer yields two notable benefits: a significant broadening of the operational bandwidth and an enhancement in the absorption efficiency within the shorter wavelength region. Within Region I, the analysis reveals that the absorption rate of the nanostructures falls below 90% across a relatively extensive bandwidth, in the absence of a single layer of molybdenum disulfide. After the introduction of single-layer molybdenum disulfide, it can be seen that due to its high absorption rate in the near-ultraviolet band, the absorption efficiency within the shorter wavelength spectrum undergoes a substantial augmentation, surpassing the 90% threshold. Furthermore, Figure 2b, depicting a magnified section of Region II, evidences a marked expansion in the bandwidth where the absorption rate surpasses 90%. Specifically, the introduction of a single layer of MoS_2 results in a broadening of this high-absorption region by approximately 310 nanometers. In a word, the introduction of single-layer molybdenum disulfide makes the $Ti-SiO_2$ cuboid structure designed for superior absorption address the challenge of insufficient absorption in the short-wavelength region, thereby effectively broadening the operational bandwidth.

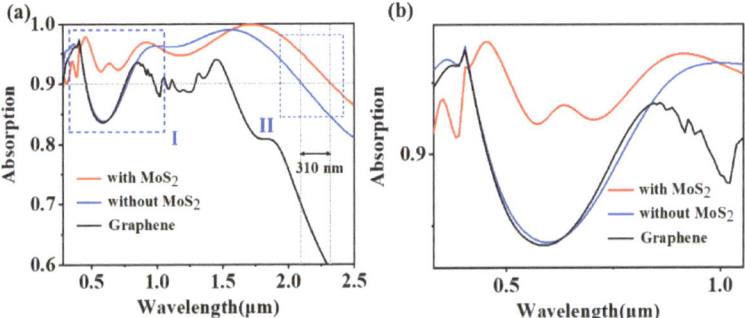

Figure 2. (a) Illustrative sketch outlining the enhancement in absorption for a monolayer MoS_2 structure; (b) Magnified section of Region I for closer inspection.

In the following research, we deeply discuss the influence of different geometric patterns on the performance of the absorber. We replaced the original cuboid structure with a cylindrical structure (B) and a circular column structure (C), respectively, and calculated their absorption efficiency under the same lighting conditions. Through careful analysis of the absorption spectrum in Figure 3a, our analysis indicates that the rectangular parallelepiped configuration exhibits a marginally reduced absorption efficiency compared to both cylindrical and circular cylindrical structures, within the spectral range spanning from 400 nm in the visible light region to 1600 nm in the near-infrared, but it shows significant advantages in the band over 1600 nm. Its wider bandwidth and higher overall absorption rate of solar full spectrum grant the cuboid structure greater potential in wide-band light absorption applications. Based on the above analysis, we finally chose the cuboid structure as the micro-nano-structure of the absorber in order to achieve better performance. This selection not only considers the evaluation of the absorber's efficacy within the visible light and near-infrared spectral bands, but also fully considers its comprehensive performance in a wider band.

In addition, from the electric field distribution diagram shown in Figure 3b–d, we can observe that the electric field is mainly concentrated on the geometric surface. This phenomenon is mainly attributed to the excitation of PSPs (surface plasmon) on the surface of the pattern layer [38,40,41]. PSPs are electromagnetic oscillations generated at the metal-medium interface; this capability allows for the conversion of light energy into thermal energy or alternative energy forms, thus achieving efficient light absorption [42–44]. Therefore, the intensified electric field distribution along the geometric surface underscores the pivotal role played by PSPs in facilitating the light absorption process.

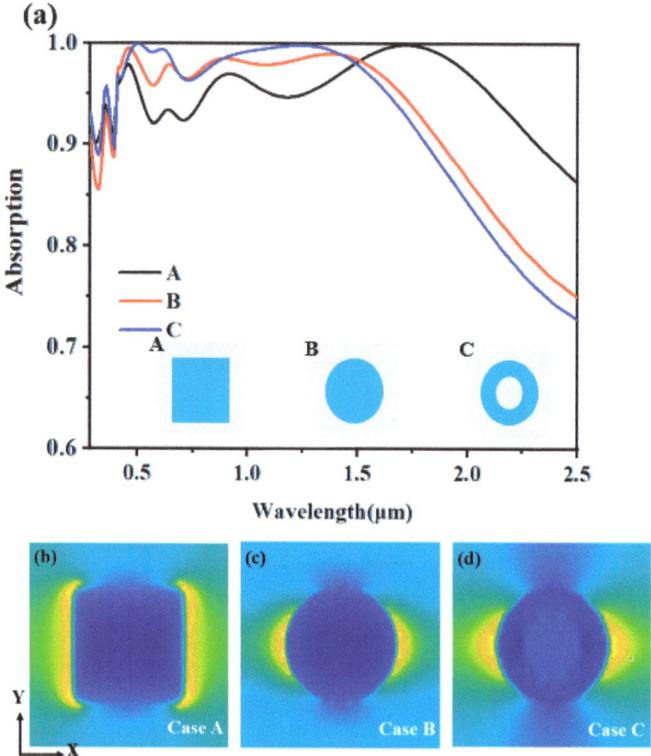

Figure 3. (**a**) is an absorption spectrum diagram of different micro-nano structures; (**b**–**d**) are the electric field intensities of different micro-nano structures in the XOY plane within a period.

2.2. Physical Mechanism Analysis of High Absorption and Wide Bandwidth of Structure

To gain deeper insight into the underlying physical mechanisms that contribute to the high absorption efficiency and extended bandwidth of the absorber, spanning from the visible light to the near-infrared region, we have conducted calculations pertaining to the electric field distributions at specific wavelengths: $\lambda 1 = 456$ nm, $\lambda 2 = 916$ nm, and $\lambda 3 = 1720$ nm, and compared them by drawing. Among them, Figure 4a–c shows the electric field distribution in the XOY direction in the next period from $\lambda 1$ to $\lambda 3$, and Figure 4d–f shows the electric field distribution in the XOZ direction in the two periods from $\lambda 1$ to $\lambda 3$.

Integrating the analysis of the electric field distribution within both the X-Y and X-Z planes, as shown in Figure 4a–f, with the increase in resonance wavelength, optical coupling goes from the edge of the metal cuboid of the pattern layer to the air cavity formed by the adjacent pattern layers, and it is obvious that excitation of surface plasmon resonance occurs within the absorber [45,46]. When the wavelength is $\lambda 1 = 456$ nm, observation of Figure 4a reveals that the electric field is predominantly concentrated on the metallic upper segment of the patterned layer. Therefore, the primary factor contributing to the perfect absorption observed at wavelength $\lambda 1$ is the excitation of surface plasmon polaritons (PSPs) on the surface of the patterned layer.

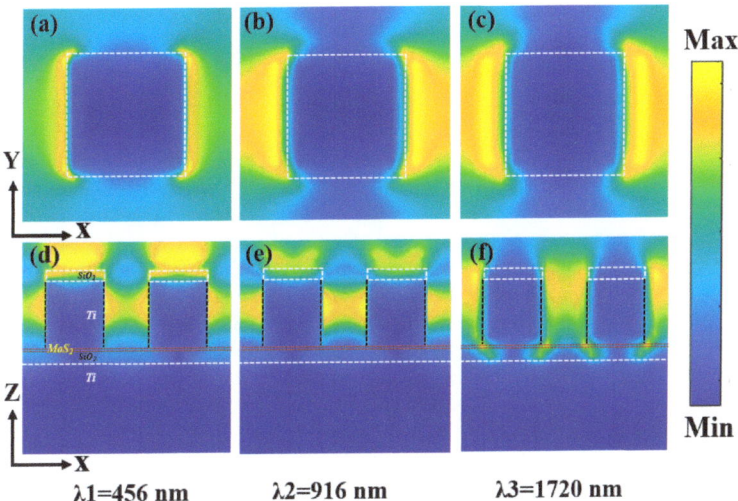

Figure 4. (**a–c**) are the electric field distributions of the XOY plane in one period at λ1, λ2, and λ3 wavelengths, respectively; (**d–f**) are the electric field distributions of two periodic XOZ planes at λ1, λ2, and λ3 wavelengths, respectively.

Different from the excitation mechanism of PSPs, the excitation of LSPs does not need specific momentum-matching conditions [47–49]. The underlying reason for this is that the incident light's wavelength far exceeds the characteristic dimensions of the metallic structures, and these nanostructures can be regarded as a kind of focused source point, which can produce diverse wave vector components, and then provide the necessary momentum matching for LSPs in the excitation structure. More specifically, the excitation intensity of local surface plasmon resonance (LSPs) is determined by the wavelength of the incoming light, the size of nanoparticles or nanostructures and the inherent characteristics of the materials [50]. Since the wavelength of incident light, λ2 = 916 nm, is much longer than the period of the absorber, 400 nm, LSPs can be generated in the metal of the pattern layer.

Illustrated in Figure 4b,e, a strong electric field appears in the adjacent pattern layer region. The electric field distribution shows that PSPs and LSPs are excited on the metal surface of the pattern layer and in the adjacent pattern air cavity, respectively; concurrently, the resonant wavelength of λ2 = 916 nm attains perfect absorption. At a wavelength of λ3 = 1720 nm, the electric field distribution in the XOY direction is similar to that when λ2 = 916 nm, as shown in Figure 4c, but different from 916 nm. An intense electric field accumulates within the dielectric spacer, localized at the interface between the cuboid structure and the MoS$_2$ layer, and the electric field intensity of the top anti-reflection layer silicon dioxide is weakened. Therefore, the perfect absorption observed in the 1720 nm band stems from the excitation of localized surface plasmons (LSPs).

2.3. Weighted Average Absorption Rate and Radiation Performance Analysis

This paper delves into the absorption and radiation properties of the proposed structure, with the objective of evaluating its performance efficiency as both a solar absorber and a thermal emitter. In this process, we use the global spectral equation under the condition of AM 1.5 incident solar energy, which is specifically expressed as [51,52]:

$$\eta_A = \frac{\int_{\lambda_{Min}}^{\lambda_{Max}} A(\omega) I_{AM1.5}(\omega) d\omega}{\int_{\lambda_{Min}}^{\lambda_{Max}} I_{AM1.5}(\omega) d\omega} \quad (1)$$

The equation of thermal emission efficiency (η_E) is [53]:

$$\eta_E = \frac{\int_{\lambda_{min}}^{\lambda_{max}} \varepsilon(\omega) \cdot I_{BE}(\omega, T) d\omega}{\int_{\lambda_{min}}^{\lambda_{max}} I_{BE}(\omega, T) d\omega} \quad (2)$$

Herein, $I_{BE}(\omega, T)$ signifies the intensity of radiation emitted by an ideal blackbody at a given frequency ω and temperature T.

Under AM 1.5 illumination conditions, as exemplified in Figure 5a, the blue curve portrays the theoretical ideal absorption spectrum, whereas the red curve represents the actual energy absorption achieved by the structure. The black area in the figure represents the main area of energy loss, mainly concentrated in the visible light band. Nevertheless, on the whole, the band absorption rate is maintained at a high level of 95.03%, while the loss rate is controlled below 5%. Further, Figure 5b divulges the thermal emission properties exhibited by the structure when subjected to a high temperature of 1500 Kelvin. The black line in the figure represents the theoretical blackbody radiation curve and serves as a benchmark, whereas the red area represents the actual thermal radiation emitted by the structure. It can be observed from the figure that within the spectral range below 1800 nanometers, the observed thermal radiation closely aligns with the theoretical blackbody radiation curve, and only slight thermal radiation loss occurs in the band above 1800 nm. However, in the whole band, the thermal radiation efficiency remains at a high level, reaching 95.96%.

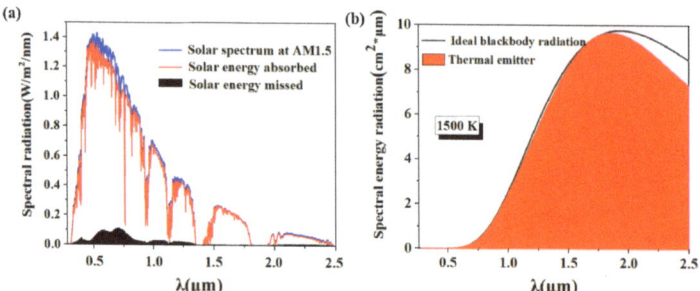

Figure 5. (a) Depicts the distribution of energy absorption and loss for solar radiation spanning from 280 nm to 2500 nm, under an atmospheric mass (AM) of 1.5; (b) Illustrates the energy emission spectrum emanating from a solar absorber operating at an elevated temperature of 1500 K.

2.4. Effect of Varying Parameters on Absorption Outcomes

To validate the optimally chosen structural parameters within this design framework, we conducted a methodical evaluation of the influence that various structural attributes exert on absorption efficiency under the premise of ensuring that other parameters remain unchanged [54,55]. Specifically, we discussed in detail the thickness H1 of top silicon dioxide (SiO_2), the thickness H2 of rectangular titanium (Ti) and its side length D, and the period P of the whole structure. In this process, we did not take into account the parameter changes in Ti and SiO_2, because according to the previous analysis, the adjustment of these parameters has no significant influence on absorption efficiency.

In Figure 6a–d, firstly, we deeply studied how the thickness H1 of the top SiO_2 affects the absorption spectrum. Especially in the ultraviolet and visible light bands, the SiO_2 layer plays a vital role [56,57]. Through careful observation, we found that with the side length d gradually increasing from 20 nm to 60 nm, the absorption efficiency first showed an upward trend, but then showed a downward trend. This discovery provides an important clue for us to further understand the role of the SiO_2 layer in light absorption. Based on this discovery, we infer that the optimal side length of a rectangle should be close to 40 nm. Then, we delved deeper into examining the effect of varying the thickness (H2) of the rectangular titanium (Ti) layer on the absorption spectrum. As H2 was incremented from 100 nm to 300 nm, we discerned a trend in the evolution of the absorption efficiency. It is particularly noteworthy that when the period p is set to 200 nm, we get the best absorption result, which not only has high absorption efficiency, but also has excellent absorption bandwidth. This discovery provides an important reference for our subsequent experiments and applications. As shown in Figure 6c, when the side length of a cuboid is increased from 180 nm to 260 nm, we can observe that the shorter the side length of the cuboid is, an enhanced absorption efficiency is observed within the short wavelength region spanning from 280 to 1500 nanometers. However, with the increase in wavelength, the absorption efficiency of the rectangular parallelepiped with longer side length is significantly enhanced and the bandwidth is wider. On the whole, when D = 220 nm, the overall absorption rate and bandwidth are better. Therefore, upon thorough analysis, we propose that a side length of D = 220 nm offers optimal performance. Figure 6d illustrates the absorption spectrum across the entire periodic variation, where it becomes evident that an increase in the periodicity corresponds to an augmentation in the absorption rate within the spectral range of 1000–1500 nanometers, but considering the overall absorption efficiency of short wave and long wave, the absorber can only exert its maximum absorption potential when P = 400 nm.

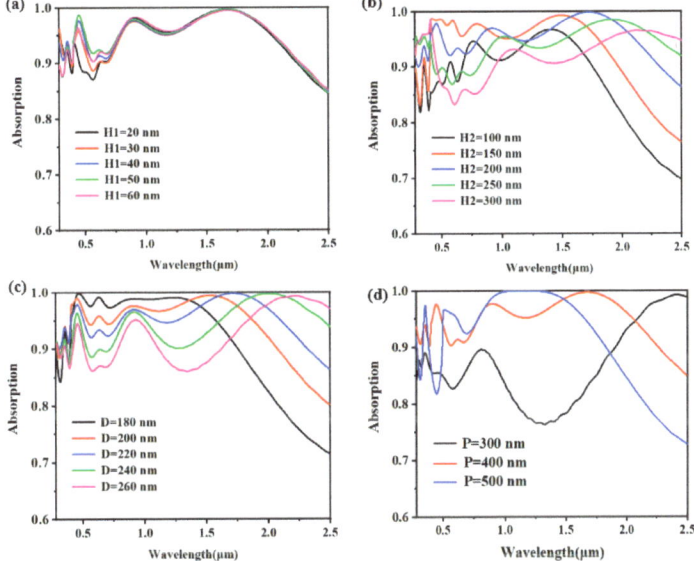

Figure 6. (a) Thickness change absorption spectrogram of top silicon dioxide; (b,c) are absorption spectrograms of thickness and side length change in rectangular titanium, respectively; (d) The absorption spectrum of structural period change.

2.5. Angle Sensitivity Analysis

Finally, in order to study the application potential of the structure, we studied the angular sensitivity of the structure [58–60]. It is acknowledged that in practical scenarios, natural light seldom strikes the solar absorber vertically, contrasting with the idealized conditions. Consequently, it is crucial to investigate the influence of varying polarization and incident angles on the performance of solar absorbers. As depicted in Figure 7, we have conducted simulations to analyze the absorption spectrum, encompassing incident angles and polarization angles ranging from 0° to 60°, respectively. As evident from Figure 7a, the designed absorber demonstrates remarkable performance within an incident angle span of 0° to 60°, exhibiting exceptional overall absorption efficiency. As the polarization angle of the incident light progresses from 0° to 60°, a marginal enhancement in visible light absorption is observed, while the near-infrared region undergoes a slight attenuation, particularly from 50%. Notably, the absorber sustains an absorption rate exceeding 90%

across a broad wavelength spectrum extending up to 2040 nm. The results of this study clearly show that the absorber we designed shows excellent insensitivity to incident angle. As shown in Figure 7b, the spectral response remains stable even when the polarization angle increases; this underscores the exceptional insensitivity of the structure towards polarization variations, thereby reinforcing its robustness. The inherent high geometric symmetry of the structure contributes to a consistently high absorption rate across the entire wavelength spectrum, exhibiting minimal variation with alterations in polarization angle [61,62]. This robustness to both oblique incidence and polarization insensitivity significantly enhances the absorber's performance, presenting substantial advantages for practical applications.

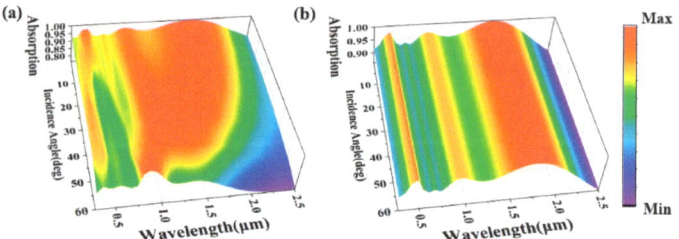

Figure 7. (**a**) Modifying the absorption spectrum by varying the incident angle within a range of 0° to 60°; (**b**) Manipulating the absorption spectrum through alterations in the polarization angle, spanning from 0° to 60°.

3. Modelling and Structural Parameters of the Micro-Nano Optical Devices

In this paper, we introduce a periodically arranged rectangular configuration consisting of Ti-SiO$_2$ layered structures, as shown in Figure 8a. The thickness of local structure is denoted as H1, H2, H3, and H4 from top to bottom. The overall width of the absorber is P = 400 nm, the bottom is made of titanium, and its thickness H4 = 300 nm. In this architecture, Ti functions as a reflective element, with its thickness (H4) set at 300 nanometers, significantly exceeding the penetration depth of electromagnetic radiation. Consequently, the transmission through this structure is effectively nullified, that is, T(ω) = 0 [63,64]. In this work, the spectral absorption efficiency is defined as:

$$A(\omega) = 1 - R(\omega) - T(\omega) \tag{3}$$

Therefore, the absorption rate can be simplified as $A(\omega) = 1 - R(\omega)$ [65,66], where R (Ω) and T (Ω) respectively represent the spectral reflection and transmission under the illumination of plane light. The second layer is dielectric silicon dioxide (SiO$_2$) with a thickness of H3 = 40 nm. In this paper, we propose a material as the supporting substrate of MoS$_2$, aiming at solving the challenge of depositing MoS$_2$ directly on the titanium layer. This support material is located under the dielectric material and above it is the two-dimensional metamaterial MoS$_2$. We set the thickness of MoS$_2$ as 0.625 nm, which is based on the standard thickness grown under most laboratory conditions, thus ensuring the accuracy and repeatability of the research. This configuration not only optimizes the growth environment of MoS$_2$, but also lays a foundation for its performance in subsequent applications. The microstructure of the surface layer is stacked by Ti-SiO$_2$, and its height is set to H1 = 40 nm and H2 = 200 nm. Non-precious metal (Ti) is chosen as the metal material here, because Ti has high loss in visible light and near-infrared, which can effectively broaden the absorption bandwidth, and the price is relatively cheap, which can reduce the manufacturing cost. Among them, the dielectric constant data of Ti and SiO$_2$ are cited and applied based on the experimental results of Palik [67]. In numerical calculations and simulations, the dielectric constant of molybdenum disulphide (MoS$_2$) is a key parameter that determines the material's response properties to light. For the monolayer MoS$_2$

in the paper, we cite the data of Ermolaev et al. [68]. Its dielectric constant is usually expressed in complex form, including real and imaginary parts, and can be expressed as n = n′ + ik, where n′ is the real part of the dielectric function, which represents the material's polarization capacity, while k is the imaginary part, which is related to the material's absorption capacity.

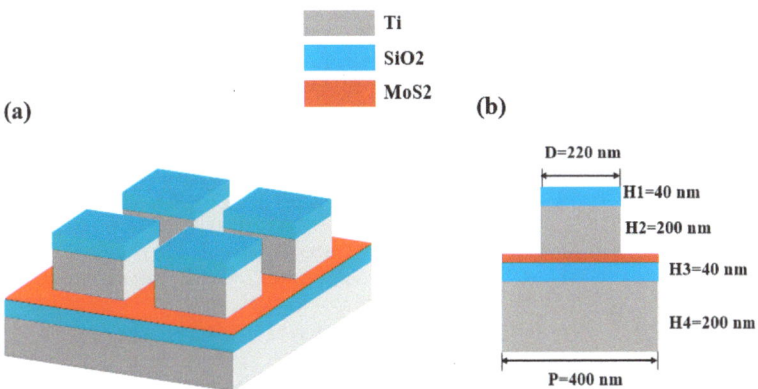

Figure 8. (**a**) Shows the three-dimensional structure of the model; (**b**) Structure diagram of XOZ plane.

In the numerical simulation calculation, we use FDTD version 2020 solutions software as an analysis tool. In this simulation, the setting of the light source is very important. We selected the plane light source from 280 nm to 2500 nm and ensured the polarization of light aligned along the X-axis by configuring the incident light to propagate vertically in the negative Z-axis direction. To uphold the rigor and fidelity of our simulation, we applied periodic boundary conditions in the x and y dimensions to simulate the infinite extended periodic structure. Along the Z-axis, we choose the perfectly matched layer (PML) as the boundary condition to effectively absorb and eliminate the possible reflected waves during the simulation, thus ensuring the accuracy of the simulation results [69].

4. Conclusions

In this paper, we have achieved the successful design of an efficient broadband solar absorber, leveraging the unique properties of MoS_2 two-dimensional materials. Through numerical calculation, our findings indicate that the absorption efficiency of the structure surpasses 90% within a wavelength spectrum exceeding 2040 nm, and the absorption rate can reach 99% or more in the wavelength range of 303 nm. It is noteworthy that an optimal thickness of 0.625 nm for the MoS_2 layer has been identified, and its absorption bandwidth is significantly extended to 310 nm, which fully verifies the superiority of two-dimensional materials in improving absorption performance. In addition, MoS_2 not only has excellent thermal stability, but also helps to reduce device size and improve overall absorption efficiency. Under the condition of AM 1.5 spectrum, our structure shows excellent weighted average absorption efficiency, reaching a high level of 95.03%. Even at an extreme temperature as high as 1500 K, its thermal radiation efficiency can be maintained at an excellent level of 95.96%. The structure demonstrates remarkable independence and resilience towards variations in both the polarization angle and the incident angle of the incoming light, and its performance remains stable even in the polarization angle range of 0 to 60. In summary, the high absorption efficiency and stability at extreme temperatures of this MoS_2-based broadband solar absorber we have designed make it ideal for solar thermal photovoltaic systems that can convert solar energy directly into electricity, thereby reducing dependence on fossil fuels and lowering greenhouse gas emissions. Additionally, by using cost-effective titanium (Ti) in place of expensive precious metals, our technology helps lower the economic barrier to solar technology, advancing its use in a wider range of applications. Our design also demonstrates insensitivity to the direction of polarization of incident light, which increases its reliability in variable environments and opens up the possibility of applying solar absorbing technologies in different geographical locations and climates.

Author Contributions: Conceptualization, W.L., F.W., Z.Y., Y.T. and Q.Z.; data curation, W.L., F.W., Z.Y., Y.T. and Q.Z.; formal analysis, W.L. and Z.Y.; methodology, W.L., F.W., Z.Y., Y.T., Y.Y., P.W. and Q.Z.; resources, W.L. and Q.Z.; software, W.L., F.W., Z.Y., Y.T., Y.Y., P.W. and Q.Z.; data curation, W.L., F.W., Z.Y., Y.T., Y.Y. and P.W.; writing—original draft preparation, W.L., F.W. and Z.Y.; writing—review and editing, W.L. and Z.Y. All authors have read and agreed to the published version of the manuscript.

Funding: The authors are grateful to the support by National Natural Science Foundation of China (No. 51606158, 11604311, 12074151); the funded by the Natural Science Foundation of Fujian Province (2022J011102, 2022H0048); the Funded by the Guangxi Science and Technology Base and Talent Special Project (No. AD21075009); the funded by the Sichuan Science and Technology Program (No. 2021JDRC0022); the funded by the Research Project of Fashu Foundation (MFK23006); the funded by the Open Fund of the Key Laboratory for Metallurgical Equipment and Control Technology of Ministry of Education in Wuhan University of Science and Technology, China (No. MECOF2022B01; MECOF2023B04); the funded by the Project supported by Guangxi Key Laboratory of Precision Navigation Technology and Application, Guilin University of Electronic Technology (No. DH202321); the funded by the Scientific Research Project of Huzhou College (2022HXKM07).

Institutional Review Board Statement: Not applicable.

Informed Consent Statement: Not applicable.

Data Availability Statement: Publicly available datasets were analyzed in this study. These data can be found here: [https://www.lumerical.com/] (accessed on 1 January 2020).

Conflicts of Interest: The authors declare no conflicts of interest.

References

1. Xiao, T.; Tu, S.; Liang, S.; Guo, R.; Tian, T.; Müller-Buschbaum, P. Solar cell-based hybrid energy harvesters towards sustainability. *Opto-Electron. Sci.* **2023**, *2*, 230011. [CrossRef]
2. Li, C.; Du, W.; Huang, Y.; Zou, J.; Luo, L.; Sun, S.; Wang, Z. Photonic synapses with ultralow energy consumption for artificial visual perception and brain storage. *Opto-Electron. Adv.* **2022**, *5*, 210069. [CrossRef]
3. Rugut, E.K.; Maluta, N.E.; Maphanga, R.R.; Mapasha, R.E.; Kirui, J.K. Structural, Mechanical, and Optoelectronic Properties of CH3NH3PbI3 as a Photoactive Layer in Perovskite Solar Cell. *Photonics* **2024**, *11*, 372. [CrossRef]
4. Han, Q.C.; Liu, S.W.; Liu, Y.Y.; Jin, J.S.; Li, D.; Cheng, S.B.; Xiong, Y. Flexible counter electrodes with a composite carbon/metal nanowire/polymer structure for use in dye-sensitized solar cells. *Solar Energy* **2020**, *208*, 469–479. [CrossRef]
5. Shioki, T.; Tsuji, R.; Oishi, K.; Fukumuro, N.; Ito, S. Designed Mesoporous Architecture by 10–100 nm TiO$_2$ as Electron Transport Materials in Carbon-Based Multiporous-Layered-Electrode Perovskite Solar Cells. *Photonics* **2024**, *11*, 236. [CrossRef]
6. Montagni, T.; Ávila, M.; Fernández, S.; Bonilla, S.; Cerdá, M.F. Cyanobacterial Pigments as Natural Photosensitizers for Dye-Sensitized Solar Cells. *Photochem* **2024**, *4*, 388–403. [CrossRef]
7. Fu, R.; Chen, K.X.; Li, Z.L.; Yu, S.H.; Zheng, G.X. Metasurface-based nanoprinting: Principle, design and advances. *Opto-Electron. Sci.* **2022**, *1*, 220011. [CrossRef]
8. Chen, Z.Y.; Cheng, S.B.; Zhang, H.F.; Yi, Z.; Tang, B.; Chen, J.; Zhang, J.G.; Tang, C.J. Ultra wideband absorption ab-sorber based on Dirac semimetallic and graphene metamaterials. *Phys. Lett. A* **2024**, *517*, 129675. [CrossRef]
9. Zhu, J.; Xiong, J.Y. Logic operation and all-optical switch characteristics of graphene surface plasmons. *Opt. Express* **2023**, *31*, 36677. [CrossRef]
10. Xiong, H.; Deng, J.H.; Yang, Q.; Wang, X.; Zhang, H.Q. A metamaterial energy power detector based on electromag-netic energy harvesting technology. *ACS Appl. Electron. Mater.* **2024**, *6*, 1204–1210. [CrossRef]
11. Wang, B.X.; Xu, C.; Duan, G.; Xu, W.; Pi, F. Review of Broadband Metamaterial Absorbers: From Principles, Design Strategies, and Tunable Properties to Functional Applications. *Adv. Funct. Mater.* **2023**, *33*, 2213818. [CrossRef]
12. Lin, H.; Sturmberg, B.C.P.; Lin, K.-T.; Yang, Y.; Zheng, X.; Chong, T.K.; de Sterke, C.M.; Jia, B. A 90-nm-thick graphene metamaterial for strong and extremely broadband absorption of unpolarized light. *Nat. Photon.* **2019**, *13*, 270–276. [CrossRef]
13. Jiang, B.; Hou, Y.; Wu, J.; Ma, Y.; Gan, X.; Zhao, J. In-fiber photoelectric device based on graphene-coated tilted fiber grating. *Opto-Electron. Sci.* **2023**, *2*, 230012. [CrossRef]
14. Li, W.; Yi, Y.; Yang, H.; Cheng, S.; Yang, W.X.; Zhang, H.; Yi, Z.; Yi, Y.; Li, H. Active Tunable Terahertz Band-width Absorber Based on single layer Graphene. *Commun. Theor. Phys.* **2023**, *75*, 045503. [CrossRef]

15. Liu, Y.; Hu, L.; Liu, M. Graphene and Vanadium Dioxide-Based Terahertz Absorber with Switchable Multifunctionality for Band Selection Applications. *Nanomaterials* **2024**, *14*, 1200. [CrossRef]
16. Yan, S.Q.; Zuo, Y.; Xiao, S.S.; Oxenløwe, L.K.; Ding, Y.H. Graphene photodetector employing double slot structure with enhanced responsivity and large bandwidth. *Opto-Electron. Adv.* **2022**, *5*, 210159. [CrossRef]
17. Gorgolis, G.; Kotsidi, M.; Messina, E.; Mazzurco Miritana, V.; Di Carlo, G.; Nhuch, E.L.; Martins Leal Schrekker, C.; Cuty, J.A.; Schrekker, H.S.; Paterakis, G.; et al. Antifungal Hybrid Graphene–Transition-Metal Dichalcogenides Aerogels with an Ionic Liquid Additive as Innovative Absorbers for Preventive Conservation of Cultural Heritage. *Materials* **2024**, *17*, 3174. [CrossRef]
18. Zeng, C.; Lu, H.; Mao, D.; Du, Y.; Hua, H.; Zhao, W.; Zhao, J. Graphene-empowered dynamic metasurfaces and metade-vices. *Opto-Electron. Adv.* **2022**, *5*, 200098. [CrossRef]
19. Guan, H.; Hong, J.; Wang, X.; Jingyuan, M.; Zhang, Z.; Liang, A.; Han, X.; Dong, J.; Qiu, W.; Chen, Z.; et al. Broadband, High-Sensitivity Graphene Photodetector Based on Ferroelectric Polarization of Lithium Niobate. *Adv. Opt. Mater.* **2021**, *9*, 2100245. [CrossRef]
20. Zhu, J.; Xiong, J.Y. Tunable terahertz graphene metamaterial optical switches and sensors based on plasma-induced transparency. *Measurement* **2023**, *220*, 113302. [CrossRef]
21. Jin, S.; Zu, H.; Qian, W.; Luo, K.; Xiao, Y.; Song, R.; Xiong, B. A Quad-Band and Polarization-Insensitive Metamaterial Absorber with a Low Profile Based on Graphene-Assembled Film. *Materials* **2023**, *16*, 4178. [CrossRef] [PubMed]
22. Cai, F.; Kou, Z. A Novel Triple-Band Terahertz Metamaterial Absorber Using a Stacked Structure of MoS$_2$ and Graphene. *Photonics* **2023**, *10*, 643. [CrossRef]
23. Ma, J.; Wu, P.H.; Li, W.X.; Liang, S.R.; Shangguan, Q.Y.; Cheng, S.B.; Tian, Y.H.; Fu, J.Q.; Zhang, L.B. A five-peaks graphene absorber with multiple adjustable and high sensitivity in the far infrared band. *Diam. Relat. Mater.* **2023**, *136*, 109960. [CrossRef]
24. Shao, M.R.; Ji, C.; Tan, J.B.; Du, B.Q.; Zhao, X.F.; Yu, J.; Man, B.; Xu, K.; Zhang, C.; Li, Z. Ferroelectrically modulate the Fermi level of graphene oxide to enhance SERS response. *Opto-Electron. Adv.* **2023**, *6*, 230094. [CrossRef]
25. He, Z.; Guan, H.; Liang, X.; Chen, J.; Xie, M.; Luo, K.; An, R.; Ma, L.; Ma, F.; Yang, T.; et al. Broadband, polarization-sensitive, and self-powered high-performance photodetection of hetero-integrated MoS$_2$ on lithium niobate. *Research* **2023**, *6*, 0199. [CrossRef] [PubMed]
26. Zhang, T.X.; Tao, C.; Ge, S.X.; Pan, D.W.; Li, B.; Huang, W.X.; Wang, W.; Chu, L.Y. Interfaces coupling defor-mation mechanisms of liquid-liquid-liquid three-phase flow in a confined microchannel. *Chem. Eng. J.* **2022**, *434*, 134769. [CrossRef]
27. Li, K.-C.; Lu, M.-Y.; Nguyen, H.T.; Feng, S.-W.; Artemkina, S.B.; Fedorov, V.E.; Wang, H.-C. Intelligent Identification of MoS$_2$ Nanostructures with Hyperspectral Imaging by 3D-CNN. *Nanomaterials* **2020**, *10*, 1161. [CrossRef]
28. Chen, S.; Li, B.; Dai, C.; Zhu, L.; Shen, Y.; Liu, F.; Deng, S.; Ming, F. Controlling Gold-Assisted Exfoliation of Large-Area MoS$_2$ Monolayers with External Pressure. *Nanomaterials* **2024**, *14*, 1418. [CrossRef]
29. Wadhwa, R.; Agrawal, A.V.; Kumar, M. A strategic review of recent progress, prospects and challenges of MoS$_2$-based photodetectors. *J. Phys. D Appl. Phys.* **2022**, *55*, 063002. [CrossRef]
30. Li, X.; Zhu, H. Two-dimensional MoS$_2$: Properties, preparation, and applications. *J. Mater.* **2015**, *1*, 33–44. [CrossRef]
31. Yue, S.; Hou, M.; Wang, R.; Guo, H.; Hou, Y.; Li, M.; Zhang, Z.; Wang, Y.; Zhang, Z. Ultra-broadband metamaterial absorber from ultraviolet to long-wave infrared based on CMOS-compatible materials. *Opt. Express* **2020**, *28*, 31844–31861. [CrossRef] [PubMed]
32. Lei, L.; Li, S.; Huang, H.; Tao, K.; Xu, P. Ultra-broadband absorber from visible to near-infrared using plasmonic metamaterial. *Opt. Express* **2018**, *26*, 5686–5693. [CrossRef] [PubMed]
33. Zhong, H.; Liu, Z.; Tang, P.; Liu, X.; Tang, C. Thermal-stability resonators for visible light full-spectrum perfect absorbers. *Sol. Energy* **2020**, *208*, 445–450. [CrossRef]
34. Lee, C.; Yan, H.; Brus, L.E.; Heinz, T.F.; Hone, J.; Ryu, S. Anomalous Lattice Vibrations of Single- and Few-Layer MoS$_2$. *ACS Nano* **2010**, *4*, 2695–2700. [CrossRef]
35. Liu, Z.; Zhang, H.; Fu, G.; Liu, G.; Liu, X.; Yuan, W.; Xie, Z. Colloid templated semiconductor meta-surface for ultra-broadband solar energy absorber. *Sol. Energy* **2020**, *198*, 194–201. [CrossRef]
36. Chen, M.; He, Y. Plasmonic nanostructures for broadband solar absorption based on the intrinsic absorption of metals. *Sol. Energy Mater. Sol. Cells* **2018**, *188*, 156–163. [CrossRef]
37. Huo, D.; Zhang, J.; Wang, Y.; Wang, C.; Su, H.; Zhao, H. Broadband Perfect Absorber Based on TiN-Nanocone Metasurface. *Nanomaterials* **2018**, *8*, 485. [CrossRef]
38. Xiong, H.; Ma, X.D.; Wang, B.X.; Zhang, H.Q. Design and analysis of an electromagnetic energy conversion device. *Sens. Actuators A Phys.* **2024**, *366*, 114972. [CrossRef]
39. Late, D.J.; Liu, B.; Matte, H.R.; Dravid, V.P.; Rao, C.N.R. Hysteresis in single-layer MoS$_2$ field effect transistors. *ACS Nano* **2012**, *6*, 5635–5641. [CrossRef]
40. Liang, S.R.; Cheng, S.B.; Zhang, H.F.; Yang, W.X.; Yi, Z.; Zeng, Q.D.; Tang, B.; Wu, P.; Ahmad, S.; Sun, T.; et al. Structural color tunable intelligent mid-infrared thermal control emitter. *Ceram. Int.* **2024**, *50*, 23611–23620. [CrossRef]

41. Gao, H.; Fan, X.H.; Wang, Y.X.; Liu, Y.C.; Wang, X.G.; Xu, K.; Deng, L.; Deng, C.; Zeng, C.; Li, T.; et al. Multi-foci metalens for spectra and polarization ellipticity recognition and reconstruction. *Opto-Electron. Sci.* **2023**, *2*, 220026. [CrossRef]
42. Cheng, S.B.; Li, W.X.; Zhang, H.F.; Akhtar, M.N.; Yi, Z.; Zeng, Q.D.; Ma, C.; Sun, T.Y.; Wu, P.H.; Ahmad, S. High sen-sitivity five band tunable metamaterial absorption device based on block like Dirac semimetals. *Opt. Commun.* **2024**, *569*, 130816. [CrossRef]
43. Li, W.; Cheng, S.; Zhang, H.; Yi, Z.; Tang, B.; Ma, C.; Wu, P.; Zeng, Q.; Raza, R. Multi-functional metasurface: Ul-trawideband/multi-band absorption switching by adjusting guided mode resonance and local surface plasmon resonance effects. *Commun. Theor. Phys.* **2024**, *76*, 065701. [CrossRef]
44. Zhang, Y.; Pu, M.; Jin, J.; Lu, X.; Guo, Y.; Cai, J.; Zhang, F.; Ha, Y.; He, Q.; Xu, M.; et al. Crosstalk-free achromatic full Stokes imaging polarimetry metasurface enabled by polarization-dependent phase optimization. *Opto-Electron. Adv.* **2022**, *5*, 220058. [CrossRef]
45. Wang, J.; Qin, X.; Zhao, Q.; Duan, G.; Wang, B.-X. Five-Band Tunable Terahertz Metamaterial Absorber Using Two Sets of Different-Sized Graphene-Based Copper-Coin-like Resonators. *Photonics* **2024**, *11*, 225. [CrossRef]
46. Zhao, H.; Wang, X.K.; Liu, S.T.; Zhang, Y. Highly efficient vectorial field manipulation using a transmitted tri-layer metasurface in the terahertz band. *Opto-Electron. Adv.* **2023**, *6*, 220012. [CrossRef]
47. Li, W.X.; Liu, Y.H.; Ling, L.; Sheng, Z.X.; Cheng, S.B.; Yi, Z.; Wu, P.H.; Zeng, Q.D.; Tang, B.; Ahmad, S. The tunable absorber films of grating structure of AlCuFe quasicrystal with high Q and refractive index sensitivity. *Surf. Interfaces* **2024**, *48*, 104248. [CrossRef]
48. Deng, J.H.; Xiong, H.; Yang, Q.; Wang, B.X.; Zhang, H.Q. Metasurface-based Microwave Power Detector for Polari-zation Angle Detection. *IEEE Sens. J.* **2023**, *23*, 22459–22465. [CrossRef]
49. Gigli, C.; Leo, G. All-dielectric χ^2 metasurfaces: Recent progress. *Opto-Electron. Adv.* **2022**, *5*, 210093. [CrossRef]
50. Liang, S.; Xu, F.; Yang, H.; Cheng, S.; Yang, W.; Yi, Z.; Song, Q.; Wu, P.; Chen, J.; Tang, C. Ultra long infrared met-amaterial absorber with high absorption and broad band based on nano cross surrounding. *Opt. Laser Technol.* **2023**, *158*, 108789. [CrossRef]
51. Zhang, Y.; Yi, Y.; Li, W.; Liang, S.; Ma, J.; Cheng, S.; Yang, W.; Yi, Y. High Absorptivity and Ultra-Wideband So-lar Absorber Based on Ti-Al2O3 Cross Elliptical Disk Arrays. *Coatings* **2023**, *13*, 531. [CrossRef]
52. Sang, T.; Mi, Q.; Yang, C.Y.; Zhang, X.H.; Wang, Y.K.; Ren, Y.Z.; Xu, T. Achieving asymmetry parameter-insensitive resonant modes through relative shift–induced quasi-bound states in the continuum. *Nanophotonics* **2024**, *13*, 1369–1377. [CrossRef]
53. Wu, L.; Yang, L.L.; Zhu, X.W.; Cai, B.; Cheng, Y.Z. Ultra-broadband and wide-angle plasmonic absorber based on all-dielectric gallium arsenide pyramid nanostructure for full solar radiation spectrum range. *Int. J. Therm. Sci.* **2024**, *201*, 109043. [CrossRef]
54. Yue, Z.; Li, J.T.; Li, J.; Zheng, C.L.; Liu, J.Y.; Lin, L.; Guo, L.; Liu, W. Terahertz metasurface zone plates with arbitrary polari-zations to a fixed polarization conversion. *Opto-Electron. Sci.* **2022**, *1*, 210014. [CrossRef]
55. Luo, J. Dynamical behavior analysis and soliton solutions of the generalized Whitham–Broer–Kaup–Boussineq–Kupershmidt equations. *Results Phys.* **2024**, *60*, 107667. [CrossRef]
56. Li, W.X.; Zhao, W.C.; Cheng, S.B.; Yang, W.X.; Yi, Z.; Li, G.F.; Zeng, L.C.; Li, H.L.; Wu, P.H.; Cai, S.S. Terahertz Se-lec-tive Active Electromagnetic Absorption Film Based on Single-layer Graphene. *Surf. Interfaces* **2023**, *40*, 103042. [CrossRef]
57. Xie, Y.D.; Liu, Z.M.; Zhou, F.Q.; Luo, X.; Gong, Y.M.; Cheng, Z.Q.; You, Y. Tunable nonreciprocal metasurfaces based on nonlinear quasi-Bound state in the Continuum. *Opt. Lett.* **2024**, *49*, 3520–3523. [CrossRef]
58. Cao, T.; Lian, M.; Chen, X.Y.; Mao, L.B.; Liu, K.; Jia, J.; Su, W.; Ren, H.; Zhang, S.; Xu, Y.; et al. Multi-cycle reconfigurable THz extraordinary optical trans-mis-sion using chalcogenide metamaterials. *Opto-Electron. Sci.* **2022**, *1*, 210010. [CrossRef]
59. Li, W.X.; Liu, M.S.; Cheng, S.B.; Zhang, H.F.; Yang, W.X.; Yi, Z.; Zeng, Q.D.; Tang, B.; Ahmad, S.; Sun, T.Y. Po-lar-ization independent tunable bandwidth absorber based on single-layer graphene. *Diam. Relat. Mater.* **2024**, *142*, 110793. [CrossRef]
60. Ha, Y.L.; Luo, Y.; Pu, M.B.; Zhang, F.; He, Q.; Jin, J.; Xu, M.; Guo, Y.; Li, X.; Ma, X.; et al. Physics-data-driven intelligent optimization for large-aperture metalenses. *Opto-Electron. Adv.* **2023**, *6*, 230133. [CrossRef]
61. Shangguan, Q.; Zhao, Y.; Song, Z.; Wang, J.; Yang, H.; Chen, J.; Liu, C.; Cheng, S.; Yang, W.; Yi, Z. High sensitivity active adjustable graphene absorber for refractive index sensing applications. *Diam. Relat. Mater.* **2022**, *128*, 109273. [CrossRef]
62. Sreekanth, K.V.; Alapan, Y.; ElKabbash, M.; Wen, A.M.; Ilker, E.; Hinczewski, M.; Gurkan, U.A.; Steinmetz, N.F.; Strangi, G. Enhancing the angular sensitivity of plasmonic sensors using hyperbolic metamaterials. *Adv. Opt. Mater.* **2016**, *4*, 1767–1772. [CrossRef] [PubMed]
63. Deng, X.; Shui, T.; Yang, W.X. Inelastic two-wave mixing induced high-efficiency transfer of optical vortices. *Opt. Express* **2024**, *32*, 16611–16628. [CrossRef] [PubMed]
64. Fan, J.X.; Li, Z.L.; Xue, Z.Q.; Xing, H.Y.; Lu, D.; Xu, G.; Gu, J.; Han, J.; Cong, L. Hybrid bound states in the continuum in terahertz metasurfaces. *Opto-Electron. Sci.* **2023**, *2*, 230006. [CrossRef]
65. Luo, J.; Zhang, J.H.; Gao, S.S. Design of Multi-Band Bandstop Filters Based on Mixed Electric and Magnetic Coupling Resonators. *Electronics* **2024**, *13*, 1552. [CrossRef]
66. Li, W.X.; Zhao, W.C.; Cheng, S.B.; Zhang, H.F.; Yi, Z.; Sun, T.Y.; Wu, P.H.; Zeng, Q.D.; Raza, R. Tunable Metamaterial Absorption Device based on Fabry–Perot Resonance as Temperature and Refractive Index Sensing. *Opt. Lasers Eng.* **2024**, *181*, 108368. [CrossRef]
67. Palik, E.D. *Handbook of Optical Constants of Solids I–III*; Academic Press: Orlando, FL, USA, 1998.

68. Ermolaev, G.A.; Stebunov, Y.V.; Vyshnevyy, A.A.; Tatarkin, D.E.; Yakubovsky, D.I.; Novikov, S.M.; Baranov, D.G.; Shegai, T.; Nikitin, A.Y.; Arsenin, A.V.; et al. Broadband Optical Properties of Monolayer and Bulk MoS$_2$. *Npj 2D Mater. Appl.* **2020**, *4*, 21. [CrossRef]
69. Liang, S.R.; Xu, F.; Li, W.X.; Yang, W.X.; Cheng, S.B.; Yang, H.; Chen, J.; Yi, Z.; Jiang, P.P. Tunable smart mid infrared thermal control emitter based on phase change material VO$_2$ thin film. *Appl. Therm. Eng.* **2023**, *232*, 121074. [CrossRef]

Disclaimer/Publisher's Note: The statements, opinions and data contained in all publications are solely those of the individual author(s) and contributor(s) and not of MDPI and/or the editor(s). MDPI and/or the editor(s) disclaim responsibility for any injury to people or property resulting from any ideas, methods, instructions or products referred to in the content.

Article

Two-Dimensional GeC/MXY (M = Zr, Hf; X, Y = S, Se) Heterojunctions Used as Highly Efficient Overall Water-Splitting Photocatalysts

Guangzhao Wang [1,*,†], Wenjie Xie [1,†], Sandong Guo [2], Junli Chang [3], Ying Chen [4], Xiaojiang Long [1], Liujiang Zhou [5], Yee Sin Ang [6,*] and Hongkuan Yuan [3,*]

[1] School of Electronic Information Engineering, Key Laboratory of Extraordinary Bond Engineering and Advanced Materials Technology of Chongqing, Yangtze Normal University, Chongqing 408100, China; xwj52739820@163.com (W.X.); longxiaojiang@yznu.edu.cn (X.L.)
[2] School of Electronic Engineering, Xi'an University of Posts and Telecommunications, Xi'an 710121, China; sandongyuwang@163.com
[3] School of Physical Science and Technology, Southwest University, Chongqing 400715, China; jlchang66@126.com
[4] School of Electronic and Information Engineering, Anshun University, Anshun 561000, China; ychenjz@163.com
[5] School of Physics, University of Electronic Science and Technology of China, Chengdu 610054, China; ljzhou86@uestc.edu.cn
[6] Science, Mathematics and Technology, Singapore University of Technology and Design, Singapore 487372, Singapore
* Correspondence: wangyan6930@yznu.edu.cn or wangyan6930@126.com (G.W.); yeesin_ang@sutd.edu.sg (Y.S.A.); yhk10@yznu.edu.cn (H.Y.)
† These authors contributed equally to this work.

Citation: Wang, G.; Xie, W.; Guo, S.; Chang, J.; Chen, Y.; Long, X.; Zhou, L.; Ang, Y.S.; Yuan, H. Two-Dimensional GeC/MXY (M = Zr, Hf; X, Y = S, Se) Heterojunctions Used as Highly Efficient Overall Water-Splitting Photocatalysts. *Molecules* **2024**, *29*, 2793. https://doi.org/10.3390/molecules29122793

Academic Editor: Sergio Navalón

Received: 26 March 2024
Revised: 16 May 2024
Accepted: 21 May 2024
Published: 12 June 2024

Copyright: © 2024 by the authors. Licensee MDPI, Basel, Switzerland. This article is an open access article distributed under the terms and conditions of the Creative Commons Attribution (CC BY) license (https://creativecommons.org/licenses/by/4.0/).

Abstract: Hydrogen generation by photocatalytic water-splitting holds great promise for addressing the serious global energy and environmental crises, and has recently received significant attention from researchers. In this work, a method of assembling GeC/MXY (M = Zr, Hf; X, Y = S, Se) heterojunctions (HJs) by combining GeC and MXY monolayers (MLs) to construct direct Z-scheme photocatalytic systems is proposed. Based on first-principles calculations, we found that all the GeC/MXY HJs are stable van der Waals (vdW) HJs with indirect bandgaps. These HJs possess small bandgaps and exhibit strong light-absorption ability across a wide range. Furthermore, the built-in electric field (BIEF) around the heterointerface can accelerate photoinduced carrier separation. More interestingly, the suitable band edges of GeC/MXY HJs ensure sufficient kinetic potential to spontaneously accomplish water redox reactions under light irradiation. Overall, the strong light-harvesting ability, wide light-absorption range, small bandgaps, large heterointerfacial BIEFs, suitable band alignments, and carrier migration paths render GeC/MXY HJs highly efficient photocatalysts for overall water decomposition.

Keywords: hybrid density functional; GeC/MXY heterojunctions; direct Z-scheme; photocatalysis; water-splitting

1. Introduction

In recent years, environmental pollution and shortages of non-renewable energy have become increasingly severe. Photocatalytic water decomposition (PCWD) for hydrogen generation is considered as an effective approach to alleviating the energy crisis and environmental pollution [1–10]. In 1972, TiO_2 was first used as a photocatalyst (PC) to decompose water into hydrogen and oxygen [11]. The PCWD process typically comprises three steps: light capture, photoinduced carrier separation/transfer, and water redox reactions occurring on the surfaces of the catalysts [12–14]. The required water-splitting PC (WSPC) must have a band edge exceeding the water redox level; specifically, the H^+/H_2

and H_2O/O_2 levels must fall between the valence band maximum (VBM) and conduction band minimum (CBM) [15–17]. Consequently, the bandgap of a WSPC for PCWD should be greater than 1.23 eV. Furthermore, considering the energy loss during the process of photoinduced carriers transferring to the catalyst's surfaces and the kinetic potentials required to drive the water redox reactions, the bandgap of the WSPC is typically required to be greater than 1.8 eV [12,18]. In addition to the bandgap requirement, the activity of WSPCs strongly depends on other factors such as photostability, light capture ability, trapping of and photoinduced carrier recombination, and the catalyst's surface reactivities towards the hydrogen/oxygen evolution reaction (HER/OER) [19]. Although researchers have developed a series of PCs; a few single WSPCs simultaneously possess the advantages of wide light response extent, good carrier mobility, high photoexcited carrier separation, spatially separated reaction sites, strong redox capacity, and lower overpotentials for the HER and OER processes. Thus, it is still urgent to explore new photocatalytic mechanisms and develop highly efficient WSPCs.

Inspired by photosynthesis in green plants, a direct Z-scheme mechanism was constructed to overcome the shortcomings of single WSPCs [20–22]. A typical direct Z-scheme WSPC is usually composed of two parts: the hydrogen production PC (HPPC) and oxygen production PC (OPPC) [23,24]. The photoinduced electrons and holes recombine at the interface between the HPPC and OPPC, resulting in remnant electrons at the HPPC and excess holes at the OPPC. This process leads to the efficient spatial separation of photogenerated carriers, thus obtaining strong redox capacity to drive water-splitting. Up to now, the direct Z-scheme mechanism has been experimentally realized in a series of composites, including $TiO_2/ZnIn_2S_4$ [25], aza-CMP/C_2N [26], $Cd_{0.5}Zn_{0.5}/BiVO_4$ [27], α-Fe_2O_3/g-C_3N_4 [28], black P/$BiVO_4$ [29], CdS/MoS_2 [30], $CdS/Co_{1-x}S$ [31], and TiO_2/CuO_2 [32]. In particular, direct Z-scheme two-dimensional (2D) van der Waals (vdW) heterojunction (HJ) PCs (HJPCs) exhibit excellent photocatalytic performance due to their highly specific surface area, abundant active sites, good carrier mobility, and tunable interfaces [26,33,34]. In addition, strong electron–hole coupling and charge transfer around such heterointerfaces have been experimentally observed [35–37] and theoretically proposed [38–42]. This facilitates interlayer carrier recombination and helps to achieve the Z-scheme photocatalytic mechanism.

The graphene-like hexagonal GeC monolayer (ML) receives considerable attention due to its excellent electronic, mechanical, magnetic, and optical properties [43–46]. Especially, it possesses lower stiffness and a bigger Poisson's ratio compared to graphene [47]. Therefore, the excellent characteristics of the GeC ML promote it to achieve device applications in the fields of electronics, optoelectronics, and photovoltaic [48]. More excitingly, GeC thin films have been experimentally synthesized by the chemical vapor deposition (CVD) [49] and laser ablation [50] methods. Since it is known that a large variety of 2D layers can be fabricated using the mechanical exfoliation and CVD [51–53] methods, we can speculate that the GeC ML may also be synthesized using similar preparation methods. The GeC ML not only has a stable plane structure, but also shares a similar honeycomb structure and lattice constants with many other 2D materials; thus, some GeC-based HJs have been designed and studied [54–63]. Although many literature works have explored the photocatalytic performance of the GeC ML and GeC-based type-II HJPCs, there are still relatively few reports on the direct Z-scheme mechanism of GeC-based HJs, which remains an open question thus far. It is interesting and meaningful to find suitable 2D materials to construct direct Z-scheme HJPCs with GeC MLs. Recently, MXY (M = Zr, Hf; X, Y = S, Se) MLs with a stable 1T phase have been demonstrated to exhibit excellent mechanical, thermal, thermoelectric, piezoelectricity, optical, and catalytic properties [64–70]. In particular, HfS_2, $HfSe_2$, ZrS_2, and $ZrSe_2$ MLs have been experimentally verified [71–75], as well as Janus MoSSe and WSSe, which have been experimentally synthesized [76–78]. We can speculate that the Janus HfSSe and ZrSSe could be potentially fabricated by selenizing HfS_2 (or $HfSe_2$) and ZrS_2 (or $ZrSe_2$) MLs, respectively, using the CVD method, which is similar to the method used for synthesizing MoSSe and WSSe MLs. Moreover, the photocatalytic properties of MXY-based HJs have also been explored [79–86]. However, the photocatalytic properties of HJs composed of GeC and MXY MLs have not been reported yet. Therefore, we expect to combine GeC MLs and MXY MLs to construct highly efficient direct Z-scheme HJPCs.

Theoretical calculation is a simple and effective way to screen and design potential direct Z-scheme WSPCs [87–91]. From density functional theory (DFT) calculations, one can determine whether a type-II HJ exhibits a direct Z-scheme or type-II photocatalytic mechanisms based on the carrier migration path, judged according to the built-in electric field (BIEF) direction [28,92]. If the BIEF promotes interlayer carrier recombination, the carrier transfer belongs to a Z-scheme mechanism. Otherwise, the type-II mechanism dominates. Herein, first-principles calculations are performed to explore the possibility of constructing GeC/MXY HJs using GeC and MXY MLs as direct Z-scheme systems. Work functions (Φ) and charge density differences (CDDs) indicate that the BIEFs promote all eight GeC/MXY HJs to form the Z-scheme photocatalytic mechanism for overall water-splitting. Furthermore, all these HJs possess strong visible light-absorption capacity and substantial near-infrared light-absorption capacity. Moreover, these HJs can provide sufficient driving forces to overcome the HER and OER overpotentials to perform overall water redox reactions. These results are expected to guide experiment progress in exploring 2D direct Z-scheme WSPCs.

2. Results and Discussion

Before constructing GeC/MXY HJs, the geometric and electronic properties of GeC and MXY MLs are first investigated. The corresponding structural models for GeC, ZrS$_2$, ZrSe$_2$, ZrSSe, HfS$_2$, HfSe$_2$, and HfSSe MLs are plotted in Figure S1. The obtained E_g values for GeC, ZrS$_2$, ZrSe$_2$, ZrSSe, HfS$_2$, HfSe$_2$, and HfSSe MLs are, respectively, 2.87, 2.02, 1.19, 1.46, 2.13, 1.33, and 1.56 eV, and the corresponding lattice parameters are, respectively, 3.235, 3.685, 3.800, 3.743, 3.645, 3.768, and 3.705 Å (see Table 1). Furthermore, the bond lengths of Ge–C in GeC, Zr–S in ZrS$_2$, Zr–Se in ZrSe$_2$, Zr–S (or Zr–Se) in ZrSSe, Hf–S in HfS$_2$, Hf–Se in HfSe$_2$, and Hf–S (or Hf–Se) in HfSSe are 1.868, 2.574, 2.706, 2.568 (or 2.713), 2.552, 2.685, and 2.550 (or 2.687) Å, respectively (see Table 1). GeC possesses a direct bandgap with both the VBM and CBM located at the K point, while all the MXY MLs are indirect bandgap semiconductors with the VBM and CBM, respectively, located at the Γ and M points (see Figure S2). All these results agree well with previous reports [54–58,64–67,93], as displayed in Table 1, indicating that our calculations are reliable.

Table 1. Lattice constants (a), bond lengths (L_B), bandgaps (E_g), dipole moments (μ), and EPDs (ΔE) between two sides for GeC and MXY MLs.

Systems	a (Å)	a (Å) (Refs.)	L_B (Å)	L_B (Å) (Refs.)	E_g (eV)	E_g (eV) (Refs.)	μ (D)	ΔE (eV)
GeC	3.235	3.26 [54,55], 3.233 [56], 3.263 [57]	1.868	1.882 [54], 1.887 [56]	2.87	3.01 [54], 2.90 [55], 2.88 [56], 2.782 [57], 2.85 [58]	0	0
ZrS$_2$	3.685	3.70 [64,65], 3.69 [67], 3.669 [93]	2.574	2.58 [67], 2.570 [93]	2.02	1.99 [67], 1.96 [93]	0	0
ZrSe$_2$	3.800	3.82 [64,65], 3.75 [67], 3.786 [93]	2.706	2.69 [67], 2.702 [93]	1.19	1.07 [67], 1.14 [93]	0	0
ZrSSe	3.743	3.73 [67]	2.568 (2.713)	2.55 (2.72) [67]	1.46	1.37 [67]	0.043	0.135
HfS$_2$	3.645	3.66 [64,65], 3.65 [66], 3.65 [67], 3.628 [93]	2.552	2.55 [66], 2.56 [67], 2.548 [93]	2.13	2.03 [67], 2.07 [93]	0	0
HfSe$_2$	3.768	3.82 [64], 3.78 [65,66], 3.72 [67], 3.751 [93]	2.685	2.69 [66], 2.68 [67], 2.681 [93]	1.33	1.16 [67], 1.26 [93]	0	0
HfSSe	3.705	3.71 [66], 3.68 [67]	2.550 (2.687)	2.55 (2.69) [66], 2.54 (2.69) [67]	1.56	1.45 [67]	0.035	0.110

Although the lattice constants of GeC and MXY are obviously different, the 2 × 2 GeC supercell could match well with the $\sqrt{3} \times \sqrt{3}$ MXY supercell. Herein, we define the lattice mismatch as $[2 \times |L_{sGeC} - L_{sMXY}|/(L_{sGeC} + L_{sMXY})] \times 100\%$, where L_{sGeC} and L_{sMXY} are the lattice constants for the GeC and MXY supercells, respectively. The calculated lattice mismatches between GeC and MXY to construct various GeC/MXY HJs are 1.38%, 1.71%,

0.20%, 0.20%, 2.47%, 0.87%, 0.82%, and 0.82%, respectively. These small lattice mismatches are favorable for the direct growth of GeC/MXY HJs by CVD or physical epitaxy [94]. Considering that the ZrSSe (or HfSSe) ML possesses two different surfaces, we loaded $\sqrt{3} \times \sqrt{3}$ ZrS$_2$, ZrSe$_2$, ZrSSe, HfS$_2$, HfSe$_2$, and HfSSe MLs onto a 2 × 2 GeC ML to construct eight different GeC/MXY HJs, i.e., GeC/ZrS$_2$, GeC/ZrSe$_2$, GeC/SZrSe, GeC/SeZrS, GeC/HfS$_2$, GeC/HfSe$_2$, GeC/SHfSe, and GeC/SeHfS. The corresponding models of GeC/MXY HJs are shown in Figure 1. Note that, here, the average of the lattice parameters for GeC and MXY is used to build GeC/MXY HJs, and the lattice parameters for GeC/MXY HJs are illustrated in Table 2.

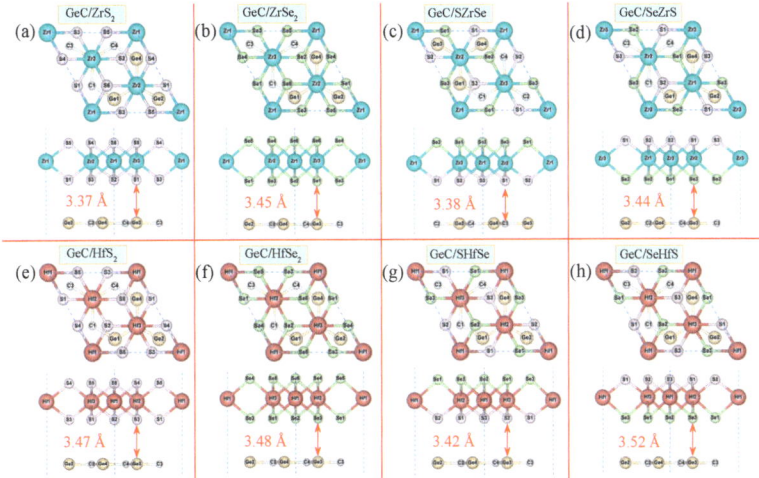

Figure 1. Top and side views for various GeC/MXY HJs.

Table 2. Lattice parameters (*a*), interlayer distances (d_i), interface formation energies (E_f), dipole moments (μ), EPDs (ΔE) between two surfaces, and charge transferred from GeC layer (ΔQ) in various GeC/MXY HJs.

Systems	a (Å)	d_i (Å)	E_f (meV/Å^2)	μ (D)	ΔE (eV)	ΔQ (e)
GeC/ZrS$_2$	6.426	3.367	−18.5	0.16	0.16	0.11
GeC/ZrSe$_2$	6.527	3.446	−28.1	0.15	0.15	0.09
GeC/SZrSe	6.477	3.376	−28.8	0.24	0.25	0.11
GeC/SeZrS	6.477	3.436	−29.4	0.04	0.05	0.09
GeC/HfS$_2$	6.392	3.468	−25.2	0.08	0.09	0.08
GeC/HfSe$_2$	6.499	3.484	−28.5	0.10	0.10	0.07
GeC/SZrSe	6.444	3.421	−27.8	0.19	0.20	0.08
GeC/SeZrS	6.444	3.519	−28.5	0.01	0.02	0.07

The thermodynamic stability of GeC/MXY HJs is assessed by calculating the interface formation energies (E_f) as follows:

$$E_f = (E^T_{GeC/MXY} - E^T_{GeC} - E^T_{MXY})/S, \quad (1)$$

where $E^T_{GeC/MXY}$, E^T_{GeC}, and E^T_{MXY}, respectively, denote the total energies of GeC/MXY HJs, GeC ML, and MXY ML. The calculated E_f values for all the considered HJs are negative, which means that the construction of all these GeC/MXY HJs release heat and tend to be thermodynamically stable. The E_f values in Table 2 range from −29.4 to −18.5 meV/Å^2, suggesting that these HJs are formed via the interaction between vdW and the MLs [95]. Moreover, the interlayer distances of GeC/MXY HJs vary from 3.367 to 3.519 Å (see Table 2), aligning with the results of some other typical vdW structures [96–100]. Consequently, all the examined GeC/MXY structures are classified as vdW HJs. The interfacial formation energy can be directly defined as: $E_f = E^T_{GeC/MXY} - E^T_{GeC} - E^T_{MXY}$. In this case, the E_f values for GeC/ZrS$_2$, GeC/ZrSe$_2$, GeC/SZrSe, GeC/SeZrS, GeC/HfS$_2$, GeC/HfSe$_2$, GeC/SHfSe, and GeC/SeHfS HJs are −0.66, −1.04, −1.05, −1.07, −0.89, −1.04, −1.00, and −1.03 eV, respectively. The E_f value of the GeC/SZrSe (or GeC/SHfSe) HJ is sightly more negative than that of the GeC/SeZrS (or GeC/SeHfS) HJ, indicating that the formation of the GeC/SZrSe (or GeC/SHfSe) HJ is energetically slightly more favorable. During experimental preparation, both GeC/SZrSe (or GeC/SHfSe) and GeC/SeZrS (or GeC/SeHfS) HJs are likely to be prepared. The difference in their preparation lies in the fact that the Janus ZrSSe (or HfSSe) ML contacts the GeC ML with different surfaces.

The band structures for various GeC/MXY HJs, computed using the HSE06 hybrid functional, are illustrated in Figure 2. The orange color denotes the contribution from the GeC layer, while the green color represents the contribution from the MXY layers. All the GeC/MXY HJs are indirect bandgap semiconductors, as their VBMs are located at the K point, while the CBMs are positioned at the M point. The corresponding bandgaps for GeC/MXY HJs are 0.45 (0.446), 0.45 (0.453), 0.55, 0.43, 0.53, 0.59, 0.66, and 0.54 eV, respectively (see Table 3), which are significantly lower than those of the corresponding MLs. Consequently, these HJs are expected to achieve high solar energy utilization. The CBMs originate from the MXY layer, whereas the VBMs come from the GeC layer, confirming the staggered type-II nature of all the examined GeC/MXY HJs. These facilitate the spatial separation of the photoinduced carriers. Furthermore, the band alignments of the GeC and MXY layers in the HJs retain the primary characteristic of their isolated MLs, suggesting that the vdW interaction at the heterointerface does not significantly influence the electronic properties of the layers.

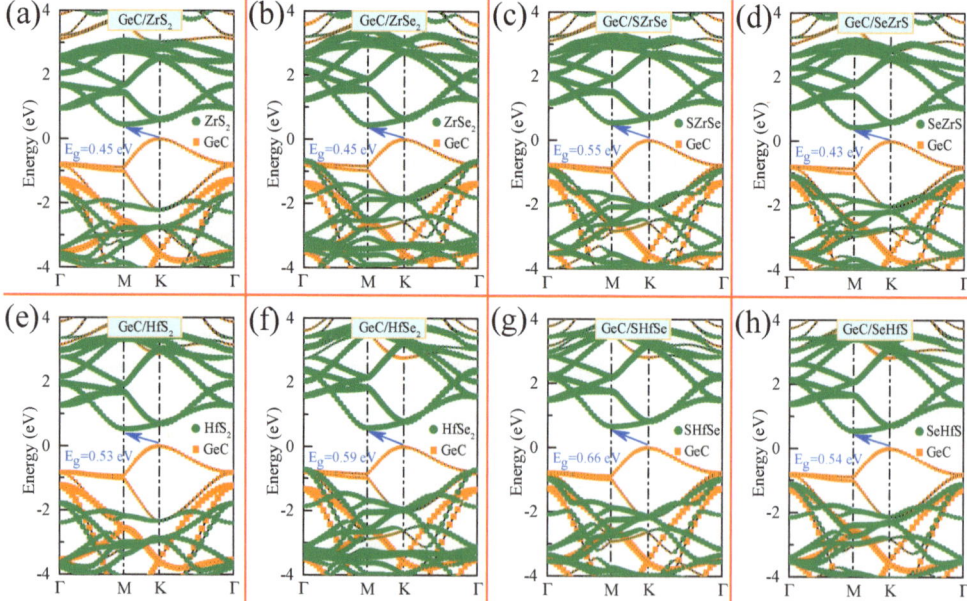

Figure 2. Band structures for various GeC/MXY HJs. The orange (or green) dots denote the contribution from the GeC (or MXY) layer.

Table 3. The bandgap (E_g), CBO, VBO, U_e, and U_h values for GeC/MXY HJs.

Systems	E_g (eV)	CBO (eV)	VBO (eV)	U_e (V)	U_h (V)
GeC/ZrS$_2$	0.45	2.59	1.70	2.14	2.75
GeC/ZrSe$_2$	0.45	2.30	0.68	1.85	1.73
GeC/SZrSe	0.55	2.19	0.92	1.84	2.08
GeC/SeZrS	0.43	2.32	1.07	1.87	1.99
GeC/HfS$_2$	0.53	2.34	1.79	2.01	2.74
GeC/HfSe$_2$	0.59	2.18	0.71	1.88	1.69
GeC/SHfSe	0.66	2.15	0.99	1.94	2.06
GeC/SeHfS	0.54	2.28	1.12	1.97	1.99

The work function (Φ) and CDD are crucial in determining the BIEF direction at the heterointerface, a factor that holds a decisive role in the design of Z-scheme PCs [101]. The Φ values can be obtained as follows:

$$\Phi = E_{\text{vac}} - E_F, \tag{2}$$

where E_{vac} and E_F refer to the vacuum and Fermi energy levels, respectively. The electrostatic potentials (EPs) of the relative MLs and HJs are depicted in Figures 3 and S3. The vacuum levels of the two surfaces in GeC, ZrS$_2$, ZrSe$_2$, HfS$_2$, and HfSe$_2$ are identical, meaning that the electrostatic potential differences (EPDs) (ΔE) between the two sides are all zero. The corresponding Φ values for these MLs are 4.80, 6.55, 5.54, 6.46, and 5.68 eV, respectively. The difference in the electronegativity between the S and Se atoms at the two opposing sides of the Janus ZrSSe and HfSSe results in inherent BIEFs perpendicular to the plane, causing the vacuum levels on both surfaces to differ. Consequently, the work functions are naturally distinct on the surfaces of both ZrSSe and HfSSe. The corresponding ΔE values are 0.13 and 0.11 eV for ZrSSe and HfSSe, respectively. The Φ values for the S-side (Se-side) for ZrSSe and HfSSe are 6.07 (5.93) and 5.95 (5.84) eV, respectively. Evidently, GeC exhibits a lower Φ value compared to the MXY MLs. Once GeC and MXY come into contact to form a GeC/MXY HJ, electrons will transfer from the material with a lower work function to the one with a higher work function until dynamic equilibrium is achieved. Consequently, a BIEF is established across the GeC/MXY heterointerface, pointing from GeC towards MXY. Additionally, the vacuum levels on both sides of the GeC/MXY HJs also differ. The calculated values ΔE for the various GeC/MXY HJs are 0.16, 0.15, 0.25, 0.05, 0.09, 0.10, 0.20, and 0.02 eV, respectively. Furthermore, the work functions for the respective GeC/MXY HJs are 5.12 (5.29), 5.13 (5.28), 5.08 (5.34), 5.12 (5.17), 5.06 (5.14), 5.05 (5.15), 5.00 (5.20), and 5.04 (5.06) eV. This indicates that, when GeC and MXY contact to form a GeC/MXY HJ, electrons migrate from GeC to MXY to reach the same Fermi level.

Moreover, we analyzed the charge transfer at the heterointerface region in GeC/MXY HJs by calculating the visual charge density difference (VCDD) based on the following relationship [102,103]:

$$\Delta \rho = \rho_{\text{GeC/MXY}} - \rho_{\text{GeC}} - \rho_{\text{MXY}}, \tag{3}$$

where $\rho_{\text{GeC/MXY}}$, ρ_{GeC}, and ρ_{MXY} represent the charge densities for the GeC/MXY HJ, GeC ML, and MXY ML, respectively. The yellow (or cyan) region denotes charge accumulation (or consumption). Additionally, the planar-averaged CDD (PACDD) along the z-direction is obtained by the following equation [103,104]:

$$\Delta \rho(z) = \int \rho_{\text{GeC/MXY}} dxdy - \int \rho_{\text{GeC}} dxdy - \int \rho_{\text{MXY}} dxdy, \tag{4}$$

where $\int \rho_{\text{GeC/MXY}} dxdy$, $\int \rho_{\text{GeC}} dxdy$, and $\int \rho_{\text{MXY}} dxdy$ represent the planar-averaged charge densities of the GeC/MXY HJ, GeC ML, and MXY ML, respectively. The positive (or negative) value indicates the charge accumulation (or consumption). It can be clearly seen from Figure 4 that the charge around the heterointerfaces of all the GeC/MXY HJs is redistributed. Charge accumulation primarily occurs at the heterointerface region near

MXY, while charge consumption mainly takes place at the heterointerface region near GeC. This further confirms that electrons migrate from GeC to MXY in all the GeC/MXY HJs. The Bader charge analysis also suggests that 0.11, 0.09, 0.11, 0.09, 0.08. 0.07, 0.08, and 0.07 e, respectively, migrate from GeC to MXY in the various GeC/MXY HJs. The charge transfer at the heterointerfaces of GeC/MXY HJs could cause the BIEF to point away from GeC toward MXY, which commonly promotes the spatial separation of carriers, thus extending the lifetime of photoexcited carriers and enhancing the photocatalytic activity.

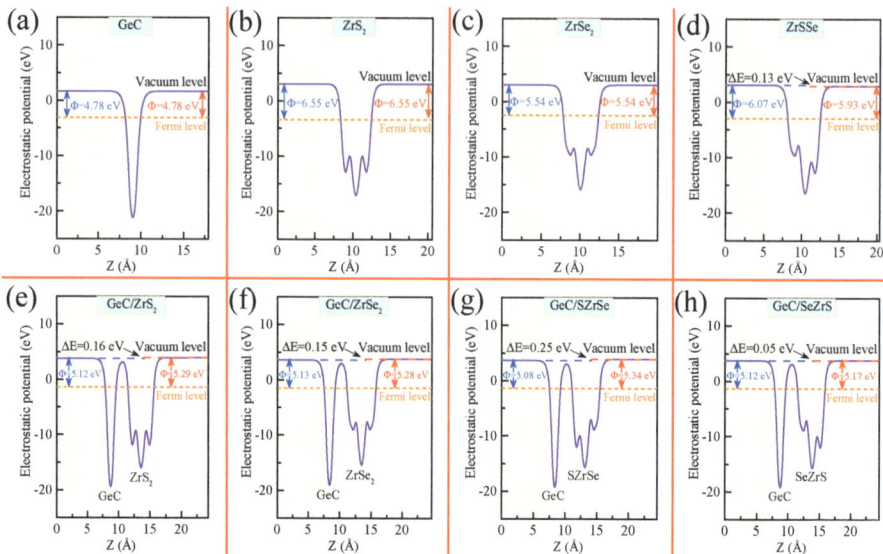

Figure 3. Electrostatic potential diagrams of (**a**) GeC, (**b**) ZrS$_2$, (**c**) ZrSe$_2$, (**d**) ZrSSe, (**e**) GeC/ZrS$_2$, (**f**) GeC/ZrSe$_2$, (**g**) GeC/SZrSe, and (**h**) GeC/SeZrS, respectively.

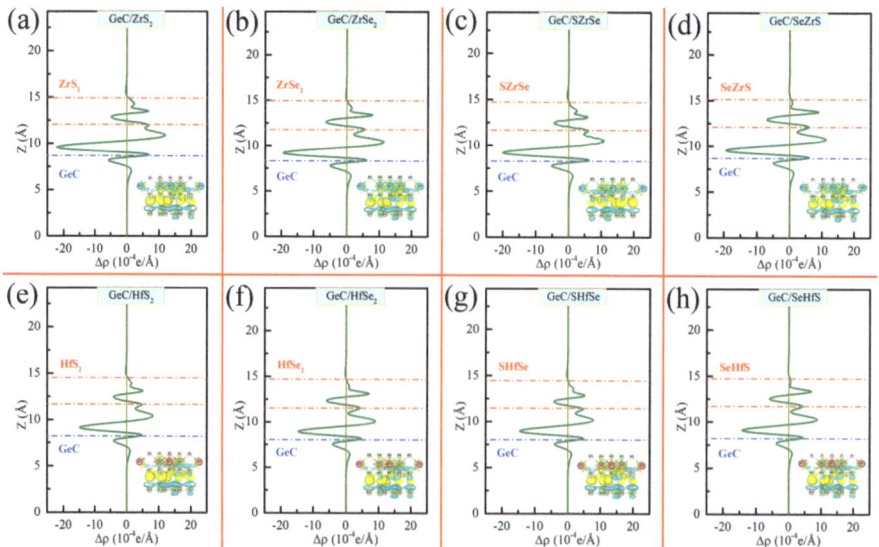

Figure 4. VCDDs and PACDDs for various GeC/MXY HJs.

As is well known, the type-II band alignment corresponds to both the type-II and direct Z-scheme photocatalytic mechanisms based on different charge transfer pathways. For a type-II HJPC, the band edges of its two components must simultaneously straddle the water redox potentials. Thus, a type-II HJPC usually does not provide sufficient driving force for water redox processes. For a direct Z-scheme HJPC, the VBM of one component should be lower than the water oxidation potential (WOP), while the CBM of the other component should be higher than the water reduction potential (WRP) [105,106]. Thus, a direct Z-scheme HJPC is usually capable of proving sufficient driving force for redox reactions. Next, we arrange the band edges of the GeC ML, MXY MLs, and GeC/MXY HJs in contrast to the water redox levels in Figures 5 and S4, in order to further determine the photocatalytic mechanisms of the considered GeC/MXY HJs. It is known that the water redox levels are determined by the electrochemical potentials relative to the vacuum level, so the difference in the vacuum energy levels on the two surfaces of PCs causes the movement of H^+/H_2 and H_2O/O_2 levels between the two surfaces. The band edges of GeC only span the WOP, indicating that GeC is only suitable for the HER. Conversely, the band edges of MXY MLs solely cross the WRP, meaning that MXY MLs only serve for the OER. Thus, neither GeC nor MXY alone could achieve overall PCWD.

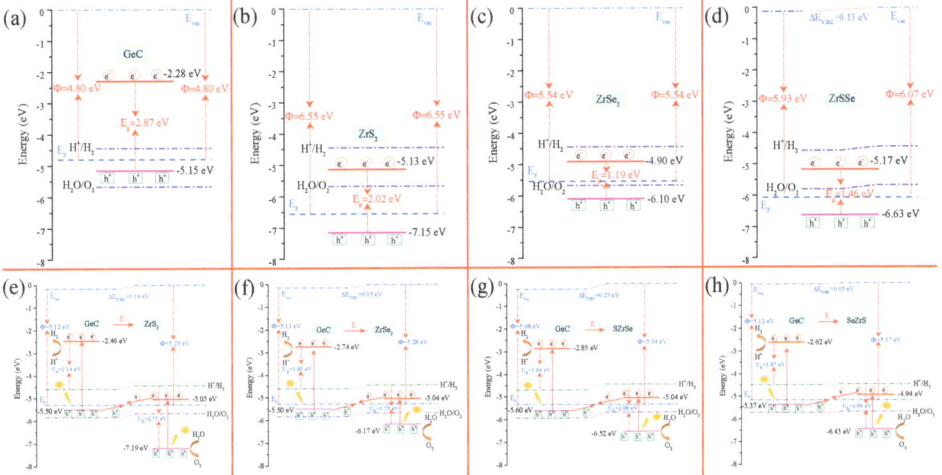

Figure 5. Band alignments for (**a**) GeC, (**b**) ZrS_2, (**c**) $ZrSe_2$, (**d**) ZrSSe, (**e**) GeC/ZrS_2, (**f**) GeC/$ZrSe_2$, (**g**) GeC/SZrSe, and (**h**) GeC/SeZrS versus vacuum level.

Since the photocatalytic mechanisms for all GeC/MXY HJs are similar, we will use GeC/ZrS_2 as an illustrative example for a detailed discussion. As the GeC ML and ZrS_2 ML approach each other to form the GeC/ZrS_2 HJ, electrons migrate from GeC to ZrS_2 due to the smaller work function of GeC compared to ZrS_2. Consequently, the GeC and ZrS_2 layers become positively and negatively charged, respectively. This results in a BIEF that is directed away from GeC toward ZrS_2 across the GeC/ZrS_2 heterointerface. Electrons in GeC are repelled by the negatively charged ZrS_2, causing GeC's bands to bend upward. Similarly, ZrS_2's bands will bend downward near the heterointerface due to the same mechanism [107,108]. To simplify the discussion, we have omitted the band bending in the band-alignment diagram of GeC/MXY HJs. When exposed to sunlight, both GeC and ZrS_2 can absorb photons with greater energy than their respective bandgaps. This stimulates electrons to transition from the valence bands (VBs) to the conduction bands (CBs), leaving holes in the VBs. However, GeC is unsuitable for the OER due to its higher VBM than the WOP, while ZrS_2 is unsuitable for the HER owing to its lower CBM than the WRP. This implies that the photoinduced holes in the VBs of GeC (or the photoexcited holes in the CBs of ZrS$_2$) cannot directly participate in the OER (or HER) process. The calculated conduction band offset (CBO) and valence band offset (VBO) are 2.59 and 1.70 eV, respectively (see Table 3). Due to the BIEF directed from GeC to ZrS$_2$, the migration of photoinduced electrons from the CBs of GeC to the CBs of ZrS$_2$ and the migration of photoinduced holes from the VBs of ZrS$_2$ to the VBs of GeC are hindered. Conversely, the photoexcited electrons are encouraged to migrate from the CBs of ZrS$_2$ to the VBs of GeC, where they recombine with the photoexcited holes. Furthermore, the interlayer bandgap of 0.45 eV is significantly smaller than both the CBO and the VBO, favoring the interlayer electron–hole recombination. Consequently, GeC (or ZrS$_2$) accumulates more photoinduced electrons (or holes). Naturally, the superfluous electrons on the CBs for GeC can achieve the HER, while the excess holes on the VBs of ZrS$_2$ can realize the OER. Evidently, the migration path of photoexcited carriers is like a "Z". Thus, the GeC/ZrS$_2$ HJ constitutes a direct Z-scheme system. The spatial separation of photoexcited electrons and holes contributes to enhancing photocatalytic efficiency. Additionally, schematic diagrams illustrating the photocatalytic mechanisms of all considered MLs and HJs versus the normal hydrogen electrode (NHE) are presented in Figures S5 and S6.

Furthermore, the sufficient kinetic potentials (U_e and U_h) provided by photoexcited electrons and holes are crucial for driving the OERs and HERs. The U_e and U_h values affect the number of active electrons and holes participating in water redox reactions, thereby influencing the photocatalytic activity. Here, U_e (or U_h) is defined as the potential difference between the CBM and the H$^+$/H$_2$ level (or between the H$^+$/H$_2$ level and the VBM). Given that the water redox levels depend on the pH values, U_e and U_h can be expressed as follows [109]:

$$U_e = U_e(pH = 0) - pH \times 0.059 \text{ V},$$
$$U_h = U_h(pH = 0) + pH \times 0.059 \text{ V}. \tag{5}$$

For the sake of simplicity, we will only discuss the U_e (or U_h) value at $pH = 0$. The calculated U_e and U_h values are 2.14 and 2.75 V, respectively, which are comparable to some previously studied Z-scheme PCs (see Figure 6) [24,40–42,89–91,109]. Consequently, GeC/ZrS$_2$ HJ emerges as a highly efficient Z-scheme WSPC.

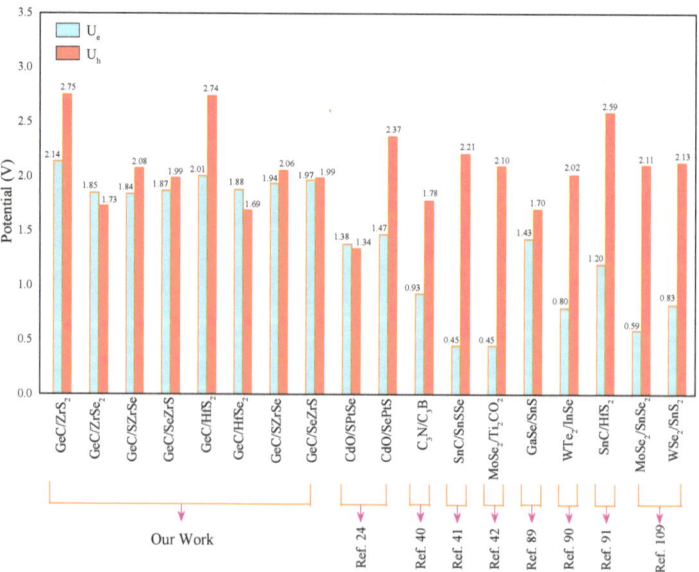

Figure 6. Comparison of U_e and U_h values of the proposed GeC/MXY HJs with some other reported HJs [24,40–42,89–91,109].

The CBO (VBO) values for GeC/ZrSe$_2$, GeC/SZrSe, GeC/SeZrS, GeC/HfS$_2$, GeC/HfSe$_2$, GeC/SHfSe, and GeC/SeHfS HJs are 2.30 (0.68), 2.19 (0.92), 2.32 (1.07), 2.34 (1.79), 2.18 (0.71), 2.15 (0.99), and 2.28 (1.12) eV, respectively (see Table 3). Obviously, the bandgaps of these HJs are smaller than their CBOs and VBOs, which is conducive to interlayer electron–hole recombination. Additionally, the BIEF direction is pointing from GeC to MXY. Similarly, the GeC/ZrS$_2$, GeC/ZrSe$_2$, GeC/SZrSe, GeC/SeZrS, GeC/HfS$_2$, GeC/HfSe$_2$, GeC/SHfSe, and GeC/SeHfS HJs are all Z-scheme WSPCs. Moreover, the obtained U_e (U_h) values are 1.85 (1.73), 1.84 (2.08), 1.87 (1.99), 2.01 (2.74), 1.88 (1.69), 1.94 (2.06), and 1.97 (1.99) V, respectively (see Table 3). These values are also close to those reported for Z-scheme WSPCs (see Figure 6) [24,40–42,89–91,109]. This indicates that the GeC/ZrSe$_2$, GeC/SZrSe, GeC/SeZrS, GeC/HfS$_2$, GeC/HfSe$_2$, GeC/SHfSe, and GeC/SeHfS HJs could supply sufficient dynamic potentials to drive HERs and OERs under light irradiation.

At the initial stage of photocatalytic water-splitting, the light absorption capacity serves as another crucial factor. For highly efficient solar utilization, a wide and intense light absorption spectrum is typically required. Therefore, we investigated the optical absorption curves of GeC, MXY, and GeC/MXY HJs using the HSE06 method. The optical absorption coefficient can be calculated using the following formula [110]:

$$\alpha(\omega) = \frac{\sqrt{2}\omega}{c}[\sqrt{\epsilon_1^2(\omega) + \epsilon_2^2(\omega)} - \epsilon_1(\omega)]^{1/2}, \quad (6)$$

where ϵ_1 (or ϵ_2) represents the real (or imaginary) part of the dielectric function and ω is the frequency of light. As shown in Figure 7, GeC/MXY HJs possess strong visible light absorption ability and non-negligible near-infrared light absorption ability. In addition, GeC/MXY HJs exhibit higher absorption coefficients in the visible and near-infrared light regions, along with a redshift of the absorption spectra, compared to the corresponding MLs. Herein, the GeC/MXY HJs demonstrate excellent light absorption capacity. Furthermore, the proper band alignments and suitable directions of the heterointerface BIEF enable these GeC/MXY HJs to form a Z-scheme photocatalytic mechanism. This facilitates the HERs and OERs to occur in different sublayers and provides sufficient driving force to spontaneously achieve water redox reactions under illumination. Generally speaking, GeC/XYs HJs are promising candidates for direct Z-scheme WSPCs.

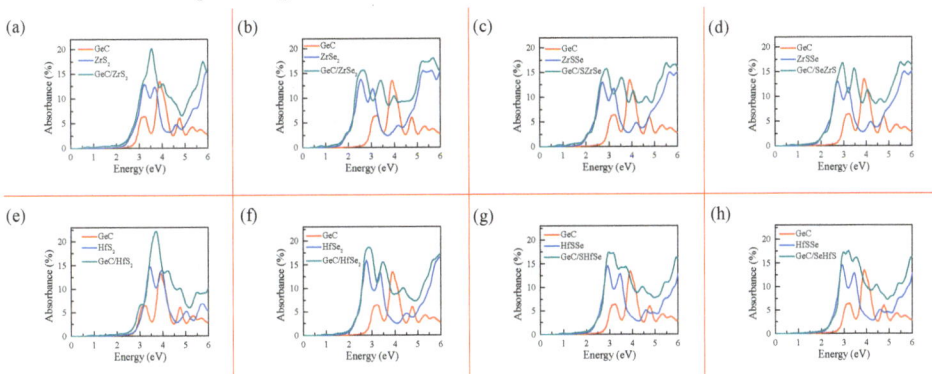

Figure 7. (**a**–**h**) Optical absorption curves of various GeC/MXY HJs compared with those of GeC and MXY MLs.

3. Computational Details

In this work, the GeC/MXY (M = Zr, Hf; X, Y = S, Se) HJs are constructed by stacking the $\sqrt{3} \times \sqrt{3}$ MXY supercell onto a 2×2 GeC supercell with an 18 Å vacuum layer to eliminate the image interaction between adjacent layers. Additionally, dipole correction is introduced along the z-direction [111]. All DFT calculations were carried out using *VASP5.4* [112,113], and the electron–ion interactions were described using the projector-enhanced wave (PAW) method [114]. The generalized gradient approximation (GGA) [115] of Perdew–Burke–Ernzerhof (PBE) [116] was employed for the exchange correlation functional. Furthermore, Grimme's DFT-D3 [117,118] method was employed to account for weak vdW interactions. The Monkhorst–Pack k-point grid for the first Brillouin zone was set to $13 \times 13 \times 1$ (or $7 \times 7 \times 1$) for MLs (or HJs). The energy cutoff was set to 500 eV, and all structures were sufficiently optimized with an energy (or force) tolerance of 10^{-5} eV (10^{-2} eV/Å). Given that GGA-PBE tends to underestimate the bandgaps [119], the Heyd–Scuseria–Ernzerhof functional (HSE06) was applied to accurately compute the electronic and optical properties [120]. The optical absorption spectra were computed based on the imaginary part of the dielectric functional, following the Kramers–Kronig dispersion relationship [110], and the band alignments of MLs and HJs were referenced to a common vacuum level.

4. Conclusions

In summary, the potential applications of GeC/MXY (M = Zr, Hf; X, Y = S, Se) HJs have been investigated through the calculation of their geometric, electronic, optical properties, band arrangement, and interface binding energies. Based on first-principles calculations, we analyzed their photocatalytic mechanism. All the considered GeC/MXY HJs, namely GeC/ZrS$_2$, GeC/ZrSe$_2$, GeC/SZrSe, GeC/SeZrS, GeC/HfS$_2$, GeC/HfSe$_2$, GeC/SHfSe, and GeC/SeHfS, were found to be direct Z-scheme photocatalytic systems with band edges spanning the water redox potentials. Charge redistribution at the heterointerface results in the formation of a BIEF pointing from GeC to MXY, enhancing the separation of the photoinduced carriers. Excitingly, the GeC/MXY HJs exhibit strong redox capacity for photocatalytic water decomposition, ensuring that the HER and OER processes occur spontaneously under light irradiation. Furthermore, the GeC/MXY HJs demonstrated strong visible light absorption and some near-infrared light absorption, guaranteeing efficient utilization of solar energy. These theoretical findings indicate that these GeC/MXY HJs are all promising WSPCs.

Supplementary Materials: The following supporting information can be downloaded at: https://www.mdpi.com/article/10.3390/molecules29122793/s1, Figure S1: Top and side view of optimized geometries of GeC, ZrS$_2$, ZrSe$_2$, ZrSSe, HfS$_2$, HfSe$_2$ and HfSSe, respectively; Figure S2: Band structures of GeC, ZrS$_2$, ZrSe$_2$, ZrSSe, HfS$_2$, HfSe$_2$ and HfSSe, respectively; Figure S3: Electrostatic potential diagrams of GeC, HfS$_2$, HfSe$_2$, HfSSe, GeC/HfS$_2$, GeC/HfSe$_2$, GeC/SHfSe and GeC/SeHfS, respectively; Figure S4: Schematic diagrams of the photocatalytic mechanisms for GeC, HfS$_2$, HfSe$_2$, HfSSe, GeC/HfS$_2$, GeC/HfSe$_2$, GeC/SHfSe and GeC/SeHfS versus vacuum level; Figure S5: Schematic diagrams of the photocatalytic mechanisms for GeC, ZrS$_2$, ZrSe$_2$, ZrSSe, GeC/ZrS$_2$, GeC/ZrSe$_2$, GeC/SZrSe and GeC/SeZrS versus NHE; Figure S6: Schematic diagrams of the photocatalytic mechanisms for GeC, HfS$_2$, HfSe$_2$, HfSSe, GeC/HfS$_2$, GeC/HfSe$_2$, GeC/SHfSe and GeC/SeHfS versus NHE; Table S1: POSCAR file for the optimized GeC/ZrS$_2$; Table S2: POSCAR file for the optimized GeC/ZrSe$_2$; Table S3: POSCAR file for the optimized GeC/SZrSe; Table S4: POSCAR file for the optimized GeC/SeZrS; Table S5: POSCAR file for the optimized GeC/HfS$_2$; Table S6: POSCAR file for the optimized GeC/HfSe$_2$; Table S7: POSCAR file for the optimized GeC/SHfSe; Table S8: POSCAR file for the optimized GeC/SeHfS.

Author Contributions: G.W., Y.S.A. and H.Y. designed the project, guided the study, and prepared the manuscript; W.X., J.C. and Y.C. carried out the calculations; S.G., X.L. and L.Z. analyzed the calculated results and produced the illustrations. All authors have read and agreed to the published version of the manuscript.

Funding: This work was supported by the National Natural Science Foundation of China under Grant No. 12304295, the GHfund B under grant No. ghfund202302023082, the Science and Technology Research Program of Chongqing Municipal Education Commission under grant No. KJQN202201405, the China Postdoctoral Science Foundation under grant No. 2022MD723798, and the Special Funding for Postdoctoral Research Projects by Chongqing Municipal Human Resources and Social Security Bureau under grant No. 2022CQBSHTB3002.

Data Availability Statement: The original contributions presented in the study are included in the article/Supplementary Material, further inquiries can be directed to the corresponding authors.

Conflicts of Interest: The authors declare no conflicts of interest.

References

1. Kudo, A.; Miseki, Y. Heterogeneous photocatalyst materials for water splitting. *Chem. Soc. Rev.* **2009**, *38*, 253–278. [CrossRef] [PubMed]
2. Zhang, X.; Zhang, Z.; Wu, D.; Zhang, X.; Zhao, X.; Zhou, Z. Computational screening of 2D materials and rational design of heterojunctions for water splitting photocatalysts. *Small Methods* **2018**, *2*, 1700359. [CrossRef]
3. Tachibana, Y.; Vayssieres, L.; Durrant, J.R. Artificial photosynthesis for solar water-splitting. *Nat. Photonics* **2012**, *6*, 511–518. [CrossRef]
4. Hisatomi, T.; Kubota, J.; Domen, K. Recent advances in semiconductors for photocatalytic and photoelectrochemical water splitting. *Chem. Soc. Rev.* **2014**, *43*, 7520–7535. [CrossRef] [PubMed]
5. Moniz, S.J.; Shevlin, S.A.; Martin, D.J.; Guo, Z.X.; Tang, J. Visible-light driven heterojunction photocatalysts for water splitting—A critical review. *Energy Environ. Sci.* **2015**, *8*, 731–759. [CrossRef]
6. Liu, J.; Liu, Y.; Liu, N.; Han, Y.; Zhang, X.; Huang, H.; Lifshitz, Y.; Lee, S.T.; Zhong, J.; Kang, Z. Metal-free efficient photocatalyst for stable visible water splitting via a two-electron pathway. *Science* **2015**, *347*, 970–974. [CrossRef]
7. Wang, W.; Xu, X.; Zhou, W.; Shao, Z. Recent progress in metal-organic frameworks for applications in electrocatalytic and photocatalytic water splitting. *Adv. Sci.* **2017**, *4*, 1600371. [CrossRef]
8. Tran, M.N.; Moreau, M.; Addad, A.; Teurtrie, A.; Roland, T.; De Waele, V.; Dewitte, M.; Thomas, L.; Levêque, G.; Dong, C.; et al. Boosting gas-phase TiO_2 photocatalysis with weak electric field strengths of volt/centimeter. *ACS Appl. Mater. Interfaces* **2024**, *16*, 14852–14863. [CrossRef]
9. Song, X.; Wei, G.; Sun, J.; Peng, C.; Yin, J.; Zhang, X.; Jiang, Y.; Fei, H. Overall photocatalytic water splitting by an organolead iodide crystalline material. *Nat. Catal.* **2020**, *3*, 1027–1033. [CrossRef]
10. Mao, J.; Ta, Q.T.H.; Tri, N.N.; Shou, L.; Seo, S.; Xu, W. 2D $MoTe_2$ nanomesh with a large surface area and uniform pores for highly active hydrogen evolution catalysis. *Appl. Mater. Today* **2023**, *35*, 101939. [CrossRef]
11. Fujishima, A.; Honda, K. Electrochemical photolysis of water at a semiconductor electrode. *Nature* **1972**, *238*, 37–38. [CrossRef] [PubMed]
12. Fu, C.F.; Wu, X.; Yang, J. Material design for photocatalytic water splitting from a theoretical perspective. *Adv. Mater.* **2018**, *30*, 1802106. [CrossRef] [PubMed]
13. Wang, G.Z.; Chang, J.L.; Tang, W.; Xie, W.; Ang, Y.S. 2D materials and heterostructures for photocatalytic water-splitting: A theoretical perspective. *J. Phys. D Appl. Phys.* **2022**, *55*, 293002. [CrossRef]
14. Ran, J.; Zhang, J.; Yu, J.; Jaroniec, M.; Qiao, S.Z. Earth-abundant cocatalysts for semiconductor-based photocatalytic water splitting. *Chem. Soc. Rev.* **2014**, *43*, 7787–7812. [CrossRef]
15. Faraji, M.; Yousefi, M.; Yousefzadeh, S.; Zirak, M.; Naseri, N.; Jeon, T.H.; Choi, W.; Moshfegh, A.Z. Two-dimensional materials in semiconductor photoelectrocatalytic systems for water splitting. *Energy Environ. Sci.* **2019**, *12*, 59–95. [CrossRef]
16. Fu, J.; Yu, J.; Jiang, C.; Cheng, B. $g-C_3N_4$-Based heterostructured photocatalysts. *Adv. Energy Mater.* **2018**, *8*, 1701503. [CrossRef]
17. Wang, G.; Tang, W.; Xie, W.; Tang, Q.; Wang, Y.; Guo, H.; Gao, P.; Dang, S.; Chang, J. Type-II CdS/PtSSe heterostructures used as highly efficient water-splitting photocatalysts. *Appl. Surf. Sci.* **2022**, *589*, 152931. [CrossRef]
18. Walter, M.G.; Warren, E.L.; McKone, J.R.; Boettcher, S.W.; Mi, Q.; Santori, E.A.; Lewis, N.S. Solar water splitting cells. *Chem. Rev.* **2010**, *110*, 6446–6473. [CrossRef] [PubMed]
19. Jiang, C.; Moniz, S.J.; Wang, A.; Zhang, T.; Tang, J. Photoelectrochemical devices for solar water splitting–materials and challenges. *Chem. Soc. Rev.* **2017**, *46*, 4645–4660. [CrossRef]
20. Zhou, P.; Yu, J.; Jaroniec, M. All-solid-state Z-scheme photocatalytic systems. *Adv. Mater.* **2014**, *26*, 4920–4935. [CrossRef]
21. Li, H.; Tu, W.; Zhou, Y.; Zou, Z. Z-Scheme photocatalytic systems for promoting photocatalytic performance: Recent progress and future challenges. *Adv. Sci.* **2016**, *3*, 1500389. [CrossRef] [PubMed]
22. Zhang, R.; Zhang, L.; Zheng, Q.; Gao, P.; Zhao, J.; Yang, J. Direct Z-scheme water splitting photocatalyst based on two-dimensional Van Der Waals heterostructures. *J. Phys. Chem. Lett.* **2018**, *9*, 5419–5424. [CrossRef] [PubMed]
23. Wang, Q.; Hisatomi, T.; Jia, Q.; Tokudome, H.; Zhong, M.; Wang, C.; Pan, Z.; Takata, T.; Nakabayashi, M.; Shibata, N.; et al. Scalable water splitting on particulate photocatalyst sheets with a solar-to-hydrogen energy conversion efficiency exceeding 1%. *Nat. Mater.* **2016**, *15*, 611–615. [CrossRef] [PubMed]
24. Wang, G.; Tang, W.; Xu, C.; He, J.; Zeng, Q.; Xie, W.; Gao, P.; Chang, J. Two-dimensional CdO/PtSSe heterojunctions used for Z-scheme photocatalytic water-splitting. *Appl. Surf. Sci.* **2022**, *599*, 153960. [CrossRef]
25. Zuo, G.; Wang, Y.; Teo, W.L.; Xian, Q.; Zhao, Y. Direct Z-scheme TiO_2–$ZnIn_2S_4$ nanoflowers for cocatalyst-free photocatalytic water splitting. *Appl. Catal. B Environ.* **2021**, *291*, 120126. [CrossRef]
26. Wang, L.; Zheng, X.; Chen, L.; Xiong, Y.; Xu, H. Van der Waals heterostructures comprised of ultrathin polymer nanosheets for efficient Z-scheme overall water splitting. *Angew. Chem. Int. Ed.* **2018**, *130*, 3512–3516. [CrossRef]

27. Zeng, C.; Hu, Y.; Zhang, T.; Dong, F.; Zhang, Y.; Huang, H. A core–satellite structured Z-scheme catalyst $Cd_{0.5}Zn_{0.5}S/BiVO_4$ for highly efficient and stable photocatalytic water splitting. *J. Mater. Chem. A* **2018**, *6*, 16932–16942. [CrossRef]
28. She, X.; Wu, J.; Xu, H.; Zhong, J.; Wang, Y.; Song, Y.; Nie, K.; Liu, Y.; Yang, Y.; Rodrigues, M.T.F.; et al. High efficiency photocatalytic water splitting using 2D α-Fe_2O_3/g-C_3N_4 Z-scheme catalysts. *Adv. Energy Mater.* **2017**, *7*, 1700025. [CrossRef]
29. Zhu, M.; Sun, Z.; Fujitsuka, M.; Majima, T. Z-scheme photocatalytic water splitting on a 2D heterostructure of black phosphorus/bismuth vanadate using visible light. *Angew. Chem. Int. Ed.* **2018**, *57*, 2160–2164. [CrossRef]
30. Yuan, Y.J.; Chen, D.; Yang, S.; Yang, L.X.; Wang, J.J.; Cao, D.; Tu, W.; Yu, Z.T.; Zou, Z.G. Constructing noble-metal-free Z-scheme photocatalytic overall water splitting systems using MoS_2 nanosheet modified CdS as a H_2 evolution photocatalyst. *J. Mater. Chem. A* **2017**, *5*, 21205–21213. [CrossRef]
31. Li, L.; Guo, C.; Shen, J.; Ning, J.; Zhong, Y.; Hu, Y. Construction of sugar-gourd-shaped $CdS/Co_{1-x}S$ hollow hetero-nanostructure as an efficient Z-scheme photocatalyst for hydrogen generation. *Chem. Eng. J.* **2020**, *400*, 125925. [CrossRef]
32. Wei, T.; Zhu, Y.N.; An, X.; Liu, L.M.; Cao, X.; Liu, H.; Qu, J. Defect modulation of Z-scheme TiO_2/Cu_2O photocatalysts for durable water splitting. *ACS Catal.* **2019**, *9*, 8346–8354. [CrossRef]
33. Zhao, D.; Wang, Y.; Dong, C.L.; Huang, Y.C.; Chen, J.; Xue, F.; Shen, S.; Guo, L. Boron-doped nitrogen-deficient carbon nitride-based Z-scheme heterostructures for photocatalytic overall water splitting. *Nat. Energy* **2021**, *6*, 388–397. [CrossRef]
34. Li, Z.; Hou, J.; Zhang, B.; Cao, S.; Wu, Y.; Gao, Z.; Nie, X.; Sun, L. Two-dimensional Janus heterostructures for superior Z-scheme photocatalytic water splitting. *Nano Energy* **2019**, *59*, 537–544. [CrossRef]
35. Hong, X.; Kim, J.; Shi, S.F.; Zhang, Y.; Jin, C.; Sun, Y.; Tongay, S.; Wu, J.; Zhang, Y.; Wang, F. Ultrafast charge transfer in atomically thin MoS_2/WS_2 heterostructures. *Nat. Nanotechnol.* **2014**, *9*, 682–686. [CrossRef] [PubMed]
36. Rivera, P.; Schaibley, J.R.; Jones, A.M.; Ross, J.S.; Wu, S.; Aivazian, G.; Klement, P.; Seyler, K.; Clark, G.; Ghimire, N.J.; et al. Observation of long-lived interlayer excitons in monolayer $MoSe_2$–WSe_2 heterostructures. *Nat. Commun.* **2015**, *6*, 6242. [CrossRef] [PubMed]
37. Chiu, M.H.; Zhang, C.; Shiu, H.W.; Chuu, C.P.; Chen, C.H.; Chang, C.Y.S.; Chen, C.H.; Chou, M.Y.; Shih, C.K.; Li, L.J. Determination of band alignment in the single-layer MoS_2/WSe_2 heterojunction. *Nat. Commun.* **2015**, *6*, 7666. [CrossRef] [PubMed]
38. Long, R.; Prezhdo, O.V. Quantum coherence facilitates efficient charge separation at a $MoS_2/MoSe_2$ van der Waals junction. *Nano Lett.* **2016**, *16*, 1996–2003. [CrossRef] [PubMed]
39. Zhou, Z.; Niu, X.; Zhang, Y.; Wang, J. Janus MoSSe/WSeTe heterostructures: A direct Z-scheme photocatalyst for hydrogen evolution. *J. Mater. Chem. A* **2019**, *7*, 21835–21842. [CrossRef]
40. Niu, X.; Bai, X.; Zhou, Z.; Wang, J. Rational design and characterization of direct Z-scheme photocatalyst for overall water splitting from excited state dynamics simulations. *ACS Catal.* **2020**, *10*, 1976–1983. [CrossRef]
41. Jiang, X.; Gao, Q.; Xu, X.; Xu, G.; Li, D.; Cui, B.; Liu, D.; Qu, F. Design of a noble-metal-free direct Z-scheme photocatalyst for overall water splitting based on a SnC/SnSSe van der Waals heterostructure. *Phys. Chem. Chem. Phys.* **2021**, *23*, 21641–21651. [CrossRef] [PubMed]
42. Fu, C.F.; Li, X.; Yang, J. A rationally designed two-dimensional $MoSe_2/Ti_2CO_2$ heterojunction for photocatalytic overall water splitting: Simultaneously suppressing electron–hole recombination and photocorrosion. *Chem. Sci.* **2021**, *12*, 2863–2869. [CrossRef] [PubMed]
43. Ji, Y.; Dong, H.; Hou, T.; Li, Y. Monolayer graphitic germanium carbide (g-GeC): The promising cathode catalyst for fuel cell and lithium–oxygen battery applications. *J. Mater. Chem. A* **2018**, *6*, 2212–2218. [CrossRef]
44. Şahin, H.; Cahangirov, S.; Topsakal, M.; Bekaroglu, E.; Akturk, E.; Senger, R.T.; Ciraci, S. Monolayer honeycomb structures of group-IV elements and III-V binary compounds: First-principles calculations. *Phys. Rev. B* **2009**, *80*, 155453. [CrossRef]
45. Pan, L.; Liu, H.; Wen, Y.; Tan, X.; Lv, H.; Shi, J.; Tang, X. First-principles study of monolayer and bilayer honeycomb structures of group-IV elements and their binary compounds. *Phys. Lett. A* **2011**, *375*, 614–619. [CrossRef]
46. Hao, A.; Yang, X.; Wang, X.; Zhu, Y.; Liu, X.; Liu, R. First-principles investigations on electronic, elastic and optical properties of XC (X = Si, Ge, and Sn) under high pressure. *J. Appl. Phys.* **2010**, *108*, 063531. [CrossRef]
47. Peng, Q.; Liang, C.; Ji, W.; De, S. A first-principles study of the mechanical properties of g-GeC. *Mech. Mater.* **2013**, *64*, 135–141. [CrossRef]
48. Ren, K.; Sun, M.; Luo, Y.; Wang, S.; Xu, Y.; Yu, J.; Tang, W. Electronic and optical properties of van der Waals vertical heterostructures based on two-dimensional transition metal dichalcogenides: First-principles calculations. *Phys. Lett. A* **2019**, *383*, 1487–1492. [CrossRef]
49. Wu, X.; Zhang, W.; Yan, L.; Luo, R. The deposition and optical properties of $Ge_{1-x}C_x$ thin film and infrared multilayer antireflection coatings. *Thin Solid Film.* **2008**, *516*, 3189–3195. [CrossRef]
50. Yuan, H.; Williams, R.S. Synthesis by laser ablation and characterization of pure germanium-carbon alloy thin films. *Chem. Mater.* **1993**, *5*, 479–485. [CrossRef]
51. Xu, M.; Liang, T.; Shi, M.; Chen, H. Graphene-like two-dimensional materials. *Chem. Rev.* **2013**, *113*, 3766–3798. [CrossRef] [PubMed]
52. Li, Z.; Lv, Y.; Ren, L.; Li, J.; Kong, L.; Zeng, Y.; Tao, Q.; Wu, R.; Ma, H.; Zhao, B.; et al. Efficient strain modulation of 2D materials via polymer encapsulation. *Nat. Commun.* **2020**, *11*, 1151. [CrossRef] [PubMed]
53. Li, J.; Chen, M.; Zhang, C.; Dong, H.; Lin, W.; Zhuang, P.; Wen, Y.; Tian, B.; Cai, W.; Zhang, X. Fractal-theory-based control of the shape and quality of CVD-grown 2D materials. *Adv. Mater.* **2019**, *31*, 1902431. [CrossRef] [PubMed]

54. Din, H.; Idrees, M.; Albar, A.; Shafiq, M.; Ahmad, I.; Nguyen, C.V.; Amin, B. Rashba spin splitting and photocatalytic properties of GeC-MSSe (M= Mo, W) van der Waals heterostructures. *Phys. Rev. B* **2019**, *100*, 165425. [CrossRef]
55. Jiang, X.; Xie, W.; Xu, X.; Gao, Q.; Li, D.; Cui, B.; Liu, D.; Qu, F. A bifunctional GeC/SnSSe heterostructure for highly efficient photocatalysts and photovoltaic devices. *Nanoscale* **2022**, *14*, 7292–7302. [CrossRef] [PubMed]
56. Wang, G.; Zhang, L.; Li, Y.; Zhao, W.; Kuang, A.; Li, Y.; Xia, L.; Li, Y.; Xiao, S. Biaxial strain tunable photocatalytic properties of 2D ZnO/GeC heterostructure. *J. Phys. D Appl. Phys.* **2020**, *53*, 015104. [CrossRef]
57. Gao, X.; Shen, Y.; Ma, Y.; Wu, S.; Zhou, Z. ZnO/g-GeC van der Waals heterostructure: Novel photocatalyst for small molecule splitting. *J. Matr. Chem. C* **2019**, *7*, 4791–4799. [CrossRef]
58. Cao, M.; Luan, L.; Wang, Z.; Zhang, Y.; Yang, Y.; Liu, J.; Tian, Y.; Wei, X.; Fan, J.; Xie, Y.; et al. Type-II GeC/ZnTe heterostructure with high-efficiency of photoelectrochemical water splitting. *Appl. Phys. Lett.* **2021**, *119*, 083101. [CrossRef]
59. Huong, P.T.; Idrees, M.; Amin, B.; Hieu, N.N.; Phuc, H.V.; Hoa, L.T.; Nguyen, C.V. Electronic structure, optoelectronic properties and enhanced photocatalytic response of GaN–GeC van der Waals heterostructures: A first principles study. *RSC Adv.* **2020**, *10*, 24127–24133. [CrossRef]
60. Yang, Z.; Wang, J.; Hu, G.; Yuan, X.; Ren, J.; Zhao, X. Strain-tunable Zeeman splitting and optical properties of $CrBr_3$/GeC van der Waals heterostructure. *Results Phys.* **2022**, *37*, 105559. [CrossRef]
61. Lou, P.; Lee, J.Y. GeC/GaN vdW heterojunctions: A promising photocatalyst for overall water splitting and solar energy conversion. *ACS Appl. Mater. Interf.* **2020**, *12*, 14289–14297. [CrossRef] [PubMed]
62. Liu, Y.L.; Shi, Y.; Yang, C.L. Two-dimensional MoSSe/g-GeC van der waals heterostructure as promising multifunctional system for solar energy conversion. *Appl. Surf. Sci.* **2021**, *545*, 148952. [CrossRef]
63. Gao, X.; Shen, Y.; Liu, J.; Lv, L.; Zhou, M.; Zhou, Z.; Feng, Y.P.; Shen, L. Boost the large driving photovoltages for overall water splitting in direct Z-scheme heterojunctions by interfacial polarization. *Catal. Sci. Technol.* **2022**, *12*, 3614–3621. [CrossRef]
64. Abdulsalam, M.; Joubert, D.P. Optical spectrum and excitons in bulk and monolayer MX_2 (M = Zr, Hf; X = S, Se). *Phys. Status Solidi B* **2016**, *253*, 705–711. [CrossRef]
65. Abdulsalam, M.; Rugut, E.; Joubert, D. Mechanical, thermal and thermoelectric properties of MX_2 (M = Zr, Hf; X = S, Se). *Mater. Today Commun.* **2020**, *25*, 101434. [CrossRef]
66. Bera, J.; Betal, A.; Sahu, S. Spin orbit coupling induced enhancement of thermoelectric performance of HfX_2 (X = S, Se) and its Janus monolayer. *J. Alloys Compd.* **2021**, *872*, 159704. [CrossRef]
67. Dimple; Jena, N.; Rawat, A.; Ahammed, R.; Mohanta, M.K.; De Sarkar, A. Emergence of high piezoelectricity along with robust electron mobility in Janus structures in semiconducting Group IVB dichalcogenide monolayers. *J. Mater. Chem. A* **2018**, *6*, 24885–24898. [CrossRef]
68. Shi, W.; Wang, Z. Mechanical and electronic properties of Janus monolayer transition metal dichalcogenides. *J. Phys. Condens. Mat.* **2018**, *30*, 215301. [CrossRef] [PubMed]
69. Som, N.N.; Jha, P.K. Hydrogen evolution reaction of metal di-chalcogenides: ZrS_2, $ZrSe_2$ and Janus ZrSSe. *Int. J. Hydrogen Energy* **2020**, *45*, 23920–23927. [CrossRef]
70. Hoat, D.; Naseri, M.; Hieu, N.N.; Ponce-Pérez, R.; Rivas-Silva, J.; Vu, T.V.; Cocoletzi, G.H. A comprehensive investigation on electronic structure, optical and thermoelectric properties of the HfSSe Janus monolayer. *J. Phys. Chem. Solids* **2020**, *144*, 109490. [CrossRef]
71. Kaur, H.; Yadav, S.; Srivastava, A.K.; Singh, N.; Rath, S.; Schneider, J.J.; Sinha, O.P.; Srivastava, R. High-yield synthesis and liquid-exfoliation of two-dimensional belt-like hafnium disulphide. *Nano Res.* **2018**, *11*, 343–353. [CrossRef]
72. Yue, R.; Barton, A.T.; Zhu, H.; Azcatl, A.; Pena, L.F.; Wang, J.; Peng, X.; Lu, N.; Cheng, L.; Addou, R.; et al. $HfSe_2$ thin films: 2D transition metal dichalcogenides grown by molecular beam epitaxy. *ACS Nano* **2015**, *9*, 474–480. [CrossRef] [PubMed]
73. Zhang, M.; Zhu, Y.; Wang, X.; Feng, Q.; Qiao, S.; Wen, W.; Chen, Y.; Cui, M.; Zhang, J.; Cai, C.; et al. Controlled synthesis of ZrS_2 monolayer and few layers on hexagonal boron nitride. *J. Am. Chem. Soc.* **2015**, *137*, 7051–7054. [CrossRef] [PubMed]
74. Tsipas, P.; Tsoutsou, D.; Marquez-Velasco, J.; Aretouli, K.; Xenogiannopoulou, E.; Vassalou, E.; Kordas, G.; Dimoulas, A. Epitaxial $ZrSe_2$/$MoSe_2$ semiconductor vd Waals heterostructures on wide band gap AlN substrates. *Microelectron. Eng.* **2015**, *147*, 269–272. [CrossRef]
75. Zeng, Z.; Yin, Z.; Huang, X.; Li, H.; He, Q.; Lu, G.; Boey, F.; Zhang, H. Single-layer semiconducting nanosheets: High-yield preparation and device fabrication. *Ang. Chem. Int. Ed.* **2011**, *123*, 11289–11293. [CrossRef]
76. Zhang, J.; Jia, S.; Kholmanov, I.; Dong, L.; Er, D.; Chen, W.; Guo, H.; Jin, Z.; Shenoy, V.B.; Shi, L.; et al. Janus monolayer transition-metal dichalcogenides. *ACS Nano* **2017**, *11*, 8192–8198. [CrossRef] [PubMed]
77. Lu, A.Y.; Zhu, H.; Xiao, J.; Chuu, C.P.; Han, Y.; Chiu, M.H.; Cheng, C.C.; Yang, C.W.; Wei, K.H.; Yang, Y.; et al. Janus monolayers of transition metal dichalcogenides. *Nat. Nanotechnol.* **2017**, *12*, 744–749. [CrossRef]
78. Lin, Y.C.; Liu, C.; Yu, Y.; Zarkadoula, E.; Yoon, M.; Puretzky, A.A.; Liang, L.; Kong, X.; Gu, Y.; Strasser, A.; et al. Low energy implantation into transition-metal dichalcogenide monolayers to form Janus structures. *ACS Nano* **2020**, *14*, 3896–3906. [CrossRef] [PubMed]
79. Singh, A.; Jain, M.; Bhattacharya, S. MoS_2 and Janus (MoSSe) based 2D van der Waals heterostructures: Emerging direct Z-scheme photocatalysts. *Nanoscale Adv.* **2021**, *3*, 2837–2845. [CrossRef]
80. Zhu, X.T.; Xu, Y.; Cao, Y.; Zou, D.F.; Sheng, W. Direct Z-scheme arsenene/HfS_2 van der Waals heterojunction for overall photocatalytic water splitting: First-principles study. *Appl. Surf. Sci.* **2022**, *574*, 151650. [CrossRef]

81. Bai, Y.; Zhang, H.; Wu, X.; Xu, N.; Zhang, Q. Two-dimensional arsenene/ZrS_2 (HfS_2) deterostructures as direct Z-Scheme photocatalysts for overall water splitting. *J. Phys. Chem. C* **2022**, *126*, 2587–2595. [CrossRef]
82. Sun, R.; Yang, C.L.; Wang, M.S.; Ma, X.G. High solar-to-hydrogen efficiency photocatalytic hydrogen evolution reaction with the $HfSe_2$/InSe heterostructure. *J. Power Sources* **2022**, *547*, 232008. [CrossRef]
83. Zhang, X.; Meng, Z.; Rao, D.; Wang, Y.; Shi, Q.; Liu, Y.; Wu, H.; Deng, K.; Liu, H.; Lu, R. Efficient band structure tuning, charge separation, and visible-light response in ZrS_2-based van der Waals heterostructures. *Energy Environ. Sci.* **2016**, *9*, 841–849. [CrossRef]
84. Cao, J.; Zhang, X.; Zhao, S.; Wang, S.; Cui, J. Mechanism of photocatalytic water splitting of 2D WSeTe/XS_2 (X = Hf, Sn, Zr) van der Waals heterojunctions under the interaction of vertical intrinsic electric and built-in electric field. *Appl. Surf. Sci.* **2022**, *599*, 154012. [CrossRef]
85. Opoku, F.; Akoto, O.; Oppong, S.O.B.; Adimado, A.A. Two-dimensional layered type-II MS_2/BiOCl (M = Zr, Hf) van der Waals heterostructures: Promising photocatalysts for hydrogen generation. *New J. Chem.* **2021**, *45*, 20365–20373. [CrossRef]
86. Ahmad, S.; Idrees, M.; Khan, F.; Nguyen, C.; Ahmad, I.; Amin, B. Strain engineering of Janus ZrSSe and HfSSe monolayers and ZrSSe/HfSSe van der Waals heterostructure. *Chem. Phys. Lett.* **2021**, *776*, 138689. [CrossRef]
87. Fu, C.F.; Luo, Q.; Li, X.; Yang, J. Two-dimensional van der Waals nanocomposites as Z-scheme type photocatalysts for hydrogen production from overall water splitting. *J. Mater. Chem. A* **2016**, *4*, 18892–18898. [CrossRef]
88. Gao, Y.; Fu, C.; Hu, W.; Yang, J. Designing direct Z-scheme heterojunctions enabled by edge-modified phosphorene nanoribbons for photocatalytic overall water splitting. *J. Phys. Chem. Lett.* **2021**, *13*, 1–11. [CrossRef] [PubMed]
89. Meng, J.; Wang, J.; Wang, J.; Li, Q.; Yang, J. β-SnS/GaSe heterostructure: A promising solar-driven photocatalyst with low carrier recombination for overall water splitting. *J. Mater. Chem. A* **2022**, *10*, 3443–3453. [CrossRef]
90. Xiong, R.; Shu, Y.; Yang, X.; Zhang, Y.; Wen, C.; Anpo, M.; Wu, B.; Sa, B. Direct Z-scheme WTe_2/InSe van der Waals heterostructure for overall water splitting. *Catal. Sci. Technol.* **2022**, *12*, 3272–3280. [CrossRef]
91. Dai, Z.N.; Cao, Y.; Yin, W.J.; Sheng, W.; Xu, Y. Z-scheme SnC/HfS_2 van der Waals heterojunction increases photocatalytic overall water splitting. *J. Phys. D Appl. Phys.* **2022**, *55*, 315503. [CrossRef]
92. Xu, Q.; Zhang, L.; Yu, J.; Wageh, S.; Al-Ghamdi, A.A.; Jaroniec, M. Direct Z-scheme photocatalysts: Principles, synthesis, and applications. *Mater. Today* **2018**, *21*, 1042–1063. [CrossRef]
93. Wang, G.; Chang, J.; Guo, S.D.; Wu, W.; Tang, W.; Guo, H.; Dang, S.; Wang, R.; Ang, Y.S. MoSSe/Hf(Zr)S_2 heterostructures used for efficient Z-scheme photocatalytic water-splitting. *Phys. Chem. Chem. Phys.* **2022**, *24*, 25287–25297. [CrossRef] [PubMed]
94. Novoselov, K.; Mishchenko, A.; Carvalho, A.; Castro Neto, A. 2D materials and van der Waals heterostructures. *Science* **2016**, *353*, aac9439. [CrossRef]
95. Björkman, T.; Gulans, A.; Krasheninnikov, A.V.; Nieminen, R.M. van der Waals bonding in layered compounds from advanced density-functional first-principles calculations. *Phys. Rev. Lett.* **2012**, *108*, 235502. [CrossRef]
96. Guo, H.; Zhang, Z.; Huang, B.; Wang, X.; Niu, H.; Guo, Y.; Li, B.; Zheng, R.; Wu, H. Theoretical study on the photocatalytic properties of 2D InX (X = S, Se)/transition metal disulfide (MoS_2 and WS_2) van der Waals heterostructures. *Nanoscale* **2020**, *12*, 20025–20032. [CrossRef] [PubMed]
97. Xu, L.; Huang, W.Q.; Hu, W.; Yang, K.; Zhou, B.X.; Pan, A.; Huang, G.F. Two-dimensional MoS_2-graphene-based multilayer van der Waals heterostructures: Enhanced charge transfer and optical absorption, and electric-field tunable Dirac point and band gap. *Chem. Mater.* **2017**, *29*, 5504–5512. [CrossRef]
98. Bafekry, A.; Obeid, M.; Nguyen, C.V.; Ghergherehchi, M.; Tagani, M.B. Graphene hetero-multilayer on layered platinum mineral jacutingaite (Pt_2HgSe_3): Van der Waals heterostructures with novel optoelectronic and thermoelectric performances. *J. Mater. Chem. A* **2020**, *8*, 13248–13260. [CrossRef]
99. Zhang, C.F.; Yang, C.L.; Wang, M.S.; Ma, X.G. Z-Scheme photocatalytic solar-energy-to-hydrogen conversion driven by the HfS_2/SiSe heterostructure. *J. Mater. Chem. C* **2022**, *10*, 5474–5481. [CrossRef]
100. Wang, G.; Li, Z.; Wu, W.; Guo, H.; Chen, C.; Yuan, H.; Yang, S.A. A two-dimensional h-BN/C_2N heterostructure as a promising metal-free photocatalyst for overall water-splitting. *Phys. Chem. Chem. Phys.* **2020**, *22*, 24446–24454. [CrossRef]
101. Zhou, F.; Yuan, S.; Wang, J. Theoretical progress on direct Z-scheme photocatalysis of two-dimensional heterostructures. *Front. Phys.* **2021**, *16*, 43203. [CrossRef]
102. He, C.; Zhang, J.; Zhang, W.; Li, T. Type-II InSe/g-C_3N_4 heterostructure as a high-efficiency oxygen evolution reaction catalyst for photoelectrochemical water splitting. *J. Phys. Chem. Lett.* **2019**, *10*, 3122–3128. [CrossRef] [PubMed]
103. Xu, L.; Huang, W.Q.; Wang, L.L.; Tian, Z.A.; Hu, W.; Ma, Y.; Wang, X.; Pan, A.; Huang, G.F. Insights into enhanced visible-light photocatalytic hydrogen evolution of g-C_3N_4 and highly reduced graphene oxide composite: The role of oxygen. *Chem. Mater.* **2015**, *27*, 1612–1621. [CrossRef]
104. Liu, J.; Cheng, B.; Yu, J. A new understanding of the photocatalytic mechanism of the direct Z-scheme g-C_3N_4/TiO_2 heterostructure. *Phys. Chem. Chem. Phys.* **2016**, *18*, 31175–31183. [CrossRef] [PubMed]
105. Bai, S.; Jiang, J.; Zhang, Q.; Xiong, Y. Steering charge kinetics in photocatalysis: Intersection of materials syntheses, characterization techniques and theoretical simulations. *Chem. Phys. Rev.* **2015**, *44*, 2893–2939.
106. Fu, C.F.; Wu, X.; Yang, J. Theoretical design of two-dimensional visible light-driven photocatalysts for overall water splitting. *Chem. Phys. Rev.* **2022**, *3*, 011310. [CrossRef]

107. Huang, Z.F.; Song, J.; Wang, X.; Pan, L.; Li, K.; Zhang, X.; Wang, L.; Zou, J.J. Switching charge transfer of $C_3N_4/W_{18}O_{49}$ from type-II to Z-scheme by interfacial band bending for highly efficient photocatalytic hydrogen evolution. *Nano Energy* **2017**, *40*, 308–316. [CrossRef]
108. Zhang, Z.; Yates, J.T., Jr. Band bending in semiconductors: Chemical and physical consequences at surfaces and interfaces. *Chem. Rev.* **2012**, *112*, 5520–5551. [CrossRef] [PubMed]
109. Fan, Y.; Wang, J.; Zhao, M. Spontaneous full photocatalytic water splitting on 2D $MoSe_2/SnSe_2$ and $WSe_2/SnSe_2$ vdW heterostructures. *Nanoscale* **2019**, *11*, 14836–14843. [CrossRef]
110. Tian, F.; Liu, C. DFT description on electronic structure and optical absorption properties of anionic S-doped anatase TiO_2. *J. Phys. Chem. B* **2006**, *110*, 17866–17871. [CrossRef]
111. Bengtsson, L. Dipole correction for surface supercell calculations. *Phys. Rev. B* **1999**, *59*, 12301. [CrossRef]
112. Kresse, G.; Furthmüller, J. Efficient iterative schemes for ab initio total-energy calculations using a plane-wave basis set. *Phys. Rev. B* **1996**, *54*, 11169. [CrossRef] [PubMed]
113. Kresse, G.; Furthmüller, J. Efficiency of ab-initio total energy calculations for metals and semiconductors using a plane-wave basis set. *Comput. Mater. Sci.* **1996**, *6*, 15–50. [CrossRef]
114. Blöchl, P.E. Projector augmented-wave method. *Phys. Rev. B* **1994**, *50*, 17953. [CrossRef] [PubMed]
115. Singh, D.; Ashkenazi, J. Magnetism with generalized-gradient-approximation density functionals. *Phys. Rev. B* **1992**, *46*, 11570. [CrossRef] [PubMed]
116. Ernzerhof, M.; Scuseria, G.E. Assessment of the Perdew–Burke–Ernzerhof exchange-correlation functional. *J. Chem. Phys.* **1999**, *110*, 5029–5036. [CrossRef]
117. Grimme, S.; Antony, J.; Ehrlich, S.; Krieg, H. A consistent and accurate ab initio parametrization of density functional dispersion correction (DFT-D) for the 94 elements H-Pu. *J. Chem. Phys.* **2010**, *132*, 154104. [CrossRef] [PubMed]
118. Moellmann, J.; Grimme, S. DFT-D3 study of some molecular crystals. *J. Phys. Chem. C* **2014**, *118*, 7615–7621. [CrossRef]
119. Kümmel, S.; Kronik, L. Orbital-dependent density functionals: Theory and applications. *Rev. Mod. Phys.* **2008**, *80*, 3. [CrossRef]
120. Heyd, J.; Scuseria, G.E.; Ernzerhof, M. Hybrid functionals based on a screened Coulomb potential. *J. Chem. Phys.* **2003**, *118*, 8207–8215. [CrossRef]

Disclaimer/Publisher's Note: The statements, opinions and data contained in all publications are solely those of the individual author(s) and contributor(s) and not of MDPI and/or the editor(s). MDPI and/or the editor(s) disclaim responsibility for any injury to people or property resulting from any ideas, methods, instructions or products referred to in the content.

Gold Nanoparticle Mesoporous Carbon Composite as Catalyst for Hydrogen Evolution Reaction

Erik Biehler, Qui Quach and Tarek M. Abdel-Fattah *

Applied Research Center at Thomas Jefferson National Accelerator Facility, Department of Molecular Biology and Chemistry at Christopher Newport University, Newport News, VA 23606, USA; erik.biehler@cnu.edu (E.B.)
* Correspondence: fattah@cnu.edu

Abstract: Increased environmental pollution and the shortage of the current fossil fuel energy supply has increased the demand for eco-friendly energy sources. Hydrogen energy has become a potential solution due to its availability and green combustion byproduct. Hydrogen feedstock materials like sodium borohydride ($NaBH_4$) are promising sources of hydrogen; however, the rate at which the hydrogen is released during its reaction with water is slow and requires a stable catalyst. In this study, gold nanoparticles were deposited onto mesoporous carbon to form a nano-composite catalyst (AuNP-MCM), which was then characterized via transmission electron microscopy (TEM), powder X-ray diffraction (P-XRD), and scanning electron microscopy/energy dispersive X-ray spectroscopy (SEM/EDS). The composite's catalytic ability in a hydrogen evolution reaction was tested under varying conditions, including $NaBH_4$ concentration, pH, and temperature, and it showed an activation of energy of 30.0 kJ mol^{-1}. It was determined that the optimal reaction conditions include high NaBH4 concentrations, lower pH, and higher temperatures. This catalyst, with its stability and competitively low activation energy, makes it a promising material for hydrogen generation.

Keywords: nanocomposite; hydrogen evolution; mesoporous carbon; gold nanoparticles; mesoporous carbon; sustainable source

1. Introduction

Hydrogen fuel is an alternative type of fuel that has potential in solving the world's energy crisis [1]. The energy generation of gasoline combustion is lower than the energy released from hydrogen combustion [2]. When hydrogen is used as a fuel, the major byproduct, in terms of its energy reaction, is water, so its environmental impact is minimal [2]. Furthermore, hydrogen can be generated from the reaction between hydrogen feedstock, such as sodium borohydrides ($NaBH_4$), and water [3], as seen in Equation (1). The widespread implementation of hydrogen as a fuel would reduce dependence on fossil fuels; however, the biggest disadvantage of this reaction is the slow rate of reaction between metal hydrides and water. As such, a catalyst is necessary before hydrogen gas can become a viable fuel source [3].

$$NaBH_4 + 2H_2O \rightarrow 4H_2 + NaBO_2 \quad (1)$$

Among the catalysts, nanoparticles have been the focus of much scientific study due to their unique properties, including their catalytic properties [4,5]; however, it is difficult to control nanoparticle performance in reactions as this depends greatly on characteristics such as shape, size, crystal structure, and texture [6–8]. The agglomeration of nanoparticles often affects their size and structure and leads to the degeneration of catalytic ability [9,10]. In order to mitigate this issue, the nanoparticles can be imbedded on a carbon template to prevent their agglomeration and improve the durability of the material in catalytic reactions [5,11].

One family of durable carbon materials that also has catalytic potential is that of mesoporous carbon materials (MCMs) [12,13]. The different types of porous carbons are characterized by their differences in pore size, with mesoporous carbon materials having a pore size 2–50 nm, while microporous and macroporous carbon have smaller and larger pore sizes, respectively [14,15]. The narrow pore size range allows for significant control of various characteristics during synthesis, which includes thermal, mechanical, and electrical stability, chemical inertness, ordered pore structure, large surface area, pore volume, and catalytic activity [16,17]. Most traditional metal and biological catalysts have several disadvantages, such as high reaction temperature, long reaction time, low conversion ratio, and low regioselectivity, so research is being conducted with MCMs to reduce these disadvantages [18]. For example, factors like cost, complexity of synthesis, and potential wasted materials can be reduced by using MCMs as opposed to traditional catalysts as MCMs are only made of ordered carbon atoms that are synthesized via well-established procedures [19].

One such method of synthesizing MCMs involves filling mesoporous silica with a carbon precursor like sucrose, which undergoes several high-temperature processes. The silica is removed with hydrofluoric acid, which leaves only carbon in the form of MCMs [20]. Unfortunately, this process involves aggressive chemicals that, along with intermediary products, become dangerous waste products. One novel method of forming MCMs has recently been developed, which involves using starches and expanding the present pores to increase the catalytic potential [21]. Starches are typically high-density substances, which makes them catalytically inert; but by expanding the pores, this activity was increased [21]. The pore expansion temperature is the key to controlling pore size and volume, with a similar process used on plant fibers resulting in similar results [21,22]. The benefits of this new method include the starting material being completely nontoxic and renewable, the lack of toxic materials needed for the reaction, and the ease of synthesis [21,23,24].

For this study, MCMs were decorated with gold nanoparticles to form a nanocomposite catalyst (AuNP-MCM) that is highly active, stable, and easily recyclable. The synthesized AuNP-MCM was characterized via TEM, P-XRD, and SEM-EDS. The catalytic ability of nanocomposite was tested in the hydrogen evolution reaction under various pHs (6, 7, 8), reactant concentrations (793 µmol, 952 µmol, 1057 µmol), and temperatures (273 K, 288 K, 295 K, 303 K). The recyclability of AuNP-MCM was also examined in reusability trials.

2. Results/Discussions

2.1. Catalyst Characterization

The gold nanoparticles for the AuNP-MCM seen in the TEM micrograph of Figure 1 vary greatly in diameter, but the particle specifically observed in Figure 1d under a smaller scale has a diameter of 20 nm. The SEM/EDS analysis in Figure 2a,b shows the gold nanoparticles on the mesoporous carbon with the gold concentration of 8.54% and carbon concentration of 10.73%. Figures 1b and 2a also show that the gold nanoparticles are evenly distributed on the MCM backbone.

The P-XRD spectrum seen in Figure 3 for the MCM showed a peak in the 20–30° range, correlating to the carbon framework, and a peak at the 40–45° range that also corresponds to the mesoporous carbon [25]. After AuNPs were deposited on the MCM, the observed peaks of the MCM were slightly shifted, but they were still within the acceptable range. The peaks at 40–45° disappeared in AuNP-MCM; the same phenomena were reported in some previous studies [26,27]. The peaks at 38°, 44°, and 64° corresponded with the (111), (200), and (220) lattice planes of gold nanoparticles [28].

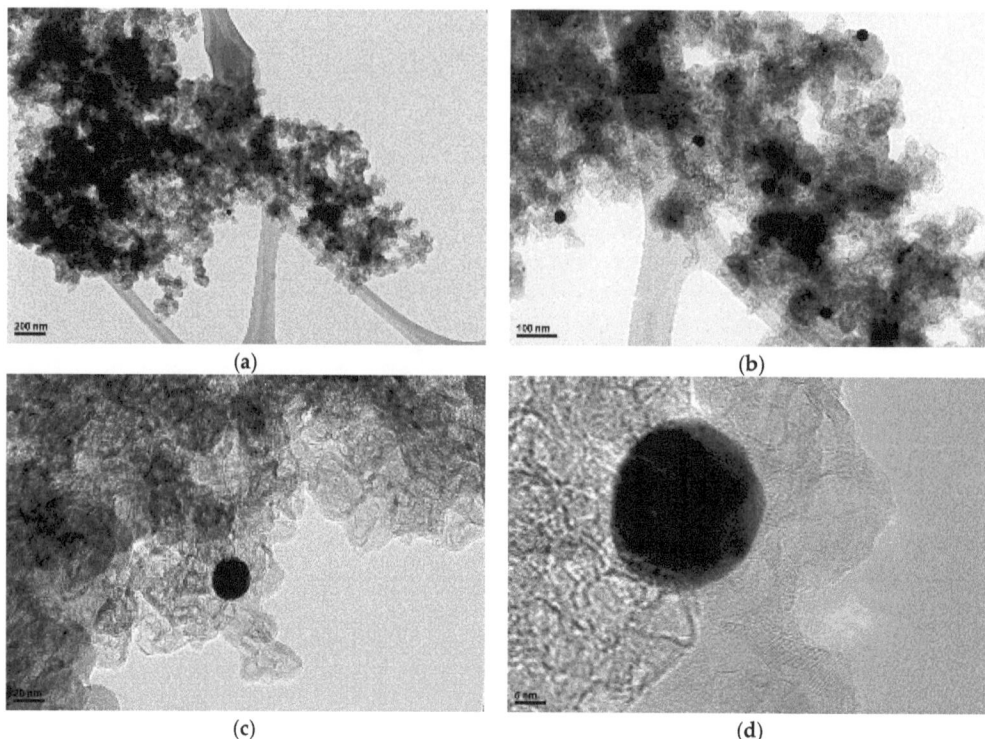

Figure 1. TEM images of the AuNP-MCM catalyst at scales of (**a**) 200 nm, (**b**) 100 nm, (c) 20 nm, and (**d**) 5 nm.

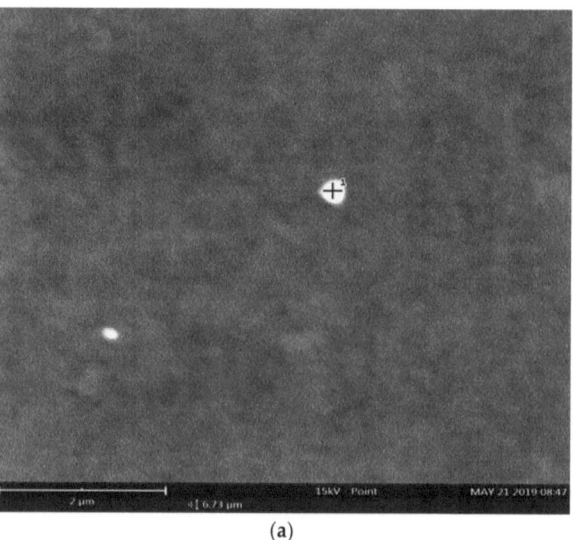

Figure 2. *Cont.*

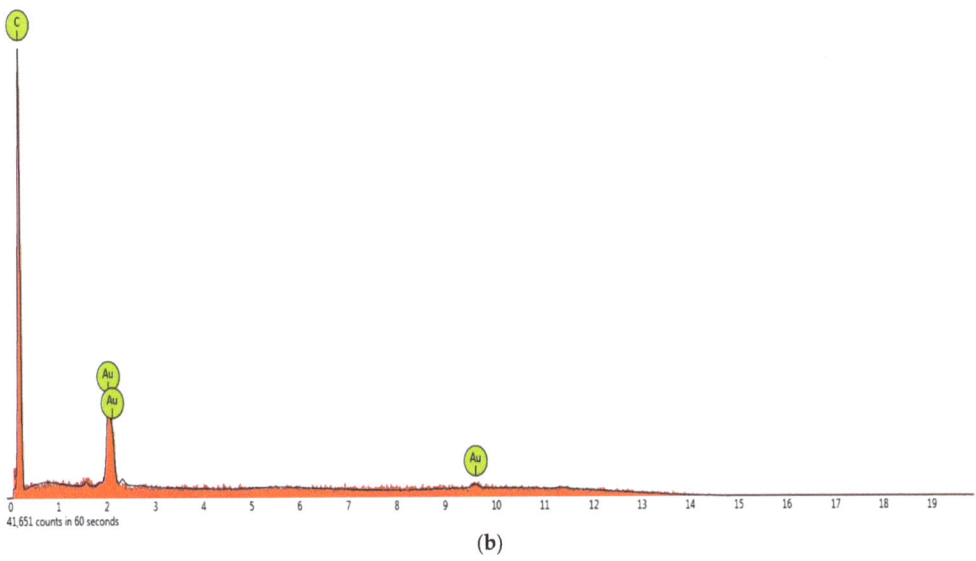

Figure 2. SEM/EDX analysis with (**a**) the SEM micrograph of the AuNP-MCM catalyst at scales of 2 μm and (**b**) the EDS spectrum correlating to the indicated gold nanoparticle.

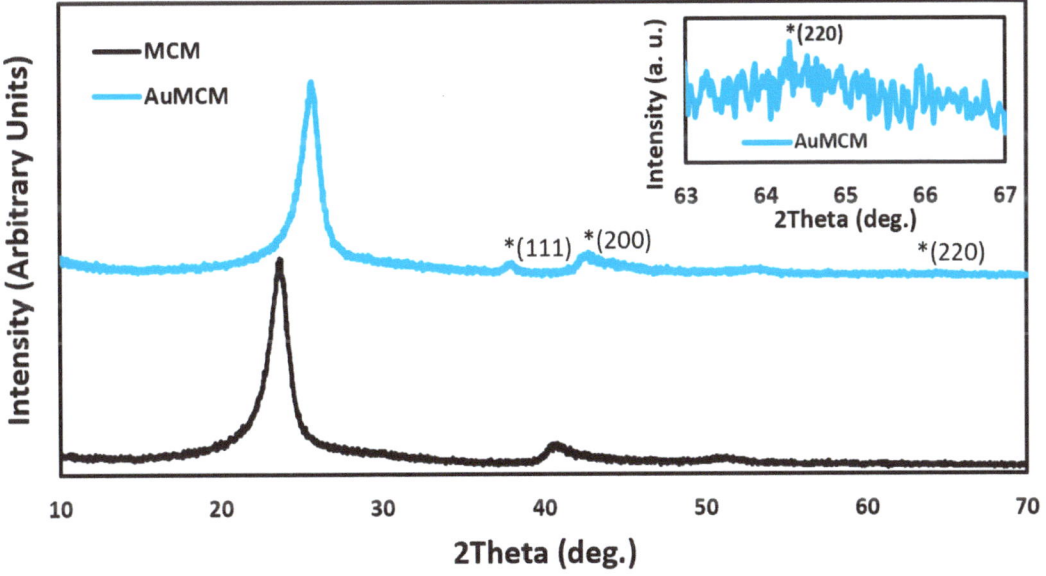

Figure 3. P-XRD spectra for the MCM and AuNP-MCM. The asterisks highlight the location of important peaks.

2.2. Catalytic Tests

Figure 4 shows the catalytic ability of AuNP-MCM at different reactant concentrations. The catalyzed reaction achieved the highest hydrogen generated rate of 0.0346 mL min^{-1} mg^{-1} at 1057 μmol, while the reaction achieved the lowest rate of 0.0140 mL min^{-1} mg^{-1} at 793 μmol. At the concentration of 952 μmol, the reaction rate was 0.0159 mL min^{-1} mg^{-1}. The hydrogen generated rate increased as the concentration of NaBH$_4$ increased. The increase in the reactant concentration shifted the equilibrium of Equation (1) and led to the formation of more products [3,29].

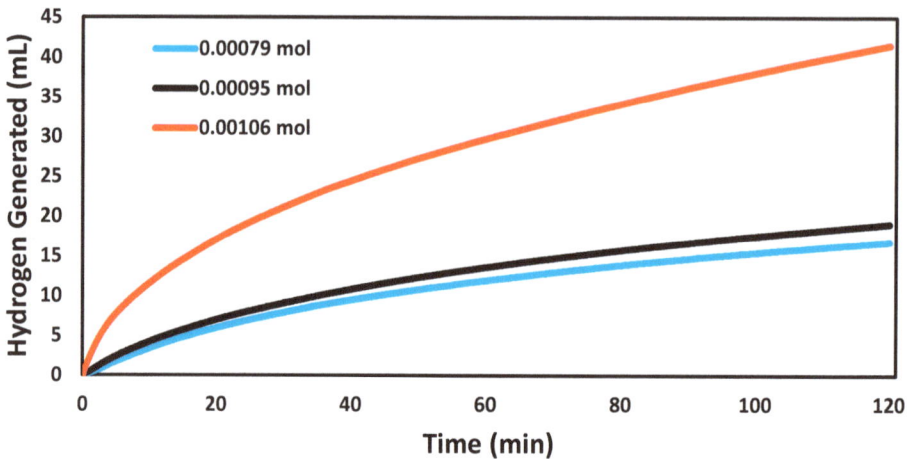

Figure 4. Volume of hydrogen generated over time in the hydrogen evolution reactions catalyzed by AuNP-MCM at different NaBH$_4$ concentrations (793 µmol, 952 µmol, and 1057 µmol).

The effect of pH on the catalyzed reactions was indicated in Figure 5. The reaction rate at pH 6 (0.0447 mL min^{-1} mg^{-1}) was higher than that of pH 7 (0.0159 mL min^{-1} mg^{-1}). The lowest reaction rate was 0.0087 mL min^{-1} mg^{-1} at pH 8. It had been stated in a previous kinetic study of sodium borohydride hydrolysis that the rate becomes lower as the pH increase due to the inhibition effect of OH$^-$ [29]. The free proton at low pH condition accelerated the hydrolysis process [29].

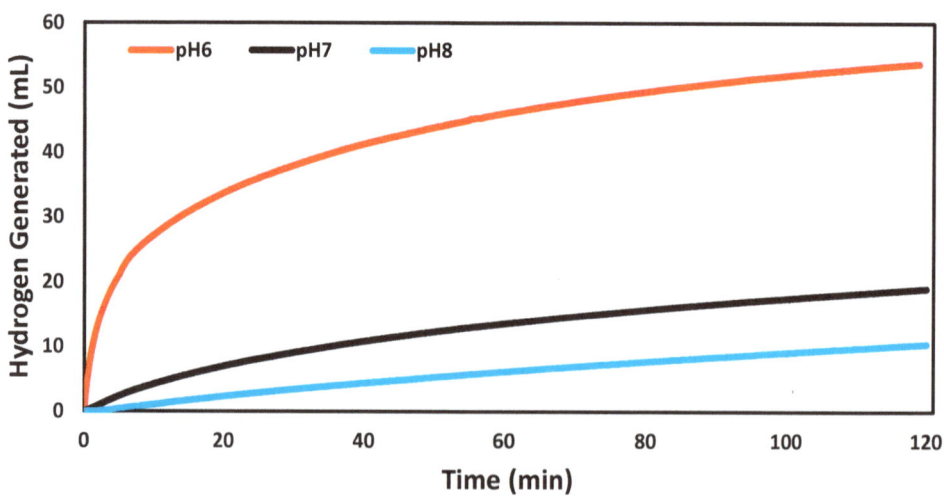

Figure 5. Volume of hydrogen generated over time in the hydrogen evolution reactions catalyzed by AuNP-MCM at different pHs (pH 6, pH 7, and pH 8).

Scheme 1 shows a proposed mechanism for the overall reaction, which could explain these results. NaBH4 reduces when hydrolyzed and forms a reversable complex with the metal nanoparticle at a catalytic site. An adjacent catalytic site receives a hydride ion and stabilizes the nanoparticle–borohydride complex, which reacts with the water to release hydrogen gas. This process repeats until there are no longer any hydride ions available and the complex breaks, releasing tetrahydroxylborate and H_2 gas. More acidic pHs increase the concentration of H^+ ions, which increase the conversion rate and supply of hydroxide ions for the conversion.

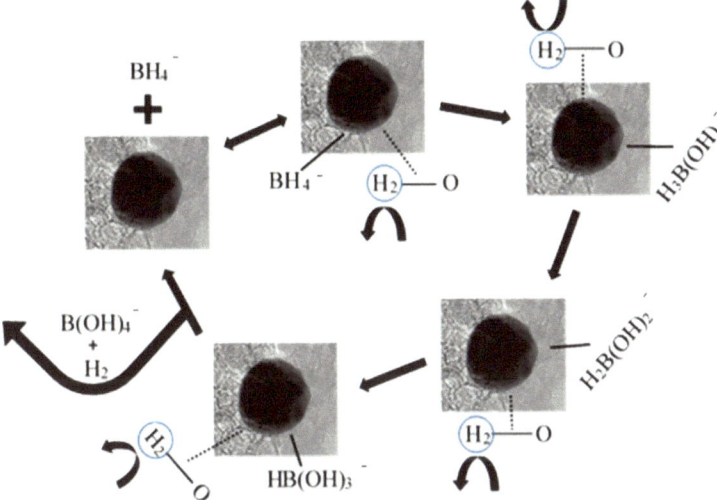

Scheme 1. Proposed mechanism for the hydrolysis of $NaBH_4$ while catalyzed by AuNP-MCMs.

Figure 6 shows that the reaction rate of catalyzed hydrogen evolution reaction improved at higher temperature. The hydrogen generation rates were 0.0062 mL min^{-1} mg^{-1}, 0.0117 mL min^{-1} mg^{-1}, 0.0159 mL min^{-1} mg^{-1}, and 0.0232 mL min^{-1} mg^{-1} at 273 K, 288 K, 295 K, and 303 K, respectively. The same pattern was observed in other studies [3,4,29].

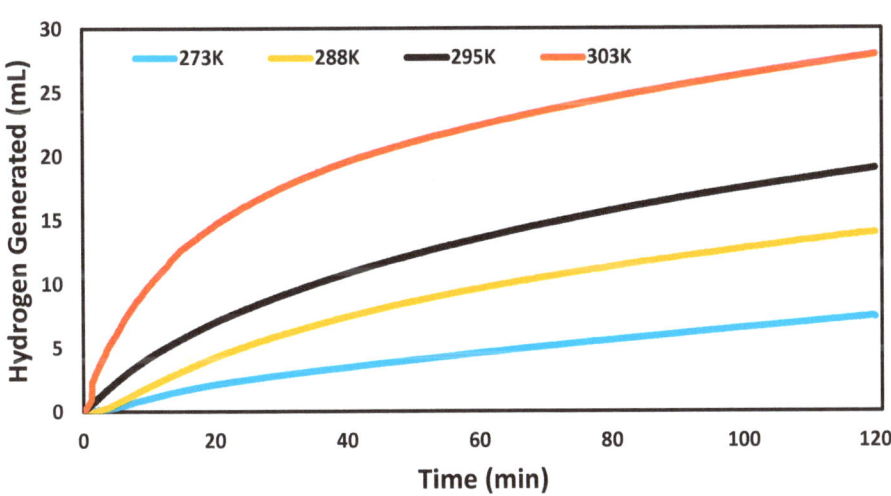

Figure 6. Volume of hydrogen generated over time in the hydrogen evolution reactions catalyzed by AuNP-MCM at different temperatures (273 K, 288 K, 295 K, and 303 K).

From the Arrhenius plot seen in Figures 7 and S2, the activation energy for the AuNP-MCM catalyst is determined to be 30 kJ mol^{-1}. The activation energy is also compared to previously recorded activation energies for other catalysts in Table 1, which shows it as one of the higher activation energies among other inorganic catalysts. AuNP-MCM has higher activation energy than other catalysts, except the gold nanoparticle supported over multiwalled carbon nanotubes (Au/MWCNTs). When compared to unsupported gold nanoparticles (BCD-AuNPs), there was a marked improvement. This indicates that the addition of MCM improves the catalytic ability of gold nanoparticles. These results highly imply that AuNP-MCM is an efficient catalyst for sodium borohydride hydrolysis.

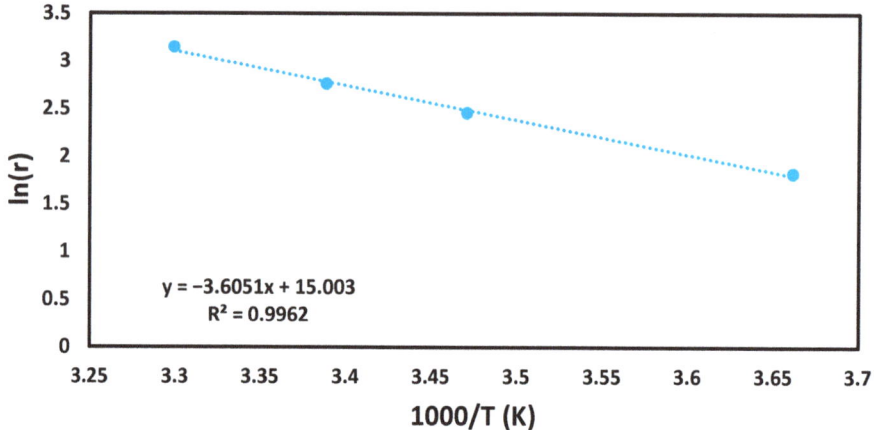

Figure 7. Arrhenius plot for calculating the activation energy of a hydrogen generation reaction with the AuNP-MCM catalyst.

Table 1. Comparison of reported activation energies for catalyzed NaBH$_4$ hydrolysis.

Catalyst	Ea (kJ mol^{-1})	Temperature (K)	Reference
CuNWs	42.48	298–333	[30]
Co/Fe$_3$O$_4$@C	49.2	288–328	[31]
Ni-Co-B	31.3	273–303	[32]
Cu-Fe-B	57	285–333	[33]
Co-Ce-B/Chi-C	33.1	266–303	[34]
Pt-MCM	37.7	298–328	[35]
Cu based catalyst	61.16	293–313	[36]
CuO/Co$_3$O$_4$	56.38	294–333	[37]
Pt/CeO$_2$-Co$_2$Ni$_2$O$_x$	47.4	298–318	[38]
NaBH$_4$@Ni	46.6	283–333	[39]
Au/MWCNTs	21.1	273–303	[20]
Ag/MWCNTs	44.45	273–303	[40]
Pt/MWCNTs	46.2	273–303	[41]
Pd/MWCNTs	62.66	273–303	[42]
AgNPs	53.3	273–295	[43]
PtNPs	39.2	273–303	[44]
Pt-MCM	37.7	273–303	[45]
Ag-MCM	15.6	273–303	[45]
Pd-MCM	27.9	273–303	[46]
BCD-AuNP	57.4	283–303	[4]
AuNP-MCM	30	273–303	This Work

2.3. Catalytic Reusability Tests

The AuNP-MCM catalyst underwent reusability trials to determine its reusability after five consecutive uses under conditions involving 952 μmol of $NaBH_4$, a pH of 7, and at 295 K. The plot in Figure 8 shows that the catalytic activity for the AuNP-MCM catalyst drastically decreased in catalytic activity after the second trial. From the third trial, the catalytic ability was increased and remained consistent in the fourth and fifth trials. This indicates that if more trials were to have been conducted, the catalytic activity may eventually remain consistent, resulting in a more ideal and durable catalyst. This would be consistent with Scheme 1, which shows the catalyst breaking away from the nanoparticle-borohydride complex to start the series of reactions again [47]. However, these reusability trials indicate that while the nanoparticles greatly increase the catalytic activity of the developed catalysts, the AuNP-MCM catalyst is generally less stable.

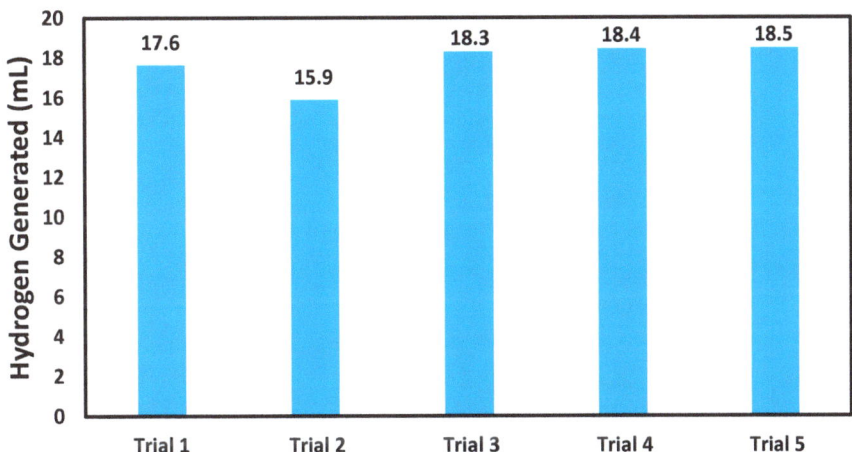

Figure 8. Testing reusability of the AuNP-MCM catalyst after five consecutive hydrogen generation reactions.

3. Experimental Section

3.1. Synthesis

The Mesoporous carbon was reported to be synthesized from starch via the Starbon synthesis method [21,22,24]. Nitrogen sorption/desorption isotherms generated at 77 K for mesoporous carbon is presented in Figure S1 and nitrogen adsorption data for mesoporous carbon is included in Table S1. Gold nanoparticles were synthesized by bringing a 1 mM aqueous chlorauric acid (Sigma Aldrich, St. Louis, MO, USA) solution to a boil and adding 1% aqueous sodium citrate (Sigma Aldrich) dropwise for five minutes with continuous stirring with a magnetic stir bar [48]. The AuNP-MCM nanocomposite was synthesized by adding 40 mL of the nanoparticle solution to 1 g of mesoporous carbon in order to functionalize the mesoporous carbon via the incipient wetness impregnation method [4,5]. The resulting precipitate was filtered and dried in a vacuum oven at 100 °C for 24 h.

3.2. Characterization

Transmission electron microscopy (TEM, JEM-2100F, JEOL, Akishima, Tokyo, Japan) was used to visualize the size of the nanoparticles in the composite and to characterize the binding between the nanoparticles and the mesoporous carbon. These samples were prepared by putting 1 μL of the nanoparticle solution onto the TEM sample grid and letting it dry overnight. X-ray diffraction (XRD, Rigaku Miniflex II, Cu Kα X-ray, nickel filters, Rigaku, Tokyo, Japan) was used to determine the crystal structure of mesoporous carbon and its composite. The sample was put onto a P-XRD slide, and a Rigaku Miniflex II was used to perform the P-XRD. Scanning electron microscopy (SEM, JEOL JSM-6060LV, JEOL, Akishima, Tokyo, Japan)/energy dispersive spectroscopy (EDS, ThermoScientific UltraDry, Thermo Fischer Scientific, Waltham, Massachusetts, USA) was used to determine the elemental composition of the nanocomposite and further confirm the presence of nanoparticles on the mesoporous carbon. The sample, in powder form, was mounted on a sample holder with carbon tape and analyzed under SEM with an EDS attachment.

3.3. Catalytic Tests

The catalytic properties of the nanocomposite were tested in the hydrogen evolution reaction between water and sodium borohydride ($NaBH_4$) as the reducing agent. The volume of generated hydrogen was determined via a gravimetric water displacement system [4,5]. Various conditions, such the concentration of $NaBH_4$ (793 µmol, 952 µmol, 1057 µmol), pH (6, 7, 8) and temperature (273 K, 288 K, 295 K, 303 K), were applied in the tests. In all trials, 0.01 g of the AuNP-MCM nanocomposite catalyst was used, and 100 mL of deionized water was used in all $NaBH_4$ solutions. During the reaction, the solution was stirred with a magnetic stir bar to maintain the dispersion of the AuNP-MCM. The water displaced during the reaction was measured via an Ohaus Pioneer Balance (Pa124) with proprietary mass logging software. Each variation was repeated in triplicate with the averages calculated.

3.4. Catalyst Reusability

In order to test for the reusability of the AuNP-MCM nanocomposites, a 952 µmol solution of $NaBH_4$ and 100 mL of deionized water at pH 7 and 295 K was made and 0.01 g of the nanocomposite was added. The same solution containing the deionized water and the catalyst was used in five reduction reactions, adding a constant amount of $NaBH_4$ for each trial.

4. Conclusions

The structure and composition of the AuNP-MCM catalyst was confirmed via transmission electron microscopy (TEM), scanning electron microscopy/energy dispersive spectroscopy (SEM/EDS), and X-ray diffraction (P-XRD). The catalytic activity for this catalyst saw its catalytic activities increase only with increasing $NaBH_4$ concentrations, increasing temperature, and lower pH. The variations in temperature allowed for the determination of an activation energy of 30.0 kJ mol^{-1}, which, when compared to previously tested inorganic catalysts, represents one of the lower activation energies, making it favorable. Catalytic reusability tests of this catalyst showed that the AuNP-MCM catalyst is stable and produces a consistent volume of hydrogen after five consecutive uses. This stability and low activation energy make this catalyst a competitive option for the hydrolysis of NBH_4.

Supplementary Materials: The following supporting information can be downloaded at: https://www.mdpi.com/article/10.3390/molecules29153707/s1. Figure S1: Nitrogen sorption/desorption isotherms generated at 77 K for Mesoporous Carbon; Figure S2: Arrhenius plot showing the Standard Deviation values for calculating the activation energy of a hydrogen generation reaction with the AuNP-MCM catalyst; Table S1: Nitrogen Adsorption data for Mesoporous Carbon;

Author Contributions: Methodology, T.M.A.-F.; Validation, T.M.A.-F.; Formal analysis, E.B., Q.Q. and T.M.A.-F.; Investigation, Q.Q.; Resources, T.M.A.-F.; Data curation, E.B.; Writing—original draft, E.B.; Writing—review & editing, Q.Q. and T.M.A.-F. All authors have read and agreed to the published version of the manuscript.

Funding: This research received no external funding.

Institutional Review Board Statement: Not applicable.

Informed Consent Statement: Not applicable.

Data Availability Statement: The raw data supporting the conclusions of this article will be made available by the authors on request.

Acknowledgments: The corresponding author acknowledges Lawrence J. Sacks' professorship in chemistry.

Conflicts of Interest: The authors declare no conflicts of interest.

References

1. Guan, D.; Wang, B.; Zhang, J.; Shi, R.; Jiao, K.; Li, L.; Wang, Y.; Xie, B.; Zhang, Q.; Yu, J.; et al. Hydrogen society: From present to future. *Energy Environ. Sci.* **2023**, *16*, 4926–4943. [CrossRef]
2. Yu, X.; Li, G.; Du, Y.; Guo, Z.; Shang, Z.; He, F.; Shen, Q.; Li, D.; Li, Y.A. comparative study on effects of homogeneous or stratified hydrogen on combustion and emissions of a gasoline/hydrogen SI engine. *Int. J. Hydrog. Energy* **2019**, *44*, 25974–25984. [CrossRef]
3. Schlesinger, H.I.; Brown, H.C.; Finholt, A.E.; Gilbreath, J.R.; Hoekstra, H.R.; Hyde, E.K. Sodium borohydride, its hydrolysis and its use as a reducing agent and in the generation of hydrogen. *J. Am. Chem. Soc.* **1953**, *75*, 215–219. [CrossRef]
4. Quach, Q.; Biehler, E.; Elzamzami, A.; Huff, C.; Long, J.M.; Abdel-Fattah, T.M. Catalytic activity of beta-cyclodextrin-gold nanoparticles network in hydrogen evolution reaction. *Catalysts* **2021**, *11*, 118. [CrossRef]
5. Huff, C.; Dushatinski, T.; Barzanji, A.; Abdel-Fattah, N.; Barzanji, K.; Abdel-Fattah, T.M. Pretreatment of gold nanoparticle multi-walled carbon Nanotube Composites for Catalytic Activity toward Hydrogen Generation Reaction. *ECS J. Solid State Sci. Technol.* **2017**, *6*, M69. [CrossRef]
6. Datta, K.K.R.; Reddy, B.V.S.; Ariga, K.; Vinu, A. Gold Nanoparticcles Embedded in a Mesoporous Carbon Nitride Stabilizer for Highly Efficent Three-Component Coupling Reaction. *Angew. Chem.* **2010**, *122*, 6097–6101. [CrossRef]
7. Gan, X.; Liu, T.; Zhong, J.; Li, G. Effect of Silver Nanoparticles on Electron Transfer Reactivity and the Catalytic Activity of Myoglobin. *ChemBioChem* **2004**, *5*, 1686–1691. [CrossRef] [PubMed]
8. Xu, R.; Wang, D.; Zhang, J.; Li, Y. Shape-Dependent Catalytic Activity of Silver Nanoparticles for the Oxidation of Styrene. *Chem. Asian J.* **2006**, *1*, 888–893. [CrossRef] [PubMed]
9. Maillard, F.; Schreier, S.; Hanzlik, M.; Savinova, E.R.; Weinkauf, S.; Stimming, U. Influence of particle agglomeration on the catalytic activity of carbon-supported Pt nanoparticles in CO monolayer oxidation. *Phys. Chem. Chem. Phys.* **2005**, *7*, 385–393. [CrossRef]
10. Comotti, M.; Pina, C.D.; Matarrese, R.; Rossi, M. The Catalytic Activity of "Naked" Gold Particles. *Angew. Chem. Int. Ed.* **2004**, *43*, 5812–5815. [CrossRef]
11. Osborne, J.; Horten, M.R.; Abdel-Fattah, T.M. Gold Nanoparticles supported over low-cost supports for hydrogen generation from a hydrogen feedstock material. *ECS J. Solid State Sci. Technol.* **2020**, *9*, 071004. [CrossRef]
12. Xia, Y.; Yang, Z.; Mokaya, R. Simultaneous Control of Morphology and Porosity in Nanoporous Carbon: Graphitic Mesoporous Carbon Nanorods and Nanotubules with Tunable Pore size. *Chem. Mater.* **2005**, *18*, 140–148. [CrossRef]
13. Horváth, E.; Puskás, R.; Rémiás, R.; Mohl, M.; Kukovecz, Á.; Kónya, Z.; Kiriesi, I. A Novel Catalyst Type Containing Noble Metal Nanoparticles Supported on Mesoporous Carbon: Synthesis, Characterization and Catalytic properties. *Top Catal.* **2009**, *52*, 1242–1250. [CrossRef]
14. McNaught, A.D.; Wilkinson, A. *Compendium of Chemical Terminology*, 2nd ed.; (the "Gold Book"); IUPAC: Malden, MA, USA, 1997.
15. Liang, C.; Li, Z.; Dai, S. Mesoporous Carbon Materials: Synthesis and Modification. *Angew. Chem. Int.* **2008**, *47*, 3696–3717. [CrossRef] [PubMed]
16. Xia, Y.; Mokaya, R. Generalized and Facile Synthesis Approach to N-Doped Highly Graphitic Mesoporous Carbon Materials. *Chem. Mater.* **2005**, *17*, 1553–1560. [CrossRef]
17. Su, F.; Zeng, J.; Bao, X.; Yu, Y.; Lee, J.Y.; Zhao, X.S. Preparation and Characterization of Highly Ordered Graphitic Mesoporous Carbon as a Pt Catalyst Support for Direct Methanol Fuel Cells. *Chem. Mater.* **2005**, *17*, 3960–3967. [CrossRef]
18. Matos, I.; Bernardo, M.; Neves, P.D.; Castanheiro, J.E.; Vital, J.; Fonseca, I.M. Mesoporous Carbon as effective and sustainable catalyst for fine chemistry. *Bol. Grupo Español Carbón* **2016**, *39*, 19–22. Available online: http://hdl.handle.net/10174/19991 (accessed on 1 January 2024).
19. Chang, H.; Joo, S.H.; Pak, C. Synthesis and characterization of mesoporous carbon for fuel cell applications. *J. Mater. Chem.* **2007**, *17*, 3078–3088. [CrossRef]
20. Ryoo, R.; Joo, S.H.; Jun, S. Synthesis of highly ordered carbon molecular sieves via template-mediated structural transformation. *J. Phys. Chem. B* **1999**, *103*, 7743–7746. [CrossRef]
21. Budarin, V.; Clark, J.H.; Hardy, J.J.E.; Luque, R.; Milkowski, K.; Tavener, S.J.; Wilson, A.J. Starbons: New Starch-Derived Mesoporous Carbonaceous Materials with Tunable Properties. *Angew. Int. Ed.* **2007**, *45*, 3782–3786. [CrossRef]
22. Shuttleworth, P.S.; Budarin, V.; White, R.J.; Gun'ko, V.M.; Luque, R.; Clark, J.H. Molecular-Level Understanding of the Carbonistation of Polysaccharides. *Chem. Eur. J.* **2013**, *19*, 9351–9357. [CrossRef]
23. Biehler, E.; Quach, Q.; Abdel-Fattah, T.M. Screening study of Different Carbon Based Materials for Hydrogen. *ECS J. Solid State Sci. Technol.* **2023**, *12*, 081002. [CrossRef]
24. Milkowski, K.; Clark, J.H.; Doi, S. New materials based on renewable resources: Chemically modified highly porous starches and their composites with synthetic monomers. *Green Chem.* **2004**, *6*, 189–190. [CrossRef]
25. Wahab, M.A.; Darain, F.; Islam, N.; Young, D.J. Nano/mesoporous carbon from rice starch for voltammetric detection of ascorbic acid. *Molecules* **2018**, *23*, 234. [CrossRef]

26. Banerjee, R.; Ghosh, D.; Satra, J.; Ghosh, A.B.; Singha, D.; Nandi, M.; Biswas, P. One Step Synthesis of a gold/ordered mesoporous carbon composite using a hard template method for electrocatalytic oxidation of methanol and colorimetric determination of glutathione. *ACS Omega* **2019**, *4*, 16360–16371. [CrossRef] [PubMed]
27. Hu, Q.L.; Zhang, Z.X.; Zhang, J.J.; Li, S.M.; Wang, H.; Lu, J.X. Ordered mesoporous carbon embedded with cu Nanoparticle materials for electrocatalytic synthesis of benzyl methyl carbonate from benzyl alcohol and carbon dioxide. *ACS Omega* **2020**, *5*, 3498–3503. [CrossRef]
28. Khalil, M.M.H.; Ismail, E.H.; El-Magdoub, F. Biosynthesis of Au nanoparticles using olive leaf extract. *Arab. J. Chem.* **2012**, *5*, 431–437. [CrossRef]
29. Yu, L.; Mathews, M.A. Hydrolysis of sodium borohydride in concentrated aqueous solution. *Int. J. Hydrog. Energy* **2011**, *36*, 7416–7422. [CrossRef]
30. Hashimi, A.S.; Nohan, M.; Chin, S.X.; Khiew, P.S.; Zakaria, S.; Chia, C.H. Copper Nanowires as Highly Efficient and Recyclable Catalyst for Rapid Hydrogen Generation from Hydrolysis of Sodium Borohydride. *Nanomaterials* **2020**, *10*, 1153. [CrossRef]
31. Chen, B.; Chen, S.; Bandal, H.A.; Appiah-Ntiamoah, R.; Jadhav, A.R.; Kim, H. Cobalt nanoparticles supported on magnetic core-shell structured carbon as a highly efficient catalyst for hydrogen generation from $NaBH_4$ hydrolysis. *Int. J. Hydrog. Energy* **2018**, *43*, 9296–9306. [CrossRef]
32. Guo, J.; Hou, Y.; Li, B.; Liu, Y. Novel Ni-Co-B hollow nanospheres promote hydrogen generation from the hydrolysis of sodium borohydride. *Int. J. Hydrog. Energy* **2018**, *43*, 15245–15254. [CrossRef]
33. Loghmani, M.H.; Shojaei, A.F.; Khakzad, M. Hydrogen generation as a clean energy through hydrolysis of sodium borohydride over Cu-Fe-B nano powders: Effect of polymers and surfactants. *Energy* **2017**, *126*, 830–840. [CrossRef]
34. Zou, Y.; Yin, Y.; Gao, Y.; Xiang, C.; Chu, H.; Qui, S.; Yan, E.; Xu, F.; Sun, L. Chitosan-mediated Co-Ce-B nanoparticles for catalyzing the hydrolysis of sodium borohydride. *Int. J. Hydrog. Energy* **2018**, *43*, 4912–4921. [CrossRef]
35. Biehler, E.; Quach, Q.; Abdel-Fattah, T.M. Application of Platinum Nanoparticles Decorating Mesoporous Carbon Derived from Sustainable Source for Hydrogen Evolution Reaction. *Catalysts* **2024**, *14*, 423. [CrossRef]
36. Balbay, A.; Saka, C. Effect of phosphoric acid addition on the hydrogen production from hydrolysis of $NaBH_4$ with Cu based catalyst. *Energy Source Part A* **2018**, *40*, 794–804. [CrossRef]
37. Xie, L.; Wang, K.; Du, G.; Asiri, A.M.; Sun, X. 3D hierarchical CuO/Co_3O_4 core-shell nanowire array on copper foam for on-demand hydrogen generation from alkaline $NaBH_4$ solution. *RSC Adv.* **2016**, *6*, 88846–88850. [CrossRef]
38. Wu, C.; Zhang, J.; Guo, J.; Sun, L.; Ming, J.; Dong, H.; Zhao, Y.; Tian, J.; Yang, X. Ceria-Induced Strategy To Tailor Pt Atomic Clusters on Cobalt-Nickel Oxide and the Synergetic Effect for Superior Hydrogen Generation. *ACS Sustain. Chem. Eng.* **2018**, *6*, 7451–7457. [CrossRef]
39. Lai, Q.; Alligier, D.; Aguey-Zinxou, K.; Demirci, U.B. Hydrogen generation from a sodium borohydride-nickel core@shell structure under hydrolytic conditions. *Nanoscale Adv.* **2019**, *1*, 2707–2717. [CrossRef] [PubMed]
40. Huff, C.; Long, J.M.; Aboulatta, A.; Heyman, A.; Abdel-Fattah, T.M. Silver Nanoparticle/Multi-Walled Carbon Nanotube Composite as Catalyst for Hydrogen Production. *ECS J. Solid State Sci. Technol.* **2017**, *6*, 115–118. [CrossRef]
41. Huff, C.; Quach, Q.; Long, J.M.; Abdel-Fattah, T.M. Nanocomposite Catalyst Derived from Ultrafine Platinum Nanoparticles and Carbon Nanotubes for Hydrogen Generation. *ECS J. Solid State Sci. Technol.* **2020**, *9*, 101008. [CrossRef]
42. Huff, C.; Long, J.M.; Heyman, A.; Abdel-Fattah, T.M. Palladium Nanoparticle Multiwalled Carbon Nanotube Composite as Catalyst for Hydrogen Production by the Hydrolysis of Sodium Borohydride. *ACS Appl. Energy Mater.* **2018**, *1*, 4635–4640. [CrossRef]
43. Huff, C.; Long, J.M.; Abdel-Fattah, F.M. Beta-Cyclodextrin-Assisted Synthesis of Silver Nanoparticle Network and Its Application in a Hydrogen Generation Reaction. *Catalysts* **2020**, *10*, 1014. [CrossRef]
44. Huff, C.; Biehler, E.; Quach, Q.; Long, J.M.; Abdel-Fattah, T.M. Synthesis of highly dispersive platinum nanoparticles and their application in a hydrogen generation reaction. *Colloid Surf. A* **2021**, *610*, 125734. [CrossRef]
45. Biehler, E.; Quach, Q.; Abdel-Fattah, T.M. Application of Silver Nanoparticles Supported over Mesoporous Carbon Produced from Sustainable Sources as Catalysts for Hydrogen Production. *Energies* **2024**, *17*, 3327. [CrossRef]
46. Biehler, E.; Quach, Q.; Abdel-Fattah, T.M. Application of Palladium Mesoporous Carbon Composite Obtained from a Sustainable Source for Catalyzing Hydrogen Generation Reaction. *J. Compos. Sci.* **2024**, *8*, 270. [CrossRef]
47. Abdel-Fattah, T.M.; Biehler, E. Carbon Based Supports for Metal Nanoparticles for Hydrogen Generation Reactions Review. *Adv. Carbon J.* **2024**, *1*, 1–19. [CrossRef]
48. Turkevich, J.; Stevenson, P.C.; Hillier, J. A Study of the Nucleation and Growth Progesses in the synthesis of Colloidal Gold. *Discuss. Faraday Soc.* **1951**, *11*, 55–75. [CrossRef]

Disclaimer/Publisher's Note: The statements, opinions and data contained in all publications are solely those of the individual author(s) and contributor(s) and not of MDPI and/or the editor(s). MDPI and/or the editor(s) disclaim responsibility for any injury to people or property resulting from any ideas, methods, instructions or products referred to in the content.

Article

Enhanced Mass Activity and Durability of Bimetallic Pt-Pd Nanoparticles on Sulfated-Zirconia-Doped Graphene Nanoplates for Oxygen Reduction Reaction in Proton Exchange Membrane Fuel Cell Applications

Maryam Yaldagard [1,*] and Michael Arkas [2,*]

1. Department of Chemical Engineering, Faculty of Engineering, Urmia University, Urmia 5766-151818, Iran
2. National Centre for Scientific Research "Demokritos", Institute of Nanoscience and Nanotechnology, 15310 Athens, Greece
* Correspondence: m.yaldagard@urmia.ac.ir (M.Y.); m.arkas@inn.demokritos.gr (M.A.)

Highlights:

- In this work, graphene nanoplates (GNPs) with a supreme medium were obtained.
- Pt particles (4.50 nm) were uniformly dispersed on the surface of S-ZrO$_2$-GNP support.
- The Pt-Pd/S-ZrO$_2$-GNPs exhibited higher ECSA than Pt-Pd/ZrO$_2$-GNPs and Pt/C.
- Pt-Pd/S-ZrO$_2$-GNPs exhibited higher ORR mass activity than other studied electrodes.
- Pt-Pd/S-ZrO$_2$-GNPs exhibited low charge transfer resistance in EIS measurements.

Citation: Yaldagard, M.; Arkas, M. Enhanced Mass Activity and Durability of Bimetallic Pt-Pd Nanoparticles on Sulfated-Zirconia-Doped Graphene Nanoplates for Oxygen Reduction Reaction in Proton Exchange Membrane Fuel Cell Applications. *Molecules* **2024**, *29*, 2129. https://doi.org/10.3390/molecules29092129

Academic Editors: Jacek Ryl, Sake Wang, Nguyen Tuan Hung and Minglei Sun

Received: 23 January 2024
Revised: 7 March 2024
Accepted: 30 April 2024
Published: 3 May 2024

Copyright: © 2024 by the authors. Licensee MDPI, Basel, Switzerland. This article is an open access article distributed under the terms and conditions of the Creative Commons Attribution (CC BY) license (https://creativecommons.org/licenses/by/4.0/).

Abstract: Developing highly active and durable Pt-based electrocatalysts is crucial for polymer electrolyte membrane fuel cells. This study focuses on the performance of oxygen reduction reaction (ORR) electrocatalysts composed of Pt-Pd alloy nanoparticles on graphene nanoplates (GNPs) anchored with sulfated zirconia nanoparticles. The results of field emission scanning electron microscopy and transmission electron microscopy showed that Pt-Pd and S-ZrO$_2$ are well dispersed on the surface of the GNPs. X-ray diffraction revealed that the S-ZrO$_2$ and Pt-Pd alloy coexist in the Pt-Pd/S-ZrO$_2$-GNP nanocomposites without affecting the crystalline lattice of Pt and the graphitic structure of the GNPs. To evaluate the electrochemical activity and reaction kinetics for ORR, we performed cyclic voltammetry, rotating disc electrode, and EIS experiments in acidic solutions at room temperature. The findings showed that Pt-Pd/S-ZrO$_2$-GNPs exhibited a better ORR performance than the Pt-Pd catalyst on the unsulfated ZrO$_2$-GNP support and with Pt on S-ZrO$_2$-GNPs and commercial Pt/C.

Keywords: Pt-Pd alloy; sulfated zirconia; graphene nanoplates; cathode; ORR; PEMFC

1. Introduction

Polymer electrolyte membrane fuel cells (PEMFCs) are increasingly gaining acceptance as a clean, efficient, and silent energy conversion technology, and are seen as a future alternative energy source [1–3]. However, the sluggish kinetics of the oxygen reduction reaction (ORR) and the instability of the platinum electrocatalysts for ORR significantly hinder the commercialization of PEMFCs [4–7]. Accordingly, a higher mass activity and longer durability/stability of the platinum electrocatalysts for ORR are required to increase the popularity of PEMFCs.

It is well understood that the specific activity of the Pt electrocatalyst in the ORR correlates with the carbon support [8]. Carbon supports such as Vulcan have excellent electrical conductivity, chemical stability in acidic solutions, and a large surface area, suitable for the dispersion of catalyst nanoparticles. On the other hand, the latest examinations have shown that the degradation of the electrodes containing carbon supports under the ORR leads to a deterioration in cell efficiency over an extended period [9]. It was found that the electrocatalysts are severely destroyed by the oxidation-induced carbon corrosion under the operating conditions of the cathode, generally due to high potentials, 0.6–1.2 V; high O_2 concentration; and high temperatures, 50–90 °C. Therefore, the support materials should be more stable to avoid the destruction of the catalyst.

Valve metals, including titanium, zirconium, tantalum, niobium, etc., are known for their high resistance to oxidization as they are passivated in a strongly acidic solution [10]. The oxides of these metals, such as Ti_4O_7 [11] and indium tin oxide [12], have been proposed as exceptionally corrosion-resistant materials. These oxides could potentially serve as support materials for cathodes in PEMFCs. In addition, the surfaces of these metal oxides are modified with sulfonic acid (SOx) to act as solid superacids [13]. Specifically, sulfated ZrO_2 (S–ZrO_2) is a solid superacid (H0 = −16.03) and is thermally stable at high temperatures of about 500 °C. In their study, Hara and Miyayama discovered that S–ZrO_2 exhibits high proton conductivity when it has a high atomic ratio of sulfur to zirconia [14]. According to their report, the high proton conductivity was attributed to the localized electrons on the oxygen in SOx and the Lewis acid sites on Zr, which can easily generate new Brønsted acid sites, resulting in a higher conductivity [14]. In the current technology, the addition of a proton conductor, such as a perfluorosulfonated ionomer (PFSI), is crucial for the construction of the wide three-phase boundaries in the back layer of the electrode. The use of a proton-conducting metal oxide as a support for the electrocatalyst would provide an innovative function for the cathode of PEMFCs due to the stability of ZrO_2. In particular, it is estimated that the amount of altered electrolyte ionomer could be reduced. Reducing ionomer content can improve gas diffusion and water transport. In addition, an increase in the utilization of platinum is expected. Since Pt nanoparticles can not penetrate the tiny pores of the carbon support, they have limited contact with a proton conductor in PFSI [15]. Consequently, the usage of Pt is reduced. S-ZrO_2 has acidic sites on its surface, even in tiny pores. Therefore, the Pt on S-ZrO_2 can exchange an H+, which increases the Pt utilization. In addition, the SO4 group on the ZrO_2 surface increases the hydrophilicity of ZrO_2 [16,17], which results in S-ZrO_2 improving the fuel cell performance in low humidity [18]. So because of the solid metal–support interaction, metal oxides improved the ORR activity of the Pt-based catalysts in most examinations [19,20] against the posing effect of the sulfate anion in PFSI in ORR kinetic activity which was recently reported by some authors [21–24].

Metal oxides generally have poor electrical conductivity. In addition, it is difficult to produce platinum in high concentrations on the metal oxide surface due to the small surface area. Therefore, metal oxides must be modified to increase electrical conductivity or they must be combined with conductive materials such as carbon [25,26]. Designing a complex structure of Pt-based alloys, metal oxides, and carbonaceous materials is a crucial challenge to fully realize their potential. When metal oxides are deposited on carbonaceous materials before the deposition of Pt-based nanoparticles, the Pt-based nanoparticles can deposit on the metal oxides, resulting in poorer ORR activity due to the non-conductivity of the metal oxides. Otherwise, when the Pt-based nanoparticles are first loaded onto the carbonaceous materials, followed by successive deposition of the metal oxide nanoparticles, the metal oxides may cover the Pt-based nanoparticles, resulting in a lower ORR activity because of the reduction in Pt active points. In several studies, metal oxides and carbon were used, but ORR activity could not be increased due to uncontrollable Pt dispersion on metal oxides and carbon [27–29]. One approach to solve this problem is to investigate a new carbon-based material as a support.

Over the past few years, graphene—a monolayer of carbon atoms in sp^2 hybridization arranged in a honeycomb matrix—has attracted considerable technical attention. This unique 2D carbon material exhibits extraordinary properties, such as exceptional electrical, thermal, and mechanical properties [30,31]. As scientists continue to explore its physical and chemical properties, graphene has become one of the most important materials in electronics, nanosensors, nanocomposites, and hydrogen storage [32–34]. Due to its large surface area, excellent electrical conductivity, and exceptional mechanical properties, graphene has the potential to serve as a catalyst support in PEMFCs.

The performance of fuel cell electrodes can be improved by using a more active electrocatalyst or by improving the structure of the catalyst layer. In addition to improving the inherent activity of Pt nanoparticles, there is also a strong interest in achieving similar activity improvements by saving Pt to reduce costs. Overcoming the sluggish reduction kinetics of molecular oxygen at low temperatures has been a major challenge in finding an effective catalyst for the cathode side of the fuel cell. Although Pt is generally considered to be the best catalyst for this reaction, successful results have also been obtained with bimetallic Pt alloys [35–39], which exhibit activity towards the ORR that is as good as that of pure Pt in an acidic solution. It is believed that bimetallic Pt alloys improve the durability of PEMFCs by reducing Pt dissolution and migration during operation [39–42]. Among the many possible combinations of Pt and other metal catalysts, Pd is one of the most commonly used catalysts for ORR in fuel cells operating with acid or alkaline electrolytes [43–48]. Moreover, except for Pt, the electrocatalytic activity of Pd due to the electronic properties very similar to Pt is one of the highest among the pure metals for ORR [49,50]. This motivates the development of Pt-Pd alloy catalysts for ORR.

As far as we know, the effectiveness of the platinum–palladium sulfated zirconia/graphene nanoplate (Pt-Pd/S-ZrO_2-GNP) electrode in the oxygen reduction reaction in PEMFCs has not been investigated. We propose a hybrid electrocatalyst combining sulfated ZrO_2 nanocrystals with GNP supports and Pt-Pd nanoparticles to achieve better proton and electron conductivity. In the present study, the superacid SO_4^{-2}-ZrO_2 was synthesized on the GNP surface by chemically linking proton-conducting sulfonic acid groups to form SO_4^{-2}-ZrO_2-GNP support for Pt-Pd catalysts in PEMFCs. Pt-Pd was dispersed on the surface of S-ZrO_2-GNPs by the electrodeposition route (Pt-Pd/S-ZrO_2-GNPs). The purpose for the choice of the electrochemical method was explained in our previous work [51]. The resulting Pt-Pd/S-ZrO_2-GNP electrode was characterized by physical and electrochemical methods, including AFM, FESEM, TEM, XRD, FTIR, CV, and LSV curves. The study investigated the impact of sulfation on the prepared electrocatalysts (Pt-Pd/S-ZrO_2-GNPs). The electrocatalysts containing sulfated zirconia oxide–graphene nanoplate (S-ZrO_2-GNP) powder outperformed the non-sulfated ZrO_2-GNP powder (Pt-Pd/ZrO_2-GNP). Additionally, the ORR mechanism behind the role of the S-ZrO_2 nanoparticles in the electrode structure has been proposed.

2. Results and Discussion

2.1. Physical Characterization

2.1.1. Topography Study of Graphene Nanoplates and S-ZrO_2-GNPs

The synthesized graphene nanoplates were studied with atomic force microscopy to specify lateral size and width. Figure 1 shows 2D and 3D AFM images of the ZrO_2-GNPs with a corresponding height profile along the GNPs derived from the chemical reduction of the exfoliated graphene oxide. The topography height of the graphene nanoplates shows that the graphene nanoplates consist of few layers. As shown in Figure 1c, twenty percent of the produced GNPs have pore size distributions ranging from 1 to 7 nm, while the remaining 80% have a distribution ranging from 8 to 100 nm. Moreover, the images of zirconia nanospheres on graphene nanosheets are visible in Figure 1a. From the assessment (by the Nanosurf Easyscan2 software (Version 2.2.1.16), specific for the used AFM device in this work), the thickness of the S-ZrO_2-doped graphene nanoplates is about 32–87 nm.

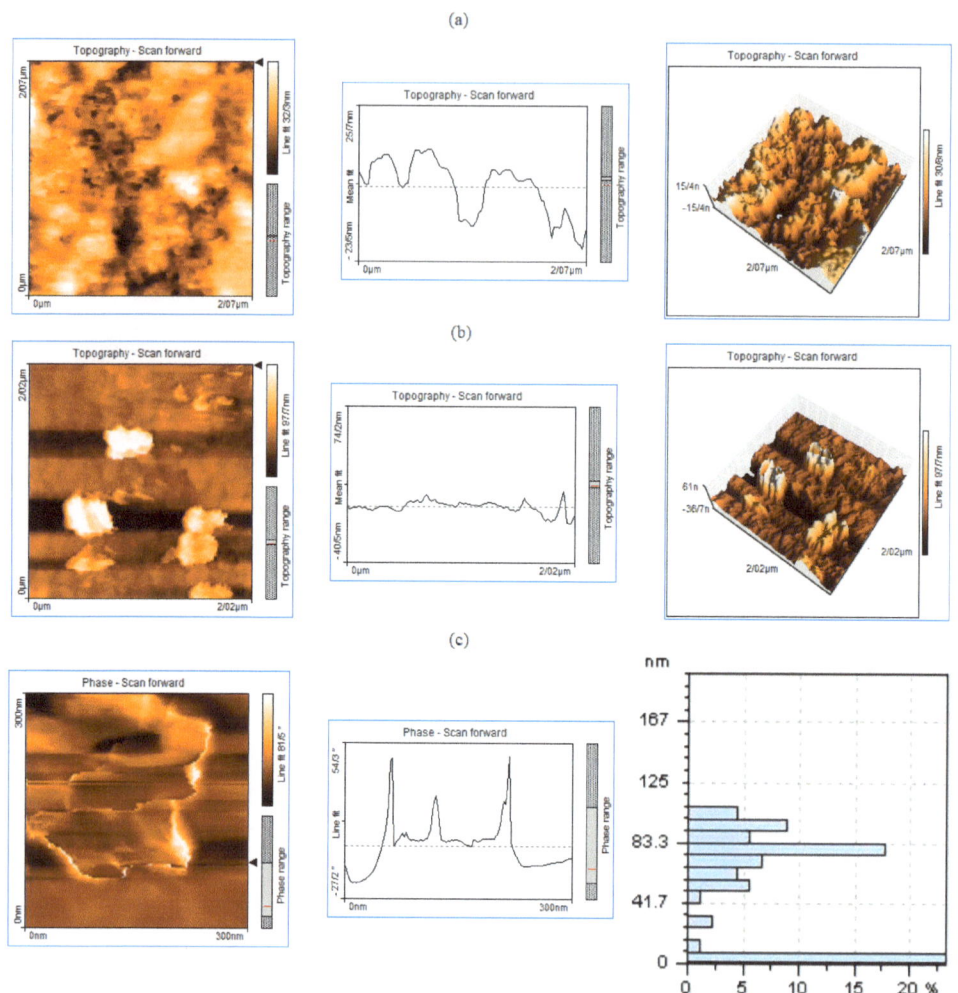

Figure 1. The 2D and 3D AFM topography images along with an equivalent height profile of the (**a**) S-ZrO$_2$/GNPs, (**b**) GNPs, and (**c**) the phase scan of GNPs and pore size distribution.

2.1.2. Morphological Study of GNPs and Pt-Pd Nanocrystals on S-ZrO$_2$-GNPs

Figure 2 shows the general FESEM and TEM illustrations of the (a) GNPs and (b) Pt-Pd/S-ZrO$_2$-GNP nanocomposites. Figure 2a reveals a high mass of the chemically prepared GNPs. Figure 2b,c reveal that the uniform spherical Pt-Pd/S-ZrO$_2$ nanoparticles were well and homogeneously distributed on the GNPs. Meanwhile, the particle size distribution of Pt-Pd on the S-ZrO$_2$–GNPs with sizes ranging from 2 to 8 nm are depicted in Figure 2d. From this histogram, it is concluded that the prepared Pt-Pd nanoparticles were highly dispersed on the support with quite a narrow particle size distribution.

It should be mentioned that to take the microscopic images and peel off the thin layer of the electrodeposit from the surface of the GC electrode, the electrodeposition process was repeated several times. Depending on the number of the repetition cycles of electrodeposition, the Pt-Pd covers the entire surface of the GNPs.

Figure 2. (a) FESEM illustrations of graphene nanoplates and (b,c) FESEM and TEM images of Pt-Pd/S-ZrO$_2$-GNPs electrocatalyst. (d) Corresponding Pt-Pd nanoparticle size histogram.

2.1.3. Structural Specifications of Synthesized Support Material

The study was conducted to identify the vibrational modes of the GNPs and ZrO$_2$-GNPs after sulfation. The G-band is a distinctive feature of the graphitic layers and represents the tangential vibration of sp2-bonded carbon atoms in a two-dimensional hexagonal lattice. Another characteristic feature of carbon materials is the presence of the D-band, which corresponds to the vibration of defective graphitic structures resulting from the doubly resonant disorder-induced mode [52–60]. The I_D/I_G intensity ratio is a measurement of the quality of a sample. A higher value of this ratio indicates a more ordered structure. In Figure 3, the Raman spectra for the GNPs and S-ZrO$_2$-GNPs are shown. The G (graphite) and D (disorder) bands and their second-order harmonic (2D band) are visible in the spectra at 1585.66 cm^{-1}, 1311.30 cm^{-1}, and 2611.92 cm^{-1} in the GNPs, respectively. The form of these bands is not impacted by doping with S-ZrO$_2$, indicating that the overall structure of the graphene sheet remains intact during sulfation treatments or doping. Moreover, the Raman spectra of the two nanostructures presented in Figure 3 show that the intensity ratio of D-band to G-band (I_D/I_G) does not significantly change with the addition of zirconia. However, the peaks of D-, G-, and 2D-band have shifted to slightly higher wavelengths. This indicates that the addition of zirconia through the adopted method does not introduce specific functional groups to the surface of the GNPs [61,62].

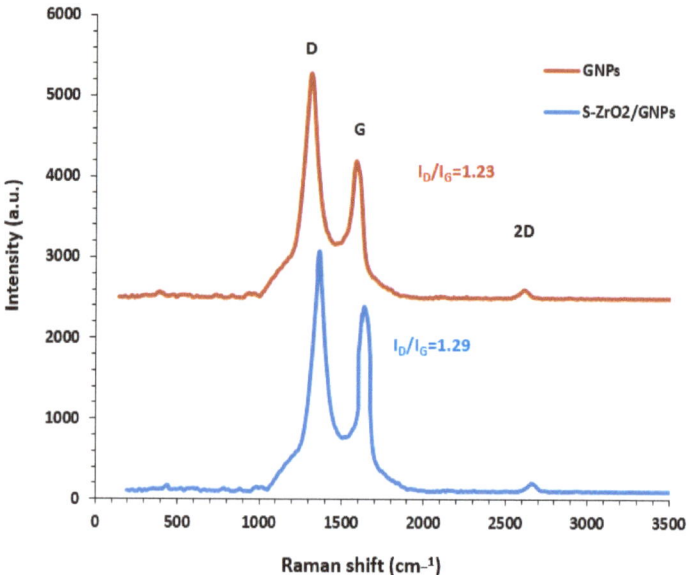

Figure 3. Raman characterization of GNP and S-ZrO$_2$-GNP support.

The FTIR results of GO and ZrO$_2$-graphene nanoplates and sulfated zirconia-GNPs were presented in our previously published work [63].

2.1.4. XRD Pattern Characterization

The crystal lattice formation of the Pt-Pd/S-ZrO$_2$-GNP electrocatalyst is shown in Figure 4, indicating the presence of carbon and platinum. The obtained nanocomposite exhibits the characteristic prominent peaks of the Pt Fcc structure at the Bragg angles of 86.44°, 82.08°, 68.05°, 46.62°, and 40.10° which relate to the (222), (311), (220), (200), and (111) diffraction peaks of the crystal surface, respectively, showing that it is a disordered Pt-Pd alloy. The formation of the bimetallic Pt-Pd nanoparticles in the Pt-Pd/S-ZrO$_2$-GNP electrocatalyst was confirmed by a small transference of the XRD spectrum to greater angles due to the reduction in the lattice compared to Pt, indicating that the interatomic space of platinum was reduced by the replacement of the smaller Pd atom in the Pt metal lattice. The sharper diffraction peaks at 2θ = 26.651° and 54.27° are the features of the parallel GNP layers in the Pt-Pd/S-ZrO$_2$-GNP composite, signifying a highly graphitic organization of the GNP in faces of (002) and (004) correspondingly. These plans show that the S-ZrO$_2$ does not affect the graphitic structure of the GNPs. The preservation of the graphitic structure of the carbon is advantageous in the preparation of an electrocatalyst because it maintains the conductivity of the support materials. Besides the characteristic peaks of the graphite and Pt construction, some additional peaks at 2θ of 29.95° and 31.69° with the surfaces of −111 and 111 [64–69] were found in the nanocomposite, which is related to the ZrO$_2$ nanocrystals used in the composition of the catalyst. The average size of the Pt-Pd nanoparticles was evaluated from the Debby–Scherrer formulation using the full width at half maximum (FWHM) of the (111) plane. This reflection was chosen for the Scherrer examination due to its high-strength level. This formula can be defined as follows [70]:

$$d = 0.9 \frac{\lambda}{B\cos\theta} \quad (1)$$

where d is the diameter of the mean particle size in Å; λ is the X-ray wavelength (1.5406 Å) for CuK $_\alpha$; θ is the Bragg angle; and B is the FWHM in radians. Based on the sample's

Pt(111) reflection, the average sizes of the Pt particles were calculated to be 4.50 nm with a d-spacing of 0.2246 nm (2.22 Å). The XRD results indicated that the innate attributes of the catalyst were not altered by the treatment of the crystalline ZrO_2 with H_2SO_4.

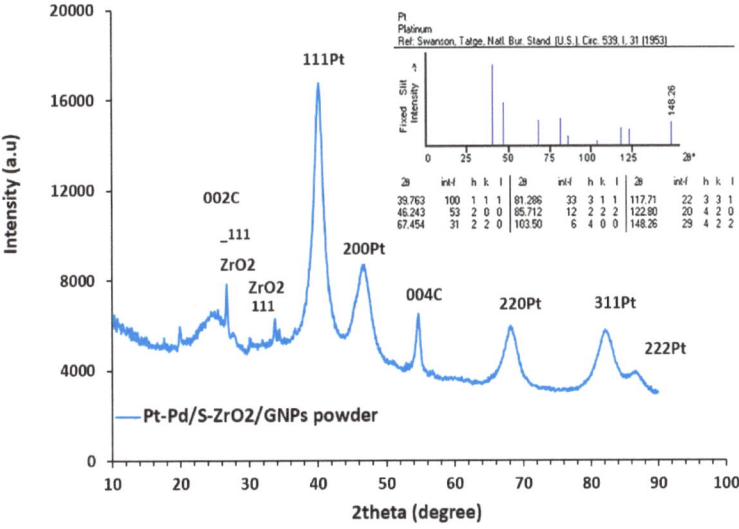

Figure 4. XRD diffraction chart of Pt-Pd/S-ZrO_2-GNP nanocomposite along with standard PDF card of Pt [71].

2.1.5. Chemical Composition of Synthesized Nanocomposite

Figure 5 illustrates the composition of the Pt-Pd/S-ZrO_2-GNP electrocatalyst. The main ingredients of the spectrum are Pt, Pd, Zr, S, and C (from the GNPs). The sulfur comes from sulfated zirconia. Additionally, the ingredients Si, Al, and K were present. The solid silicon and Al peaks are likely due to the Si and Al substratum used in the FESEM examination. A small quantity of potassium observed in the spectrum may be caused by the $KMnO_4$ used in graphene oxide production. A relatively trivial amount of Na and Cl observed in the EDS image is essentially from the NaCl electrolyte used in the plating bath. Table 1 provides further information on the chemical composition of the synthesized nanocomposite resulting from the energy-dispersive X-ray (EDX) spectrum. The Pt-Pd (with the atomic ratio of 0.68%/0.61%) loading on the electrocatalyst containing the graphene nanoplates was quantitatively computed as 18.24%, which is close to the theoretic extent of 20 wt%.

Table 1. Chemical composition of Pt-Pd/S-ZrO_2-GNP nanocomposite (quantitative values).

Element	Line	wt%
C	Kα	52.84
O	Kα	11.27
Na	Kα	0.95
Al	Kα	2.43
Si	Kα	1.77
S	Kα	10.30
Cl	Kα	0.28
K	Kα	0.57
Zr	Lα	1.35
Pt	Lα	10.03
Pd	Lα	8.21
		100

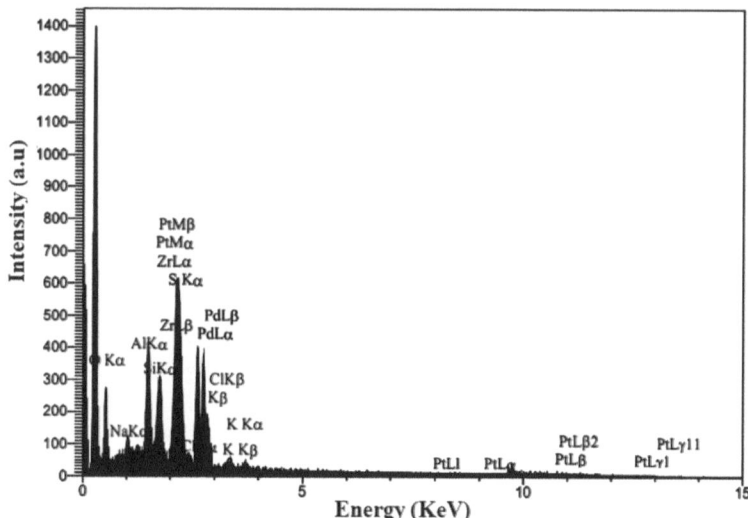

Figure 5. EDX pattern of Pt-Pd/S-ZrO$_2$-GNP electrocatalyst.

2.2. Electrochemical Measurements

2.2.1. Electrochemical Surface Area (ECSA) of Electrodes

Cyclic voltammetry is a commonly used electroanalytical technique to study electroactive species and electrode surfaces. It is usually the first electrochemical study performed. This technique is used to measure the ECSA of electrocatalysts using the hydrogen desorption method in a three-electrode system [72]. In Figure 6, the steady-state cyclic voltammograms of the different working electrodes, such as Pt-Pd/S-ZrO$_2$-GNPs, Pt-Pd/ZrO$_2$-GNPs, Pt/S-ZrO$_2$-GNPs, and Pt/C, are presented. These electrodes were submerged in a 0.1 M HClO$_4$ solution saturated by 99.9995% N$_2$ in the voltage range between -0.2 and 1.2 V vs. Ag/AgCl (saturated KCl) at the scan rate of 50 mVs^{-1}. All given potentials were converted and reported in the reversible hydrogen electrode (RHE) scale using the following: $E_{RHE} = E_{Ag/AgCl} + 0.205$ V ($=E_{RHE} = E_{Ag/AgCl} + \varnothing_{Ag/AgCl} + 0.0591(V) \times pH$) [73,74].

The figure shows clear hydrogen absorption/desorption peaks ranging from 0.04 to 0.3 V vs. RHE, along with distinct surface oxidation and reduction peaks. The electrochemical surface area (ECSA) of the electrocatalysts was evaluated by computing the hydrogen adsorption/desorption area after double-layer correction, using a conversion factor of 210 μCcm^{-2} for the polycrystalline Pt. The ECSA (m^2 g$_{metal}$$^{-1}$) was calculated as follows [75,76]:

$$\text{ECSA}\left(m^2 g_{metal}^{-1}\right) = \left[\frac{Q_{H-adsorption}(C)}{210 \mu Ccm^{-2} \times L_{Pt}(mgcm^{-2}) \times A_s(cm^{-2})}\right] 10^5 \quad (2)$$

where L_{Pt} is the Pt loading (mg cm^{-2}) on the surface of an electrode, Q_H is the hydrogen adsorption charge (mCcm^{-2}), and A_s is the electrode surface area.

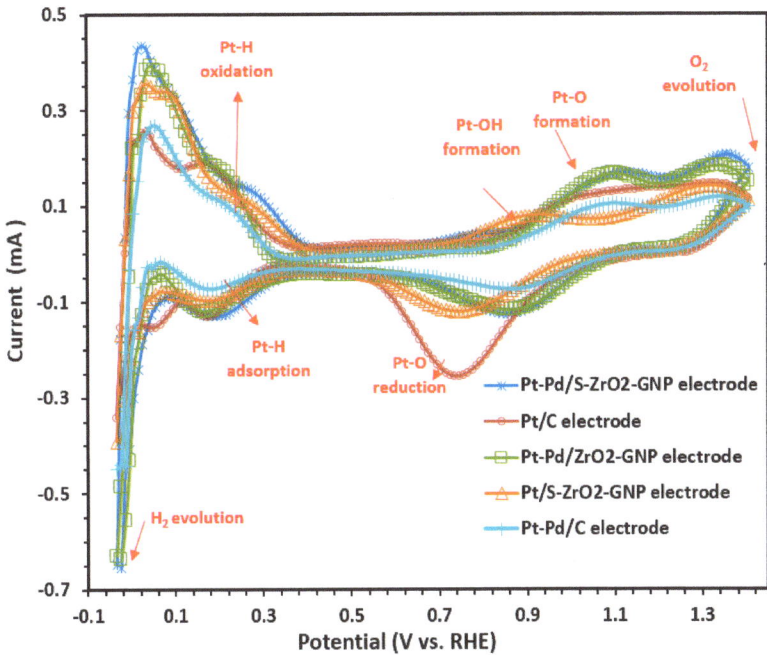

Figure 6. CVs of Pt-Pd/S-ZrO$_2$-GNPs, Pt-Pd/ZrO$_2$-GNPs, Pt/S-ZrO$_2$-GNPs, Pt-Pd/C, and Pt/C in 0.1 M HClO$_4$ solution scan rate: 50 mVs^{-1} at 25 °C under N$_2$ flux.

The comparison of the cyclic voltammograms of the Pt-Pd/S-ZrO$_2$-GNP, Pt-Pd/ZrO$_2$-GNP, Pt/S-ZrO$_2$-GNP, Pt-Pd/C and commercial Pt/C electrodes in Figure 6 indicates that the double-layer of the Pt-Pd/S-ZrO$_2$-GNP electrode is a little thinner than that of the Pt/S-ZrO$_2$-GNP electrode due to the contribution of Pd. Additionally, the current density of the H$_2$ release area extends when Pd is added, indicating a variation in the ECSA value of the catalyst due to Pd addition. Furthermore, adding Pd to platinum caused a slightly positive shift in the start of the anode O$_2$ chemisorption (oxide development) and the reduction peak voltage of the oxide. For instance, the start of the anode oxygen chemisorption for the Pt-Pd/S-ZrO$_2$-GNP, Pt-Pd/ZrO$_2$-GNP, and Pt-Pd/C electrodes occurred at approximately 0.87 V, 0.83 V, and 0.82 (vs. RHE), in comparison to 0.79 V for Pt/C (this value for Pt/S-ZrO$_2$-GNPs was 0.75 V). The contrast in the O$_2$ chemisorption and oxide formation between the Pt/S-ZrO$_2$-GNPs and Pt/C relative to the Pt-Pd/S-ZrO$_2$-GNP, Pt-Pd/ZrO$_2$-GNP, and Pt-Pd/C electrodes is due to the transition metal Pd. It suggests that the addition of Pd (alloying process) hinders the chemisorption of OH on the platinum sites at high voltages by altering the electronic effects (increase Pt d-band vacancy) and geometric (decrease in the Pt-Pt bond distance) factors [77,78]. This might facilitate the O$_2$ adsorption at low overpotential and, therefore, enhance the oxygen reduction reaction kinetic performance. Also, the cathodic current peaks associated with the reduction of platinum oxide positively shift for the Pt-Pd/S-ZrO$_2$-GNP, Pt-Pd/ZrO$_2$-GNP, and Pt-Pd/C electrodes as compared to the Pt/S-ZrO$_2$-GNP and Pt/C electrodes. This implies that the desorption of the oxygenate species (e.g., OH) from the surfaces of the alloy particles is easier than from the surface of pure Pt, i.e., the oxygenate species have lower adsorption energy on the Pt-Pd alloy catalysts. Since the adsorption of OH or other oxygenate species on the platinum surface can inhibit its catalytic activity toward oxygen reduction reaction, the weak adsorption of the oxygenated species would increase the surface active sites for ORR. Table 2 lists the ECSA for the Pt-Pd/S-ZrO$_2$-GNP, Pt-Pd/ZrO$_2$-GNP, Pt/S-ZrO$_2$-GNP, Pt-Pd/C, and Pt/C electrodes. According to the table, the ECSA of Pt-Pd/S-ZrO$_2$-GNPs is more significant than that of Pt/C, indicating that a more significant quantity of active sites is available for H$_2$ adsorption and desorption reactions. This could be due to the structural modifications caused by alloying, the high specific surface area (SSA) of GNPs as catalyst support as well as the positive effects of ZrO$_2$ as a promoter. Moreover, comparing the ECSA value of the Pt-Pd/S-ZrO$_2$-GNPs, Pt-Pd/ZrO$_2$-GNPs show that the catalyst layer containing sulfated ZrO$_2$-GNP support utilizes a significant amount of Pt, suggesting that the sulfating increases the efficiency of PEMFCs. Furthermore, from Table 2, it is seen that the bimetallic Pt-Pd/C electrode had a relatively lower Pt surface area than the monometallic Pt/C electrode. This was expected because some Pt surface atoms are covered by Pd atoms. The smaller ECSA value of the Pt-Pd/C electrode in comparison to the Pt/C electrode also can be explained by the obtained XRD results. The reduced Pt active surface area of the Pt-Pd/C electrode is consistent with its higher particle size produced by alloying.

Table 2. Electrochemical surface area of the Pt-Pd/S-ZrO$_2$-GNP, Pt-Pd/ZrO$_2$-GNP, Pt/S-ZrO$_2$-GNP, Pt-Pd/C, and Pt/C electrodes with a catalyst loading of 0.1 mgcm^{-2} in a N$_2$-saturated solution of 0.1 M HClO$_4$.

Electrode	Crystallite Size (XRD) (nm)	Q_H(C)	ECSA ($\frac{m^2}{g\,metal}$)
Pt-Pd/S-ZrO$_2$-GNPs	4.50	14.443 × 10^{-4}	97.32 **
Pt-Pd/ZrO$_2$-GNPs	4.54	14.011 × 10^{-4}	94.51 **
Pt/S-ZrO$_2$-GNPs	4.31	12.338 × 10^{-4}	83.21 *
Pt-Pd/C	4.39	9.940 × 10^{-4}	67.02 **
Pt/C(20 wt%)	4.20	10.205 × 10^{-4}	68.83 *

* Per unit weight of Pt. ** per unit weight of metals including both Pt and Pd.

It can be seen that the Pt-Pd/S-ZrO$_2$-GNP electrode had a higher Pt-based metal active surface area (97.32 m$^2 \cdot$ g^{-1}metal^{-1}) compared to the Pt/C electrode (68.83 m$^2 \cdot$ g^{-1}Pt^{-1}). The combination of sulfation, alloying, and GNPs helped maintain the catalytic activity of the Pt-Pd/S-ZrO$_2$-GNP composite.

2.2.2. Mechanism of Reaction

The activity of the Pt catalyst towards oxygen reduction is improved by the presence of a Pd alloy structure on S-ZrO$_2$-treated GNPs. It is proposed that the presence of S-ZrO$_2$ in the catalyst promotes the activity of ORR by PtPd which is explained by the flowing reactions [79,80].

$$Pt + O_2 \rightarrow Pt - O_2 \tag{3}$$

$$Pt - O_2 + H^+ + e^- \rightarrow Pt - HO_{2ad} \tag{4}$$

$$Pt + Pt - HO_2 \rightarrow Pt - HO + Pt - O \tag{5}$$

$$ZrO_2 + H_2O \rightarrow ZrO_2 - OH_{ad} + H^+ + e^- \tag{6}$$

$$Pt - O + ZrO_2 - OH_{ad} + 3H^+ + 3e^- \rightarrow Pt + ZrO_2 + 2H_2O \tag{7}$$

$$Pt - HO + ZrO_2 - OH_{ad} + 2H^+ + 2e^- \rightarrow Pt + ZrO_2 + 2H_2O \tag{8}$$

It is well known that a catalyst for oxygen reduction needs to break the O-O bonds of oxygen and facilitate the reaction of the ORR. During this reaction, some formed intermediates like OH and O species are strongly adsorbed on the Pt surface. The surface oxygen of

S-ZrO$_2$ helps reduce the adsorbed HO and/or O on the Pt surface [81]. So, the enhancement of the Pt-Pd/S-ZrO$_2$-GNP activity towards ORR is likely to occur at the interface of Pt and S-ZrO$_2$ in the reactions (7) and (8). The importance of the surface oxygen source for the effective O$_2$ reduction in fuel cell electrocatalysts has been highlighted in the case of carbon-coated tungsten oxides [82]. ZrO$_2$ is a well-known reducible oxide that shows high oxygen storage capacity and strong metal support interaction between Pt and ZrO$_2$ [63,83–85]. The overall reaction mechanism for oxygen reduction on the Pt–Pd/S-ZrO$_2$-GNP electrode can be proposed as follows:

$$PtPd \ldots S - ZrO_{2-2x}(\text{interface}) + 2xO_2 + 4xH^+ + 4xe^- \rightarrow PtPd \ldots S - ZrO_2(\text{interface}) + 2xH_2O \quad (9)$$

The enhancement of oxygen reduction is likely to occur at the interface of PtPd and S-ZrO$_2$ in the supported system. In the presence of S-ZrO$_2$ as a promoter, O$_2$ adsorption starts at a higher potential at the interface of PtPd... S-ZrO$_2$ thereby decreasing the anodic overpotentials. The Pt–O and Pt-H species are reduced to H$_2$O at the PtPd... S-ZrO$_2$ interface [81].

2.2.3. ORR Kinetics

The prepared electrodes were evaluated in two ways. Firstly, the kinematic current and the number of exchanged electrons were calculated in the experiments conducted on the rotating disk electrode (RDE). Secondly, the mass transfer corrected Tafel plots constructed from the RDE data and the kinetic parameters of i_0 were determined through the linear sweep voltammetry (LSV) analysis.

Estimation of Electrocatalyst Efficiency Using the RDE Instrument

In the present study, the RDE method was used to investigate the oxygen reduction reaction (ORR) of the five types of electrodes Pt/C, Pt-Pd/C, Pt/S-ZrO$_2$-GNPs, Pt-Pd/ZrO$_2$-GNPs, and Pt-Pd/S-ZrO$_2$-GNPs. The electrodes were prepared and tested in 0.1 M HClO$_4$ at room temperature in a three-electrode cell. The ORR experiments were examined by flowing 99.999% O$_2$ at a rate of 100 mL/min through the cell for 30 min using an MFC (model ALIGAT, SCIENTIFIC SCinTIFic, Tucson, AZ 85743, USA). The scanning voltage ranged from 1.1 to 0 V versus RHE, with a scan rate of 5 mV/s at a rotation speed of 1600 rpm. The results (Figure 7) showed that the onset potential of the Pt-Pd/S-ZrO$_2$-GNPs was shifted by 30 mV, 50 mV, 70 mV, and 90 mV to a positive direction when compared to the Pt-Pd/ZrO$_2$-GNPs, Pt/ZrO$_2$-GNPs, Pt-Pd/C, and Pt/C, respectively (from 1.01 for Pt/C to 1.1 V for Pt-Pd/S-ZrO$_2$-GNPs). It was observed that the electrochemical reaction occurred under mixed control, which was a combination of kinetic and diffusion control, ranging from 1.1 V to 0.7 V (vs. RHE). The diffusion limiting currents were achieved in the potential region under 0.7 V. The half-wave potential ($E_{1/2}$, the potential where the current is half of its limiting value, the point of half-way between zero current and the diffusion-limited current density plateau) of the Pt-Pd/S-ZrO$_2$-GNPs was found to be 960 mV, which was higher than that of the Pt/C, Pt-Pd/C, Pt/ZrO$_2$-GNPs, and Pt-Pd/ZrO$_2$-GNPs, which were 890 mV, 900 mV, 910 mV, and 940 mV, respectively. This indicates that the Pt-Pd/S-ZrO$_2$-GNP catalyst has higher activity and is more promising than the other catalysts for PEMFC cathode applications [74,76,86].

The loading amount of Pt and ECSA was normalized to obtain the mass and specific activities, correspondingly. The current values of mass and specific activity at 0.9 V (where influences of mass transport are negligible) for the Pt-Pd/S-ZrO2-GNPs were found to be as high as 45.43 mA mgPt^{-1} and 0.0466 mA mgPt^{-1}, respectively. In contrast, the values for Pt-Pd/ZrO$_2$-GNP, Pt/S-ZrO$_2$-GNP, Pt-Pd/C, and Pt/C electrodes were 40.75 mA mgPt^{-1} and 0.0431 mA mgPt^{-1}; 33.29 mA mgPt^{-1} and 0.0401 mA mgPt^{-1}; 28.67 mA mgPt^{-1} and 0.0402 mA mgPt^{-1}; and 23.54 mA mgPt^{-1} and 0.0342 mA mgPt^{-1}, respectively. The improved mass activity of the Pt-Pd/S-ZrO$_2$-GNPs can be attributed to the bifunctional mechanism of the Pt-Pd alloy nanoparticles, as well as the interface effect between the Pt-Pd nanoparticles and O_2 vacancy-rich ZrO_2 nanoparticles and GNPs.

The hydrodynamic behavior of the Pt-Pd/S-ZrO_2-GNP electrode in the ORR was studied through rotating disk electrode experiments at different rotation speeds in an oxygen-saturated 0.1 M $HClO_4$ solution at a scan rate of 5 mV s^{-1}. The results of the experiments are presented in Figure 8. The ORR current was measured using the Koutechy–Levich equation, a standard metric for comparing various electrocatalysts [74,76,87,88].

$$\frac{1}{I} = \frac{1}{I_k} + \frac{1}{B\omega^{\frac{1}{2}}} \tag{10}$$

$$B = 0.62nFD^{\frac{2}{3}}\vartheta^{\frac{-1}{6}}C_{O_2} \tag{11}$$

where D, ϑ, and C are the diffusion coefficient, kinematic viscosity, and dissolved oxygen concentration, respectively, in a 0.1 M $HCLO_4$ solution [89]. ω is the rotation speed in radians, and F is Faraday's constant. I_k and B are defined as the kinematic current and Levich parameter, respectively.

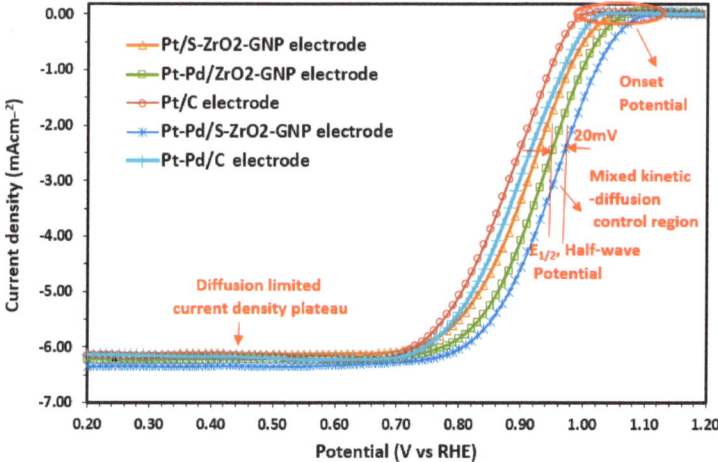

Figure 7. Polarization curves for ORR on Pt-Pd/S-ZrO_2-GNP, Pt-Pd/ZrO_2-GNP, Pt/S-ZrO_2-GNP, Pt-Pd/C, and Pt/C electrodes in 0.1 M $HClO_4$ solution scan rate: 5 mV s^{-1} rotation rate: 1600 rpm.

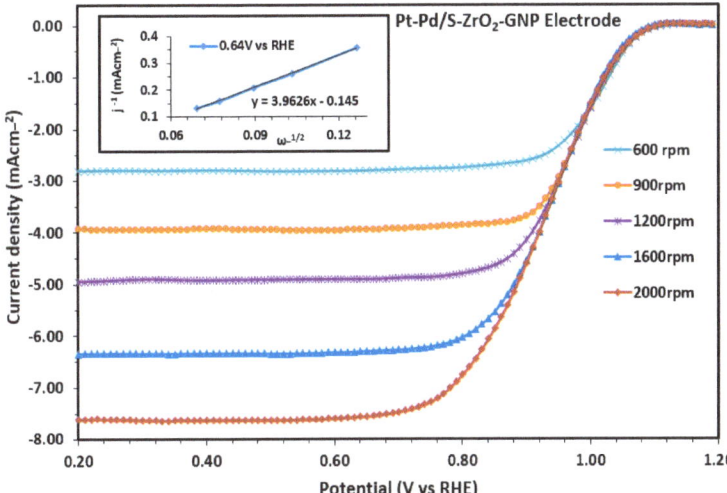

Figure 8. Disk current density achieved for the duration of the ORR in the cathodic scanning from the Pt-Pd/S-ZrO$_2$-GNP electrode at various RDE rotation speeds (scan rate: 5 mV s^{-1}). The inset displays the Koutechy–Levich plot attained at the diffusion flat terrain at 0.64 V.

The calculations were performed using Equations (10) and (11), with the following values: $D_{O2} = 1.9 \times 10^{-5}$ cm^2 s^{-1}, $\vartheta = 9.87 \times 10^{-3}$ cm^2 s^{-1}, and $C_{O2} = 1.6 \times 10^{-6}$ mol cm^{-3}. The Faraday constant is 96,485 C mol^{-1}. The calculated number of exchanged electrons with the kinematic current value of 3.13 mAcm^{-2} is 3.95 (n $\approx$ 4), which is in accordance with most scientific publications [90–93]. Consequently, oxygen reduction proceeds via the overall 4-electron path on the Pt surface of the Pt-Pd/S-ZrO$_2$-GNP electrode. A similar result was also obtained from the intercept of the inset curve of Figure 8. The inset in Figure 8 shows the current region for the Pt-Pd/S-ZrO$_2$-GNP catalyst.

Determination of Kinetic Parameters of b and i$_0$

The kinetic parameters were obtained after the measured currents were corrected for diffusion to give the kinetic currents in the mixed activation–diffusion region calculated based on the following equation:

$$i_k = \frac{i \times i_d}{i_d - i} \tag{12}$$

where $\frac{i_d}{i_d - i}$ is the mass transfer correction.

The Tafel plots (Figure 9) for Pt-Pd/S-ZrO$_2$-GNPs, Pt-Pd/ZrO$_2$-GNPs, Pt/ZrO$_2$-GNPs, Pt-Pd/C, and Pt/C were examined by plotting log($-i_k$) against overpotential. According to the Tafel equation, the relation between the electrode overpotential (η) and current density is specified as follows [94,95]:

$$\eta = -\frac{2.3RT}{n(1 - \alpha)F}\log i_0 + \frac{2.3RT}{n(1 - \alpha)F}\log(-i_k) \tag{13}$$

where $\eta = (E - E^0)$ is the difference between the applied voltage and the Nerst voltage, R is gas constant, T is absolute temperature, n is the number of exchanged electrons, α is the transfer coefficient, F is Faraday's constant, and i$_k$ and i$_0$ are the kinetic current density and exchange current density, respectively [95,96]. The equation can be simplified as follows:

$$\eta = a + b\log(-i_k) \tag{14}$$

where b = $\frac{2.3RT}{n(1-\alpha)F}$ and is called Tafel slope and a = $-\frac{2.3RT}{n(1-\alpha)F}$ logi_0. This equation implies that in a certain current density range, overpotential is linearly dependent on the logarithm of current density. The exchange current density can be obtained from the intercept (at $\eta = 0$) at the current density axis. In the higher Tafel slope, the kinetics of the ORR reaction is lower [96].

In Figure 9, the mass-transfer corrected Tafel plots from the RDE data are presented. Even though it is clear that the Tafel slope for the ORR is changing continuously on the potential range examined, in this figure, two Tafel regions with characteristic slopes near −120 and −60 mVdec^{-1} are distinguished, a transition in slope occurring at potentials between 0.95 and 1.02 V (vs. RHE, from 0.95 for Pt/C to 1.02 V for Pt-Pd/S-ZrO$_2$-GNPs). The value of −120 mVdec^{-1} indicates that the rate-determining step is the transfer of the first electron to the O$_2$ molecule. As suggested previously, the change in the slope is not related to the change in the reaction mechanism, but it has been attributed to the potential-dependent coverage of the surface oxides that inhibit the adsorption of oxygen molecules and reaction intermediates [97–100].

The experimental data were fitted to the two Tafel slope regions at low (E > 0.9 V, region1) and high overpotentials (E < 0.85 V, region2). The kinetic parameters were obtained using the Tafel equation and are presented in Table 3. They are in good agreement with the Tafel slopes for the carbon-supported Pt catalysts [88,101–103], with values around $-\frac{2.3RT}{n(1-\alpha)F}$ at low overpotentials and values of ca. $-2 \times \frac{2.3RT}{n(1-\alpha)F}$ at high overpotentials.

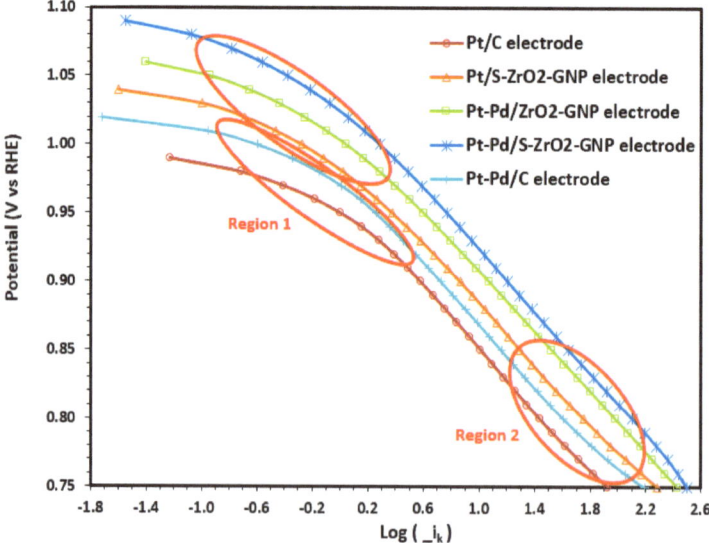

Figure 9. Mass transfer corrected Tafel plots for the ORR on Pt-Pd/S-ZrO$_2$-GNP, Pt-Pd/ZrO$_2$-GNP, Pt/S-ZrO$_2$-GNP, Pt-Pd/C, and Pt/C electrodes obtained from the RDE rotation speeds of 1600 rpm (sweep rate: 5 mV s^{-1}) in an O$_2$-saturated solution of 0.1 M HClO$_4$.

Table 3. Kinetic parameters of Pt-Pd/S-ZrO$_2$-GNP, Pt-Pd/ZrO$_2$-GN, Pt/S-ZrO$_2$-GNPs, Pt-Pd/C, and Pt/C electrodes in O$_2$-saturated solution of 0.1 M HClO$_4$.

Electrode	Tafel Slope: b($\frac{mV}{dec}$) in (E > 0.9 V)	Tafel Slope: b($\frac{mV}{dec}$) in (E < 0.85 V)	$i_0(\frac{A}{cm^2})$ For (E < 0.85 V)
Pt-Pd/S-ZrO$_2$-GNPs	−56	−106	1.662×10^{-3}
Pt-Pd/ZrO$_2$-GNPs	−57	−107	1.610×10^{-3}
Pt/S-ZrO$_2$-GNPs	−59	−111	1.021×10^{-3}
Pt-Pd/C	−58	−113	1.020×10^{-3}
Pt/C (20 wt%)	−61	−122	1.018×10^{-3}

Based on Figure 9, it is clear that the Pt-Pd/S-ZrO$_2$-GNP electrode by itself offers a substantial improvement over the Pt/C electrode. Obvious from this figure is the higher activity for ORR for the Pt-Pd/C electrode than the monometallic Pt/C electrode. The other Pt-Pd/S-ZrO$_2$-GNPs and Pt-Pd/ZrO$_2$-GNPs are very similar in activity as can be seen in Figure 9. This can be concluded also from the Tafel slopes obtained at the two electrodes of Pt-Pd/S-ZrO$_2$-GNPs and Pt-Pd/ZrO$_2$-GNPs in Table 3. This suggests that the same rate-determining step is occurring at two electrodes. The results in Figure 9 and Table 3 showed that the kinetic current was higher in the Pt-Pd/S-ZrO$_2$-GNPs compared to the Pt-Pd/ZrO$_2$-GNPs, Pt/ZrO$_2$-GNPs, Pt-Pd/C, and Pt/C. This indicates that charge transfer is faster on the Pt-Pd/S-ZrO$_2$-GNPs than on the others. These results are compatible with data reported in the references [73,104–107] for other Pt-alloy components. The improvement in the catalytic properties of Pt-alloys is attributed to various factors, such as the structural alterations caused by alloying, an increase in Pt d-band vacancy due to electronic factors, which weakens the strength of the metal-O band, and a reduction in the Pt-Pt bond space due to geometric factors [77,78]. GNPs are considered a favorable supporting material for enhancing the catalytic efficiency of Pt-based materials in fuel cell electrodes. Additionally, the use of sulfated metal oxides like sulfated zirconia as a Co-catalyst with Pt helps to enhance the electron and proton conductivity of S-ZrO$_2$-GNP nanocomposites.

2.2.4. Electrochemical Impedance Spectroscopy (EIS) Studies of the Electrodes

To seek the influence of carbon corrosion on electrode resistance changes and learn the interface process of electrodes, EIS analysis is utilized. The EIS examination of the electrodes was conducted at 0.05 V versus SCE by scanning frequencies ranging from 0.1 to 100 KHz with 67 decades and a changing sinusoidal signal of 0.05 V peak-to-peak superimposed on the DC potential. During the examinations, the electrodes were kept in an oxygen-saturated aqueous solution of 0.1 M *HClO$_4$*, and all EIS measurements were completed below 25 °C. The corresponding Nyquist plots (imaginary against actual impedance) are presented in Figure 10a. The impedance spectra of all the electrodes exhibit similar characteristics in the Nyquist plots, i.e., a broken-down semi-circle at the high-frequency section, whose diameter corresponds to the charge transfer resistance demonstrating the catalytic activity for ORR, and relatively straight lines, approximately 45° (Warburg) in the low-frequency area which relates to the diffusion-limiting of the electrocatalyst (the impedance chart of the commercial Pt/C electrode displayed a half-circle at high-frequency region). In these plots, the actual axis intercept at the high-frequency region relates to the unadjusted resistance of the electrolyte solution in bulk. As can be seen, charge transfer and ohmic resistance (IR) increased harshly in the Pt-Pd/C and Pt/C after substantial carbon corrosion. However, no major fluctuations were seen for the Pt-Pd/S-ZrO$_2$-GNP, Pt-Pd/ZrO$_2$-GNP, and Pt/S-ZrO$_2$-GNP samples (with sulfated zirconia-graphene nanoplate supports) suggesting simple oxygen reduction reactions on the surface of these electrodes. In fact, the charge transfer resistance (Rct) of the Pt-Pd/S-ZrO$_2$-GNP, Pt-Pd/ZrO$_2$-GN, and Pt/S-ZrO$_2$-GNP electrodes is considerably lower than that of Pt-Pd/C and Pt/C electrodes due to the well-known conductivity of the GNPs, showing a high activity for ORR. The finest results were achieved for the Pt-Pd/S-ZrO$_2$-GNP electrode, which exhibited a faster electron transfer and an improved mass diffusion kinetics after the sulfate treatment. Sulfated zirconia is responsible for less IR losses due to providing sufficient proton conductivity in the catalyst layer.

The contribution of each component to the cell resistance can be better understood by simplifying the circuits. In this experiment, the impedance data were fitted to appropriate equations, and the best equivalent circuits (EC) were selected, as displayed in the inset of Figure 10a. This circuit contains the sum of the electrode and electrolyte ohmic resistance (R1) with the charge transfer resistance (R2), which controls the electron transfer kinetics of electroactive kinds at the electrode boundary. The circuit also includes mass-transfer Warburge (W1) of the electrode parallel to the constant phase element, which belongs to the double-layer capacity (C1). In other words, the model R1 + (R2 + W1) × C1 corresponds to the CDC of R1(R2W1)C1.

In addition, the corresponding phase plots and bode diagrams in the real and imaginary parts of the impedance spectra of the studied electrodes are presented in Figure 10b–d and demonstrate that the ohmic resistance in all the electrodes did not change much during the ORR. However, in these diagrams, the lowest values of Rs+Rct and −1/Rct Cdl were found with the Pt-Pd/S-ZrO$_2$-GNP electrode emphasizing that the O$_2$ reduction reaction measured on the Pt-Pd/S-ZrO$_2$-GNP electrode was significantly faster than the others. This may be an explanation for the higher performance in the Pt-Pd/S-ZrO$_2$-GNP electrode in the polarization curve.

The orthogonal form of the Nyquist plots is also presented in Figure 10e.

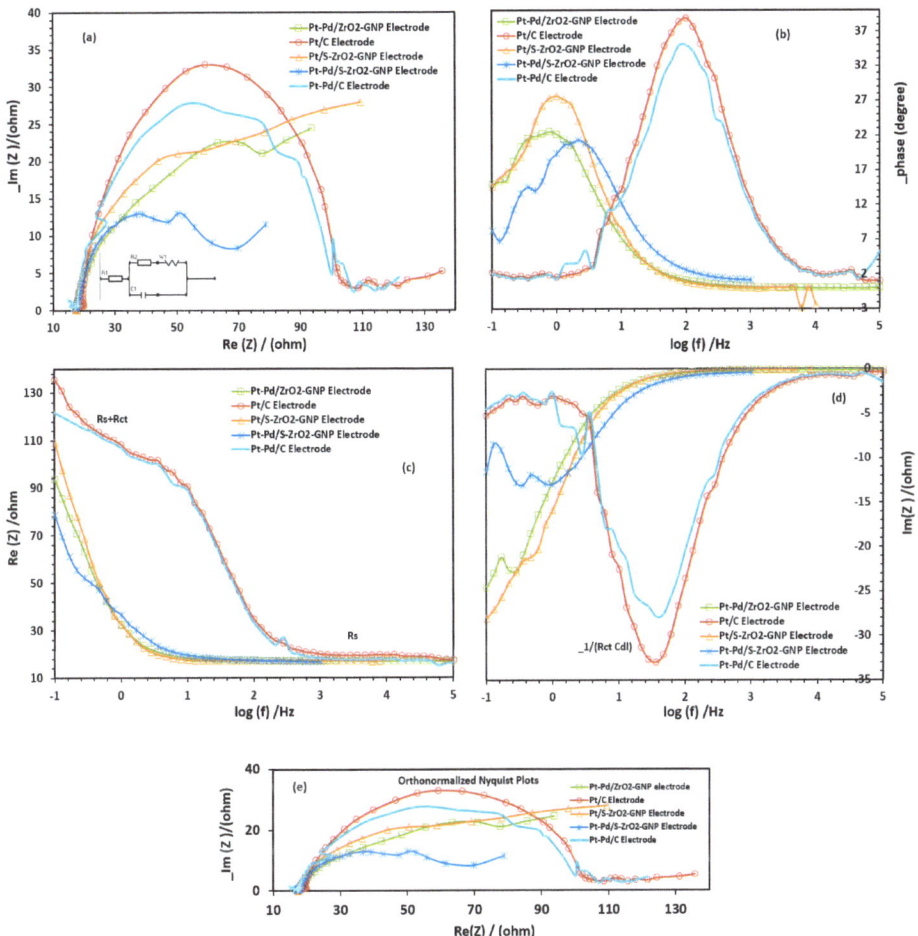

Figure 10. EIS chart in (**a**) Nyquist form, (**b**) bode phase plots, (**c**,**d**) bode magnitude diagrams, and (**e**) orthonormalized form of Nyquist plots of Pt-Pd/S-ZrO$_2$-GNP, Pt-Pd/ZrO$_2$-GNP, Pt/S-ZrO$_2$-GNP, Pt-Pd/C, and Pt/C electrodes in 0.1 M HClO$_4$ at 25 °C under O$_2$ flux.

2.2.5. Long-Term Activity and Durability

The effectiveness of the prepared electrocatalysts was evaluated through various indicators. The electrocatalysts were evaluated based on durability through successive voltage cycling in CV scanning and RDE. An accelerated durability test (ADT) was then performed on the four studied electrocatalysts. For CV scanning, first a working electrode with an electrodeposited catalyst in a plating bath covered with a Nafion-bonded catalyst layer dried in a vacuum oven at 80 °C (for the Pt/C electrode, the ink way prepared electrode, as described in Section 3.2.3, was used). The ADT involved cycling between potential −0.2 to 1.2 V (vs. Ag/AgCl reference electrode potential of 0.205 V stable in dilute acidic solution) for 500 cycles in a 0.1 M HClO$_4$ solution saturated by high purity (99.9995%) nitrogen with a scan rate of 50 mV/s. Then, the ORR mass activity of the samples was examined in RDE configuration in an O$_2$-saturated 0.1 M HClO$_4$ solution at a scan rate of 5 mV s^{-1} with a rotation speed of 1600 rpm. The corresponding mass and specific activity based on the RDE experiments, as well as the ECSA value after the ADT (also before the ADT), are presented in Table 4 and Figure 11.

Table 4. Mass and specific activity based on RDE experiments at 0.9 V as well as ECSA values of Pt-Pd/S-ZrO$_2$-GNP, Pt-Pd/ZrO$_2$-GNP, Pt/S-ZrO$_2$-GNP, Pt-Pd/C, and Pt/C electrodes before and after ADT test.

Test	Electrode	Mass Activity at 0.9 V (vs. RHE) (mA/mg metal)	Specific Activity (mA/mg metal)	ECSA ($\frac{m^2 Pt}{g metal}$)
Before ADT	Pt-Pd/S-ZrO2-GNPs	45.43	0.0466	97.32
	Pt-Pd/ZrO2-GNPs	40.75	0.0431	94.51
	Pt/S-ZrO2-GNPs	33.29	0.0401	83.21
	Pt-Pd/C	28.67	0.0402	67.02
	Pt/C	23.54	0.0342	68.83
After ADT	Pt-Pd/S-ZrO2-GNPs	17.20	0.0233	73.54
	Pt-Pd/ZrO2-GNPs	14.19	0.0213	66.49
	Pt/S-ZrO2-GNPs	9.03	0.0181	49.77
	Pt-Pd/C	5.85	0.0189	30.83
	Pt/C	3.48	0.0132	26.38

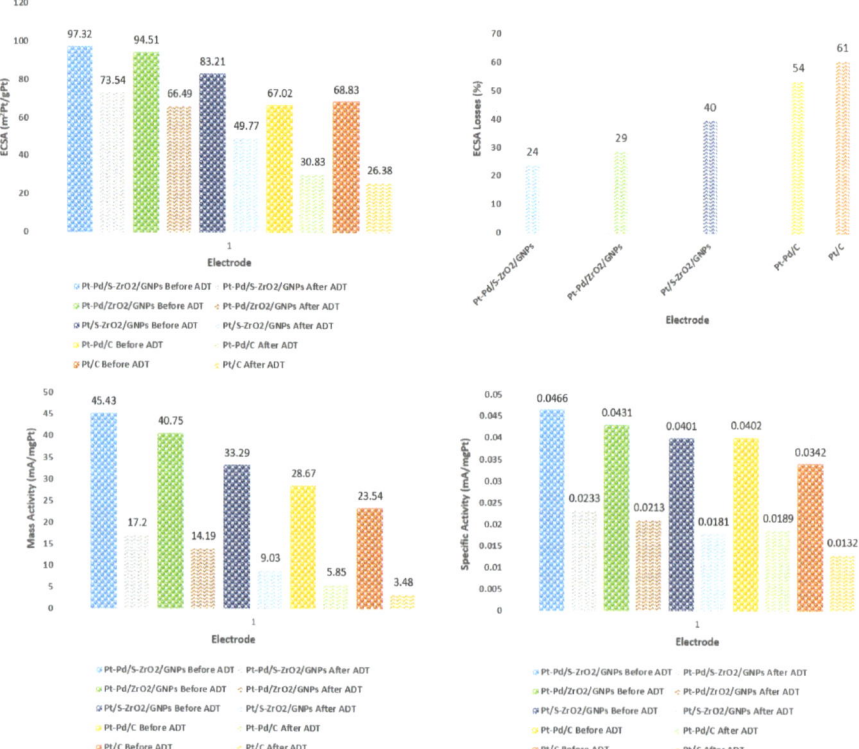

Figure 11. ECSA, mass, and specific activity based on RDE experiments at 0.9 V as well as ECSA values of Pt-Pd/S-ZrO$_2$-GNP, Pt-Pd/ZrO$_2$-GNP, Pt/S-ZrO$_2$-GNP, Pt-Pd/C, and Pt/C electrodes before and after ADT test.

The catalytic activity and stability of the Pt-Pd/S-ZrO$_2$-GNPs, Pt-Pd/ZrO$_2$-GNPs, Pt/S-ZrO$_2$-GNPs, Pt-Pd/C, composite electrocatalysts, and Pt/C were evaluated from the ECSA measurement after the ADTs and the results are presented in Figure 12a–f. With cycling, a reduction in the H$_{UPD}$ peak is observed for all electrocatalysts indicating an increase in the metal particle size. The loss of electrochemical surface area was observed on four catalysts during potential cycling caused by Pt dissolution and cluster formation. It is clear from Figure 12a that the ECSA for the Pt-Pd/S-ZrO$_2$-GNP electrode is more significant than that of the Pt-Pd/ZrO$_2$-GNPs, Pt/S-ZrO$_2$-GNPs, Pt-Pd/C, and commercial Pt/C after the CV cycling, indicating both of the sulfating metal oxide/GNPs and alloying effects can preserve the electrocatalytic activity and stability of the composite. The ECSA values of the Pt-Pd/S-ZrO$_2$-GNP electrode decreased by 24% after 500 cycles, while that of the Pt-Pd/ZrO$_2$-GNP, Pt/S-ZrO$_2$-GNP, Pt-Pd/C, and Pt/C electrodes decreased by 29, 40, 54 and 61%, respectively. The results disclosed that the Pt-Pd/S-ZrO$_2$-GNP electrode was more electrochemically stable than the Pt-Pd/ZrO$_2$-GNP, Pt/S-ZrO$_2$-GNP, Pt-Pd/C, and Pt/C electrodes. Another important point that can be seen from Table 4 is that although the electrochemically active surface area of the bimetallic Pt-Pd/C electrode in the first state is lower than the monometallic Pt/C electrode, its ECSA value is higher than that of the Pt/C after the ADT tests, and this can be explained by the alloying effects, which prevent the increase in ECSA losses.

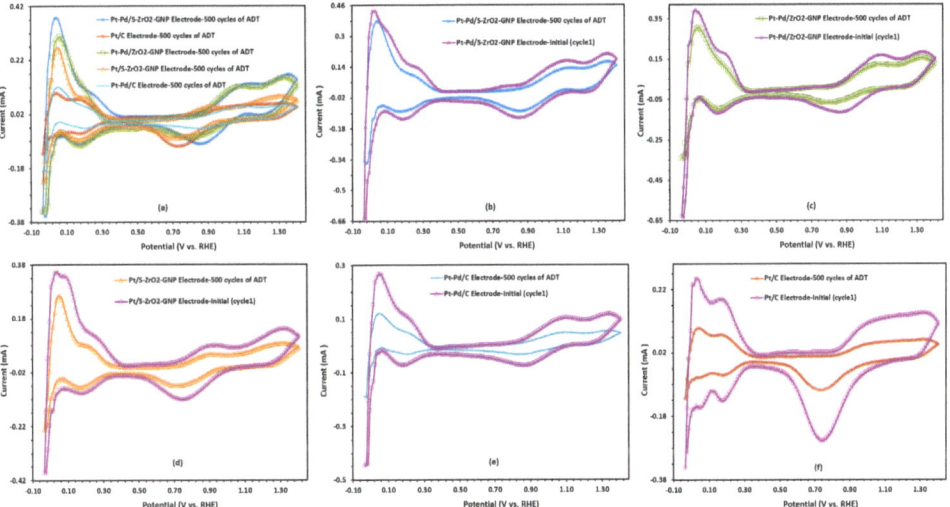

Figure 12. The CV curves (**a**–**f**) after 500 cycles of consecutive voltage scanning of the Pt-Pd/S-ZrO$_2$-GNP, Pt-Pd/ZrO$_2$-GNP, Pt/S-ZrO$_2$-GNP, Pt-Pd/C, and Pt/C electrodes in a N$_2$-saturated 0.1 M HClO$_4$ solution, scan rate: 50 mVs^{-1}.

Also, the durability of each electrode was further examined by the RDE experiments in an O$_2$-saturated 0.1M HClO$_4$ solution at a scan rate of 5 mV s^{-1} with a rotation speed of 1600 rpm, and the results are presented in Figure 13. According to Table 4, the ORR mass activity of Pt-Pd/S-ZrO$_2$-GNP composite electrocatalyst after 500 cycles at 0.9 V is 17.20 mA mg^{-1}Pt^{-1}, which is higher than that of the Pt-Pd/ZrO$_2$-GNPs (14.19), Pt/S-ZrO$_2$-GNPs (9.03), Pt-Pd/C (5.85), and Pt/C (3.48) after the ADTs. The results confirm once again that the Pt-Pd/S-ZrO$_2$-GNP composite electrocatalyst has a high tolerance to electrochemical corrosion and premiere electrocatalytic performance in ORR.

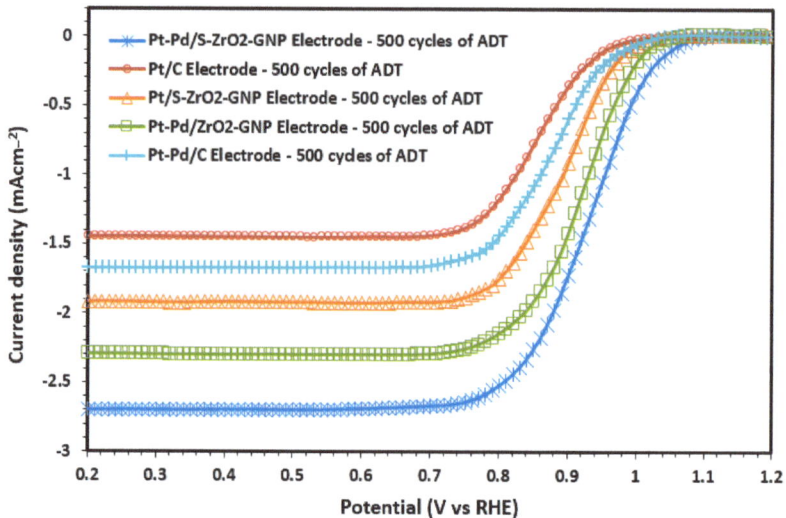

Figure 13. Polarization curves for ORR on Pt-Pd/S-ZrO$_2$-GNP, Pt-Pd/ZrO$_2$-GNP, Pt/S-ZrO$_2$-GNP, Pt-Pd/C, and Pt/C electrodes after 500 cycles of ADTs in O$_2$-saturated 0.1 M HClO$_4$ solution, scan rate: 5 mV s^{-1} and rotation speed: 1600 rpm.

3. Materials and Methods

3.1. Materials

Graphite powder (Gr) of 99.9999% purity was bought from Alfa Aesar, Heysham, Lancashire LA3 2XY, UK. Chemical materials, including H$_2$PtCl$_6$·6H$_2$O (40%), ZrOCl$_2$·8H$_2$O, Na$_2$NO$_3$, KMnO$_4$, H$_2$SO$_4$, 2-propanol, ethanol, (CH$_2$OH)$_2$, AgNO$_3$, NaCL, and H$_2$O$_2$, were obtained from Sigma-Aldrich, Chemie GmbH, Eschenstr. 5, Taufkirchen, Germany. PdCl$_2$ (59 wt.%) obtained from Scharlau chemie S.A., European Union, Pt/C (20 wt%) and Nafion® solution (5 wt%, Dupont, Wilmington, Delaware) were purchased from Fuel Cell Earth Company, Woburn, MA, USA. O$_2$ (99.999%) and N$_2$ (99.9995%) gases were provided by Canadian Sigma Inc. 2149 Winston Park Drive, Oakville, ON L6H 6J8, Canada. Milli-Q water was utilized during the electrodeposition and electrochemical analysis. Polishing kits and glassy carbon (GC) electrode (d = 3 mm) were procured from Bio-Analytical System (BASi Corporate Headquarters, 2701 Kent Avenue. West Lafayette, IN 47906, USA).

3.2. Methods

3.2.1. Synthesis of Graphene Nanoplates

To convert the pristine Gr to graphite oxide, the method of Hummer and Offenman was modified under a usual oxidation synthesis method [108]. In a container containing 90 mL of concentrated sulfuric acid (95%), 1 g of NaNO$_3$ and 2 g of graphite were added to an ice bath. Gradually, 6 g KMnO$_4$ was added and the mixture was agitated at 30 ± 5 °C for around 10 h. To the contents, 300 mL of deionized (DI) water was added and then (after about 30 min) diluted with 500 mL DI water. An amount of 5% H$_2$O$_2$ was added dropwise to the solution until the brown slurry turned yellow. The mixture was filtered, and the residue was dispersed in DI water through an ultrasonic system. By performing several centrifugation processes at 11,000 rpm for 25 min, the residue was washed out with 1:30 hydrochloric acid dilution and water to a pH value of 7. The filtrate was then dried in a vacuum furnace at 80 °C for 20 h. To generate graphene nanoplates, the resultant was dispersed in DI water and exfoliated through sonication via a welding horn. After decanting the mixture into a flask, NH$_2$NH$_2$. H$_2$O (hydrazine monohydrate) was added as a reducing agent to the brown GO nanosheet dispersion. The solution was then refluxed at about 100 °C for 3 h which caused the color to repeatedly change into dark black due to the advent of the GNP dispersal floating on the surface of the solution. A small amount of the precipitate was removed through centrifugation for 10 min at 3500 rpm. The GNPs' dispersion supernatants were directly dried in a vacuum oven to obtain the bulk of the graphene nanoplate powder.

3.2.2. Synthesis of S-ZrO$_2$-GNP Support

For synthesizing S-ZrO$_2$-GNP nanocomposite, GNPs, ZrCl$_2$ · 8H$_2$O, H$_2$SO$_4$, and NH$_3$ were used as the beginning material, sulfating agent, and precipitating agent. The blend was adjusted to a pH of 10 by regularly dripping 28% ammonia solution into zirconium oxychloride octahydrate solution (0.20 M) with proper amounts of GNPs and agitated for 36 h. The obtained ZrO$_2$ · nH$_2$O sol was washed with double distilled water by a centrifuge until chloride ions were not noticed by 0.1 N AgNO$_3$, dried at 130 °C for 6 h, and powdered. The ZrO$_2$ · 8H$_2$O and GNP admixture was added to 0.50 M H$_2$SO$_4$ with strong stirring for 25 min, filtrated, and dried at 120 °C. Following that, the resulting powder was calcinated at 600 °C under airflow for 40 min, and the obtained support was regarded as S-ZrO$_2$-GNPs. For comparison, the ZrO$_2$-GNP powder was also prepared in the same way except that the sulfation step was not performed.

3.2.3. Electrodeposition of Pt-Pd Nanoparticles on the S-ZrO$_2$-GNP-Contained Glassy Carbon

A two-electrode cell was used to carry out the galvanostatic pulse co-electrodeposition of Pt-Pd. To begin, 10 mg of S-ZrO$_2$-graphene powder was mixed with 1000 μL of isopropanol solution and sonicated for 30 min to create a soft slurry. Next, 10 μL of the slurry was micro-pipetted onto the top of the glassy carbon electrode and dried at 30 °C. The experiment used an S-ZrO$_2$/graphene ink on glassy carbon as a cathode and Pt wire as an anode electrode. The electrodeposition process was carried out in a plating cell containing a solution of 4 mM H$_2$PtCl$_6$ · 6H$_2$O + 4 mM PdCl$_2$ dissolved in 0.5 M NaCl [109]. The pH of the solution was brought to 2.8 by adding HCl and NaOH. A peak current density of 300 mAcm^{-2} was applied with an on/off time of 10/100 ms, resulting in a duty cycle of 10% for the studied electrode. The solution was gently stirred during electrodeposition to pass new metal ions to the cathode and remove any gas bubbles produced. A total of 4 mM concentration of Pt and Pd metal precursors were calculated to reach 20% of the electrocatalyst value based on the theoretical assumption, and the step of applying the S-ZrO$_2$/graphene slurry to the working electrode before electrodeposition was repeated to reach the support weight of 80% and was controlled by weighing. Before the Pt-Pd electrodeposition, the active surface of the glassy carbon containing sulfated-zirconia-doped graphene nanoplates was impregnated with a solution of 0.05% Nafion in ethanol. Since the S-ZrO$_2$ nanocrystals act as a proton conductor in the catalyst layer, only a small amount of Nafion ionomer was required in the catalyst layer composition of the electrodes in comparison to the preparation of the conventional Pt/C electrodes. After the electrodeposition, the electrode was instantly taken away from the electrolyte, carefully washed with DI water, and dried out under an IR lamp. The total amount of charge passed through the electrode was 200 mCcm^{-2}, corresponding to approximately 0.1 mgcm^{-2} of the Pt-Pd alloy catalyst according to a formula presented in our previously published article [51]. The amount of Pt-Pd deposited increased with an increase in charge passed. The metal deposition time, assuming a faradic yield of 100% is approximately 36,630 ms (37 s) with t_{off} = 100 ms and t_{on} = 10 ms.

The experimental methodology to synthesize the Pt-Pd/S-ZrO$_2$-GNP electrode is depicted in Figure 14. Also, the current(potential)–time charts of Pt-Pd/S-ZrO$_2$-GNP electrode under the electrodeposition condition at the end of the electrodeposition process and the quantity of Pt-Pd electrodeposited on the GC surface have been shown in Figure 15a,b. A conventional Pt/C electrode was made up by spraying a slurry of commercial 20 wt% Pt/C nanoparticles in ink preparation way onto a GC electrode with a Pt/C loading of

0.1 mg/cm² and an optimal Nafion load of 0.28 mg/cm² according to the formula below which was introduced in ref. [110].

$$\text{Nafion} \left(\text{mg cm}^{-2} \right) \cong 56 \frac{L_{Pt}}{P_{Pt}} \quad (15)$$

where L_{Pt} is the platinum loading (mg cm^{-2}) and P_{Pt} the weight percentage of the metal supported on carbon (Pt/C).

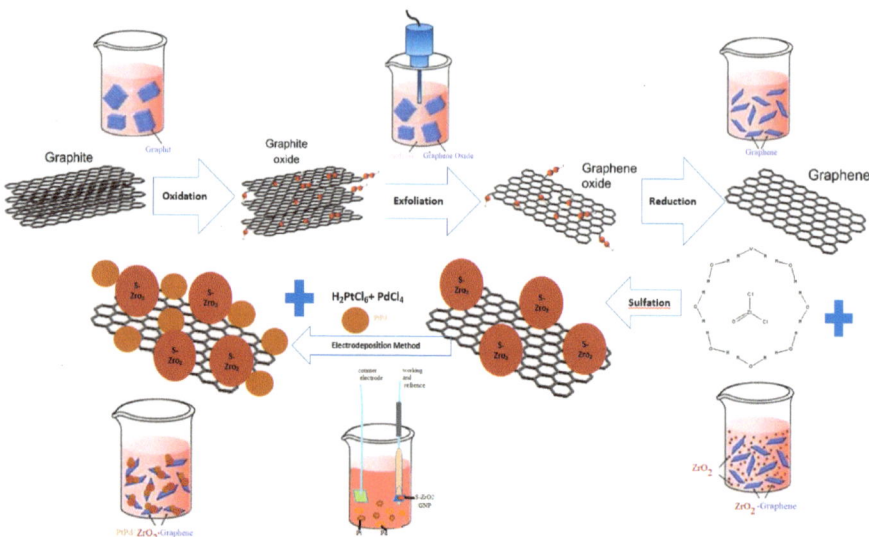

Figure 14. Schematic of experimental methodology for synthesizing S-ZrO$_2$-GNPs and fabricating Pt-Pd/S-ZrO$_2$-GNPs.

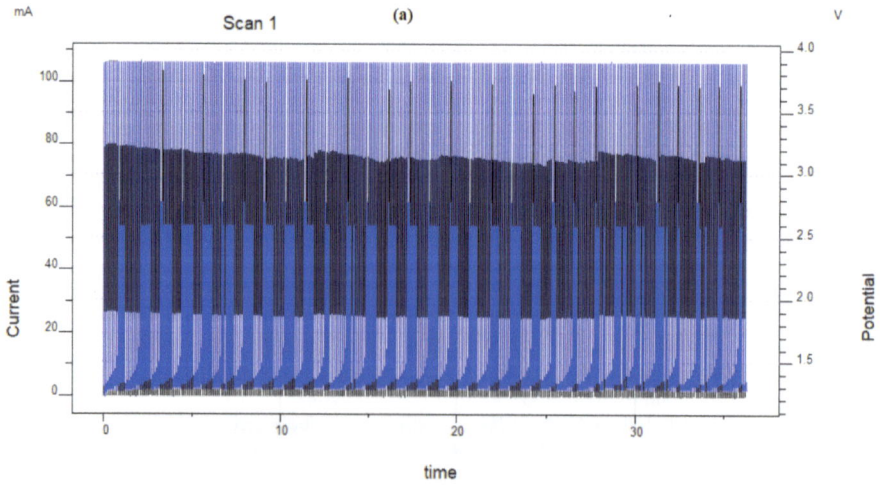

Figure 15. *Cont.*

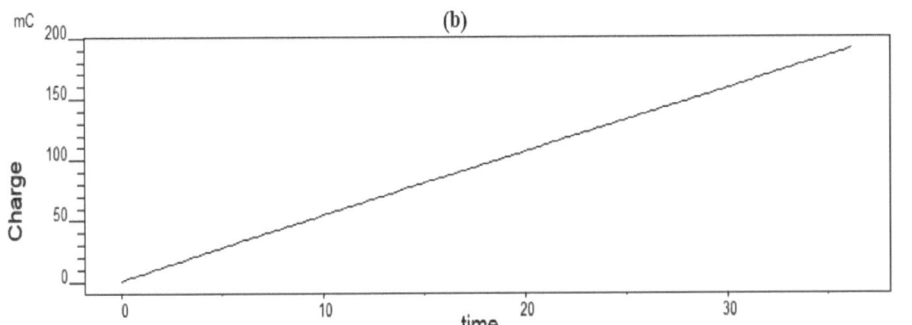

Figure 15. Presentation of the (**a**) current(potential)–time charts of the Pt-Pd/S-ZrO$_2$-GNP electrode under the electrodeposition condition at the end of the electrodeposition process and (**b**) charge–time chart of the amount of Pt-Pd electrodeposited on the S-ZrO$_2$-GNPs.

3.2.4. Characterization and Analysis

The Pt-Pd/S-ZrO$_2$-GNPs were analyzed using an Equinox 3000 spectrometer (IENL France, Head Quarters, INEL, Paris, France). The analysis employed Cu Kα λ = 0.15406 nm radiation yielded at 40 kV and 30 mA with a resolution of $\leq$0.1°. The AFM (model Nanosurf Easyscan2, Nanosurf AG, Gräubernstrasse 12, 4410 Liestal, Switzerland) was used in contact mode to study the physical structure of graphene nanosheets. A field emission scanning electron microscopy (FESEM) depiction and EDX spectroscopy joined with SEM MAG100.00kx with a silicon detector were performed at 15 kV. The particle morphology and size of Pt-Pd nanoparticles deposited on the support are characterized by transition electron microscopy (TEM, EM208, 1 − −100 kV, Philips, Eindhoven, The Neterlands). The FTIR spectra of GO, ZrO$_2$/GNPs, and S-ZrO$_2$-GNP composites were obtained using a WQF-510A/520 FTIR spectrometer (No.160 Beiqing Road, Haidian District, Beijing 100095, China). The Pt-Pd electrodeposition and electrochemical measurements as well as the ORR experiments were conducted on a conventional 2- and 3-electrode cell using Iviumstat potentiostat/galvanostat (Vertex, De Zaale 11, 5612 AJ Eindhoven, The Netherlands), respectively. A Pt foil was used as a counter electrode, Ag/AgCl saturated KCl as the reference electrode, and a glassy carbon disk as the working electrode. Due to safety considerations in the laboratory, instead of the Pt/H$_2$/H$^+$ standard dynamic hydrogen electrode, the most popular Ag/AgCl/Cl$^-$ reference electrode was used to record data in an acidic solution of 0.1 M HClO$_4$. Before use, the electrode surface was polished via alumina suspension with successively reduced particle sizes between 1 and 0.05 μm on polishing mats. The electrode was then ultrasonicated in C$_2$H$_5$OH and DI water for 15 min to eliminate contamination.

4. Conclusions

The corrosion of carbon supports and the need for more durable materials have led to the exploration of alloy- and ceramic-based support materials for Pt catalysts. Non-precious metal oxides are promising options due to their resistance to corrosion in severe fuel cell environments. However, their low surface area and electric conductivity prevent them from being used as primary support materials. In this paper, Pt-Pd/S-ZrO$_2$-GNP nanocomposite electrode was prepared and compared with Pt-Pd/ZrO$_2$-GNP, Pt/S-ZrO$_2$-GNP, Pt-Pd/C, and Pt/C electrodes in terms of the electrochemical activity and durability for ORR using CV, RDE, polarization curves (Tafel plots) and EIS in acidic solutions. All the results showed that the Pt-Pd/S-ZrO$_2$/GNPs gave a higher catalytic activity and durability for ORR. The Pt-Pd/ZrO$_2$-GNP electrode had a higher electrochemical surface area with a positive shift peak potential for ORR than other studied electrodes. The ECSA of the Pt-Pd/S-ZrO$_2$/GNPs was 97.32 m^2/gPt, which was higher than that of the Pt-

Pd/ZrO$_2$-GNPs, Pt/S-ZrO2-GNPs, Pt-Pd/C, and Pt/C with values of 94.51, 83.21, 67.02, and 68.83 m^2/gPt, respectively. Furthermore, the mass activity of the Pt-Pd/S-ZrO$_2$/GNPs for ORR at 0. 9 V versus RHE was 45.43 mA mg^{-1}Pt^{-1}, which was 1.92 times higher than commercial Pt/C (23.54 mA mg^{-1}Pt^{-1}) based on RDE experiments. The electrocatalyst also demonstrated high activity for ORR cathode operation after 500 cycles of durability testing. The electrochemical surface area of the Pt-Pd/S-ZrO$_2$/GNP electrode remained at 76% of its initial value. The charge transfer resistance of the Pt-Pd/S-ZrO$_2$/GNPs was smaller than that of the Pt-Pd/ZrO$_2$-GNPs, Pt/S-ZrO$_2$-GNPs, Pt-Pd/C, and Pt/C, indicating an increase in reaction kinetics. The enhanced mass activity and durability of the Pt-Pd/S-ZrO$_2$/GNPs could be attributed to the synergistic effect between Pt-Pd alloy NPs, oxygen vacancy-rich ZrO$_2$ NPs, sulfation effect, and high conductive GNPs. The function of each component in the Pt-based electrocatalyst was analyzed to determine their impact on electrochemical surface area, electron, and proton conductivity. It was suggested that a collaborative effect exists between metal oxides (ZrO$_2$), deposited Pt-Pd NPs, and GNP sublayer, which improves the electron transfer rate. In addition, another effect related to the attractive feature of the S-ZrO$_2$/GNP composite increases electron and proton conductivity. Based on the results of this paper, the Pt-Pd/S-ZrO$_2$-GNPs can be selected as one of the best electrodes with excellent ORR for PEMFC application. The application of sulfated metals, i.e., S-ZrO$_2$ as a Co-catalyst of Pt-Pd, seems to be a promising oxygen reduction cathode catalyst.

Author Contributions: Conceptualization, M.Y.; investigation, M.Y. and M.A.; original draft preparation, M.Y.; writing, review, and editing, M.Y. and M.A. All authors have read and agreed to the published version of the manuscript.

Funding: This research received no external funding.

Institutional Review Board Statement: Not applicable.

Informed Consent Statement: Not applicable.

Data Availability Statement: Data is contained within the article.

Conflicts of Interest: The authors declare no conflicts of interest.

References

1. Tian, N.; Lu, B.-A.; Yang, X.-D.; Huang, R.; Jiang, Y.-X.; Zhou, Z.-Y.; Sun, S.-G. Rational design and synthesis of low-temperature fuel cell electrocatalysts. *Electrochem. Energy Rev.* **2018**, *1*, 54–83. [CrossRef]
2. Hu, B.; Yuan, J.; Zhang, J.; Shu, Q.; Guan, D.; Yang, G.; Zhou, W.; Shao, Z. High activity and durability of a Pt–Cu–Co ternary alloy electrocatalyst and its large-scale preparation for practical proton exchange membrane fuel cells. *Compos. Part B Eng.* **2021**, *222*, 109082. [CrossRef]
3. Xu, X.; Wang, W.; Zhou, W.; Shao, Z. Recent advances in novel nanostructuring methods of perovskite electrocatalysts for energy-related applications. *Small Methods* **2018**, *2*, 1800071. [CrossRef]
4. Li, Y.; Gui, F.; Wang, F.; Liu, J.; Zhu, H. Synthesis of modified, ordered mesoporous carbon-supported Pt3Cu catalyst for enhancing the oxygen reduction activity and durability. *Int. J. Hydrog. Energy* **2021**, *46*, 37802–37813. [CrossRef]
5. Zhao, W.; Ye, Y.; Jiang, W.; Li, J.; Tang, H.; Hu, J.; Du, L.; Cui, Z.; Liao, S. Mesoporous carbon confined intermetallic nanoparticles as highly durable electrocatalysts for the oxygen reduction reaction. *J. Mater. Chem. A* **2020**, *8*, 15822–15828. [CrossRef]
6. Han, X.-F.; Batool, N.; Wang, W.-T.; Teng, H.-T.; Zhang, L.; Yang, R.; Tian, J.-H. Templated-assisted synthesis of structurally ordered intermetallic Pt3Co with ultralow loading supported on 3D porous carbon for oxygen reduction reaction. *ACS Appl. Mater. Interfaces* **2021**, *13*, 37133–37141. [CrossRef] [PubMed]
7. Lin, R.; Zheng, T.; Chen, L.; Wang, H.; Cai, X.; Sun, Y.; Hao, Z. Anchored Pt-Co nanoparticles on honeycombed graphene as highly durable catalysts for the oxygen reduction reaction. *ACS Appl. Mater. Interfaces* **2021**, *13*, 34397–34409. [CrossRef] [PubMed]
8. Britto, P.J.; Santhanam, K.S.; Rubio, A.; Alonso, J.A.; Ajayan, P.M. Improved charge transfer at carbon nanotube electrodes. *Adv. Mater.* **1999**, *11*, 154–157. [CrossRef]
9. Tada, T. *Handbook of Fuel Cells: Fundamentals, Technology, and Applications*; Vielstich, W., Lamm, A., Gasteiger, H., Eds.; John Wiley & Sons: New York, NY, USA, 2003; Volume 3.
10. Palit, G.; Elayaperumal, K. Passivity and pitting of corrosion resistant pure metals Ta, Nb, Ti, Zr, Cr and Al in chloride solutions. *Corros. Sci.* **1978**, *18*, 169–179. [CrossRef]
11. Ioroi, T.; Siroma, Z.; Fujiwara, N.; Yamazaki, S.-i.; Yasuda, K. Sub-stoichiometric titanium oxide-supported platinum electrocatalyst for polymer electrolyte fuel cells. *Electrochem. Commun.* **2005**, *7*, 183–188. [CrossRef]

12. Chhina, H.; Campbell, S.; Kesler, O. An oxidation-resistant indium tin oxide catalyst support for proton exchange membrane fuel cells. *J. Power Sources* **2006**, *161*, 893–900. [CrossRef]
13. Arata, K.i.; Hino, M. Preparation of superacids by metal oxides and their catalytic action. *Mater. Chem. Phys.* **1990**, *26*, 213–237. [CrossRef]
14. Hara, S.; Miyayama, M. Proton conductivity of superacidic sulfated zirconia. *Solid State Ion.* **2004**, *168*, 111–116. [CrossRef]
15. Uchida, M.; Fukuoka, Y.; Sugawara, Y.; Eda, N.; Ohta, A. Effects of microstructure of carbon support in the catalyst layer on the performance of polymer-electrolyte fuel cells. *J. Electrochem. Soc.* **1996**, *143*, 2245. [CrossRef]
16. Wu, Z.; Sun, G.; Jin, W.; Hou, H.; Wang, S.; Xin, Q. Nafion® and nano-size TiO_2–SO_4^{2-} solid superacid composite membrane for direct methanol fuel cell. *J. Membr. Sci.* **2008**, *313*, 336–343. [CrossRef]
17. Suzuki, Y.; Ishihara, A.; Mitsushima, S.; Kamiya, N.; Ota, K.-i. Sulfated-zirconia as a support of Pt catalyst for polymer electrolyte fuel cells. *Electrochem. Solid-State Lett.* **2007**, *10*, B105. [CrossRef]
18. Choi, P.; Jalani, N.H.; Datta, R. Thermodynamics and Proton Transport in Nafion: III. Proton Transport in Nafion/Sulfated Nanocomposite Membranes. *J. Electrochem. Soc.* **2005**, *152*, A1548. [CrossRef]
19. Ma, Z.; Li, S.; Wu, L.; Song, L.; Jiang, G.; Liang, Z.; Su, D.; Zhu, Y.; Adzic, R.R.; Wang, J.X. NbO_x nano-nail with a Pt head embedded in carbon as a highly active and durable oxygen reduction catalyst. *Nano Energy* **2020**, *69*, 104455. [CrossRef]
20. Liu, Y.; Ishihara, A.; Mitsushima, S.; Kamiya, N.; Ota, K.-i. Zirconium oxide for PEFC cathodes. *Electrochem. Solid-State Lett* **2005**, *8*, A400. [CrossRef]
21. Kodama, K.; Shinohara, A.; Hasegawa, N.; Shinozaki, K.; Jinnouchi, R.; Suzuki, T.; Hatanaka, T.; Morimoto, Y. Catalyst poisoning property of sulfonimide acid ionomer on Pt (111) surface. *J. Electrochem. Soc.* **2014**, *161*, F649. [CrossRef]
22. Shinozaki, K.; Morimoto, Y.; Pivovar, B.S.; Kocha, S.S. Suppression of oxygen reduction reaction activity on Pt-based electrocatalysts from ionomer incorporation. *J. Power Sources* **2016**, *325*, 745–751. [CrossRef]
23. Subbaraman, R.; Strmcnik, D.; Paulikas, A.P.; Stamenkovic, V.R.; Markovic, N.M. Oxygen Reduction Reaction at Three-Phase Interfaces. *ChemPhysChem* **2010**, *11*, 2825–2833. [CrossRef] [PubMed]
24. Yarlagadda, V.; Carpenter, M.K.; Moylan, T.E.; Kukreja, R.S.; Koestner, R.; Gu, W.; Thompson, L.; Kongkanand, A. Boosting fuel cell performance with accessible carbon mesopores. *ACS Energy Lett.* **2018**, *3*, 618–621. [CrossRef]
25. Ho, V.T.T.; Pan, C.-J.; Rick, J.; Su, W.-N.; Hwang, B.-J. Nanostructured $Ti0.7Mo0.3O2$ support enhances electron transfer to Pt: High-performance catalyst for oxygen reduction reaction. *J.Am. Chem. Soc.* **2011**, *133*, 11716–11724. [CrossRef] [PubMed]
26. Jiang, Z.-Z.; Wang, Z.-B.; Chu, Y.-Y.; Gu, D.-M.; Yin, G.-P. Ultrahigh stable carbon riveted Pt/TiO_2–C catalyst prepared by in situ carbonized glucose for proton exchange membrane fuel cell. *Energy Environ. Sci.* **2011**, *4*, 728–735. [CrossRef]
27. Sun, W.; Sun, J.; Du, L.; Du, C.; Gao, Y.; Yin, G. Synthesis of Nitrogen-doped Niobium Dioxide and its co-catalytic effect towards the electrocatalysis of oxygen reduction on platinum. *Electrochim. Acta* **2016**, *195*, 166–174. [CrossRef]
28. Orilall, M.C.; Matsumoto, F.; Zhou, Q.; Sai, H.; Abruna, H.D.; DiSalvo, F.J.; Wiesner, U. One-pot synthesis of platinum-based nanoparticles incorporated into mesoporous niobium oxide–carbon composites for fuel cell electrodes. *J. Am. Chem. Soc.* **2009**, *131*, 9389–9395. [CrossRef] [PubMed]
29. Xu, C.; Yang, J.; Liu, E.; Jia, Q.; Veith, G.M.; Nair, G.; DiPietro, S.; Sun, K.; Chen, J.; Pietrasz, P. Physical vapor deposition process for engineering Pt based oxygen reduction reaction catalysts on NbO_x templated carbon support. *J. Power Sources* **2020**, *451*, 227709. [CrossRef]
30. Geim, A.K.; Novoselov, K.S. The rise of graphene. *Nat. Mater.* **2007**, *6*, 183–191. [CrossRef]
31. Novoselov, K.S.; Jiang, D.; Schedin, F.; Booth, T.; Khotkevich, V.; Morozov, S.; Geim, A.K. Two-dimensional atomic crystals. *Proc. Natl. Acad. Sci. USA* **2005**, *102*, 10451–10453. [CrossRef]
32. Stankovich, S.; Dikin, D.A.; Dommett, G.H.; Kohlhaas, K.M.; Zimney, E.J.; Stach, E.A.; Piner, R.D.; Nguyen, S.T.; Ruoff, R.S. Graphene-based composite materials. *Nature* **2006**, *442*, 282–286. [CrossRef] [PubMed]
33. Liang, X.; Fu, Z.; Chou, S.Y. Graphene transistors fabricated via transfer-printing in device active-areas on large wafer. *Nano Lett.* **2007**, *7*, 3840–3844. [CrossRef]
34. Eda, G.; Fanchini, G.; Chhowalla, M. Large-area ultrathin films of reduced graphene oxide as a transparent and flexible electronic material. *Nat. Nanotechnol.* **2008**, *3*, 270–274. [CrossRef] [PubMed]
35. Gasteiger, H.A.; Kocha, S.S.; Sompalli, B.; Wagner, F.T. Activity benchmarks and requirements for Pt, Pt-alloy, and non-Pt oxygen reduction catalysts for PEMFCs. *Appl. Catal. B* **2005**, *56*, 9–35. [CrossRef]
36. Antolini, E. Formation of carbon-supported PtM alloys for low temperature fuel cells: A review. *Mater. Chem. Phys.* **2003**, *78*, 563–573. [CrossRef]
37. Yu, X.; Ye, S. Recent advances in activity and durability enhancement of Pt/C catalytic cathode in PEMFC: Part I. Physico-chemical and electronic interaction between Pt and carbon support, and activity enhancement of Pt/C catalyst. *J. Power Sources* **2007**, *172*, 133–144. [CrossRef]
38. Lopes, T.; Antolini, E.; Colmati, F.; Gonzalez, E.R. Carbon supported Pt–Co (3:1) alloy as improved cathode electrocatalyst for direct ethanol fuel cells. *J. Power Sources* **2007**, *164*, 111–114. [CrossRef]
39. Neyerlin, K.; Srivastava, R.; Yu, C.; Strasser, P. Electrochemical activity and stability of dealloyed Pt–Cu and Pt–Cu–Co electrocatalysts for the oxygen reduction reaction (ORR). *J. Power Sources* **2009**, *186*, 261–267. [CrossRef]
40. Koh, S.; Toney, M.F.; Strasser, P. Activity–stability relationships of ordered and disordered alloy phases of Pt_3Co electrocatalysts for the oxygen reduction reaction (ORR). *Electrochim. Acta* **2007**, *52*, 2765–2774.

41. Maillard, F.; Dubau, L.; Durst, J.; Chatenet, M.; André, J.; Rossinot, E. Durability of Pt_3Co/C nanoparticles in a proton-exchange membrane fuel cell: Direct evidence of bulk Co segregation to the surface. *Electrochem. Commun.* **2010**, *12*, 1161–1164. [CrossRef]
42. Yu, X.; Ye, S. Recent advances in activity and durability enhancement of Pt/C catalytic cathode in PEMFC: Part II: Degradation mechanism and durability enhancement of carbon supported platinum catalyst. *J. Power Sources* **2007**, *172*, 145–154. [CrossRef]
43. Li, H.; Sun, G.; Li, N.; Sun, S.; Su, D.; Xin, Q. Design and preparation of highly active Pt−Pd/C catalyst for the oxygen reduction reaction. *J. Phys. Chem. C* **2007**, *111*, 5605–5617. [CrossRef]
44. Wu, Z.-P.; Caracciolo, D.T.; Maswadeh, Y.; Wen, J.; Kong, Z.; Shan, S.; Vargas, J.A.; Yan, S.; Hopkins, E.; Park, K. Alloying-realloying enabled high durability for Pt–Pd-3d-transition metal nanoparticle fuel cell catalysts. *Nat. Commun.* **2021**, *12*, 859. [CrossRef] [PubMed]
45. Ye, H.; Crooks, R.M. Effect of elemental composition of PtPd bimetallic nanoparticles containing an average of 180 atoms on the kinetics of the electrochemical oxygen reduction reaction. *J. Am. Chem. Soc.* **2007**, *129*, 3627–3633. [CrossRef] [PubMed]
46. Hoshi, N.; Nakamura, M.; Kondo, S. Oxygen reduction reaction on the low index planes of palladium electrodes modified with a monolayer of platinum film. *Electrochem. Commun.* **2009**, *11*, 2282–2284. [CrossRef]
47. Lim, B.; Jiang, M.; Camargo, P.H.; Cho, E.C.; Tao, J.; Lu, X.; Zhu, Y.; Xia, Y. Pd-Pt bimetallic nanodendrites with high activity for oxygen reduction. *Science* **2009**, *324*, 1302–1305. [CrossRef]
48. Peng, Z.; Yang, H. Synthesis and oxygen reduction electrocatalytic property of Pt-on-Pd bimetallic heteronanostructures. *J. Am. Chem. Soc.* **2009**, *131*, 7542–7543. [CrossRef]
49. Rego, R.; Oliveira, C.; Velázquez, A.; Cabot, P.-L. A new route to prepare carbon paper-supported Pd catalyst for oxygen reduction reaction. *Electrochem. Commun.* **2010**, *12*, 745–748. [CrossRef]
50. Shao, M. Palladium-based electrocatalysts for hydrogen oxidation and oxygen reduction reactions. *J. Power Sources* **2011**, *196*, 2433–2444. [CrossRef]
51. Yaldagard, M.; Seghatoleslami, N.; Jahanshahi, M. Preparation of Pt-Co nanoparticles by galvanostatic pulse electrochemical codeposition on in situ electrochemical reduced graphene nanoplates based carbon paper electrode for oxygen reduction reaction in proton exchange membrane fuel cell. *Appl. Surf. Sci.* **2014**, *315*, 222–234. [CrossRef]
52. Park, S.; Ruoff, R.S. Chemical methods for the production of graphenes. *Nat. Nanotechnol.* **2009**, *4*, 217–224. [CrossRef] [PubMed]
53. Stankovich, S.; Piner, R.D.; Chen, X.; Wu, N.; Nguyen, S.T.; Ruoff, R.S. Stable aqueous dispersions of graphitic nanoplatelets via the reduction of exfoliated graphite oxide in the presence of poly (sodium 4-styrenesulfonate). *J. Mater. Chem.* **2006**, *16*, 155–158. [CrossRef]
54. Stankovich, S.; Dikin, D.A.; Piner, R.D.; Kohlhaas, K.A.; Kleinhammes, A.; Jia, Y.; Wu, Y.; Nguyen, S.T.; Ruoff, R.S. Synthesis of graphene-based nanosheets via chemical reduction of exfoliated graphite oxide. *Carbon* **2007**, *45*, 1558–1565. [CrossRef]
55. Park, S.; An, J.; Jung, I.; Piner, R.D.; An, S.J.; Li, X.; Velamakanni, A.; Ruoff, R.S. Colloidal suspensions of highly reduced graphene oxide in a wide variety of organic solvents. *Nano Lett.* **2009**, *9*, 1593–1597. [CrossRef] [PubMed]
56. Hassan, H.M.; Abdelsayed, V.; Abd El Rahman, S.K.; AbouZeid, K.M.; Terner, J.; El-Shall, M.S.; Al-Resayes, S.I.; El-Azhary, A.A. Microwave synthesis of graphene sheets supporting metal nanocrystals in aqueous and organic media. *J. Mater. Chem.* **2009**, *19*, 3832–3837. [CrossRef]
57. Ferrari, A.C. Raman spectroscopy of graphene and graphite: Disorder, electron–phonon coupling, doping and nonadiabatic effects. *Solid State Commun.* **2007**, *143*, 47–57. [CrossRef]
58. Berciaud, S.; Ryu, S.; Brus, L.E.; Heinz, T.F. Probing the intrinsic properties of exfoliated graphene: Raman spectroscopy of free-standing monolayers. *Nano Lett.* **2008**, *9*, 346–352. [CrossRef] [PubMed]
59. Dresselhaus, M.S.; Jorio, A.; Hofmann, M.; Dresselhaus, G.; Saito, R. Perspectives on carbon nanotubes and graphene Raman spectroscopy. *Nano Lett.* **2010**, *10*, 751–758. [CrossRef] [PubMed]
60. Yang, G.; Zhou, Y.; Pan, H.-B.; Zhu, C.; Fu, S.; Wai, C.M.; Du, D.; Zhu, J.-J.; Lin, Y. Ultrasonic-assisted synthesis of Pd–Pt/carbon nanotubes nanocomposites for enhanced electro-oxidation of ethanol and methanol in alkaline medium. *Ultrason. Sonochem.* **2016**, *28*, 192–198. [CrossRef]
61. Yıldırım, A.; Seçkin, T. In situ preparation of polyether amine functionalized MWCNT nanofiller as reinforcing agents. *Adv. Mater. Sci. Eng.* **2014**, *2014*, 1–6. [CrossRef]
62. Chaudhary, B.; Panwar, V.; Roy, T.; Pal, K. Thermomechanical behaviour of zirconia–multiwalled carbon nanotube-reinforced polypropylene hybrid composites. *Polym. Bull.* **2019**, *76*, 511–521. [CrossRef]
63. Yaldagard, M.; Shahbaz, M.; Kim, H.W.; Kim, S.S. Ethanol Electro-Oxidation on Catalysts with $S-ZrO_2$-Decorated Graphene as Support in Fuel Cell Applications. *Nanomaterials* **2022**, *12*, 3327. [CrossRef]
64. Mangla, O.; Roy, S. Monoclinic zirconium oxide nanostructures having tunable band gap synthesized under extremely non-equilibrium plasma conditions. *Proceedings* **2019**, *3*, 10. [CrossRef]
65. Ding, S.; Zhao, J.; Yu, Q. Effect of zirconia polymorph on vapor-phase ketonization of propionic acid. *Catalysts* **2019**, *9*, 768. [CrossRef]
66. Manoharan, D.; Loganathan, A.; Kurapati, V.; Nesamony, V.J. Unique sharp photoluminescence of size-controlled sonochemically synthesized zirconia nanoparticles. *Ultrason. Sonochem.* **2015**, *23*, 174–184. [CrossRef]
67. Mkhize, N.; Vashistha, V.K.; Pullabhotla, V.S.R. Catalytic Oxidation of 1, 2-Dichlorobenzene over Metal-Supported on ZrO_2 Catalysts. *Top. Catal.* **2023**, *67*, 409–421. [CrossRef]

68. Asencios, Y.J.; Yigit, N.; Wicht, T.; Stöger-Pollach, M.; Lucrédio, A.F.; Marcos, F.C.; Assaf, E.M.; Rupprechter, G. Partial Oxidation of Bio-methane over Nickel Supported on MgO–ZrO. *Top. Catal.* **2023**, *66*, 1532–1552. [CrossRef] [PubMed]
69. Abdelaziz, O.Y.; Clemmensen, I.; Meier, S.; Bjelić, S.; Hulteberg, C.P.; Riisager, A. Oxidative Depolymerization of Kraft Lignin to Aromatics Over Bimetallic V–Cu/ZrO$_2$ Catalysts. *Top. Catal.* **2023**, *66*, 1369–1380. [CrossRef]
70. Gregory, N. Elements of X-Ray Diffraction. *J. Am. Chem. Soc.* **1957**, *79*, 1773–1774. [CrossRef]
71. Swanson, H.E.; Tatge, E. *Standard X-ray Diffraction Powder Patterns*; US Department of Commerce, National Bureau of Standards: Washington, DC, USA, 1953; Volume 1, p. 31.
72. Sharma, R.; Gyergyek, S.; Andersen, S.M. Critical thinking on baseline corrections for electrochemical surface area (ECSA) determination of Pt/C through H-adsorption/H-desorption regions of a cyclic voltammogram. *Appl. Catal. B* **2022**, *311*, 121351. [CrossRef]
73. Zaman, S.; Su, Y.Q.; Dong, C.L.; Qi, R.; Huang, L.; Qin, Y.; Huang, Y.C.; Li, F.M.; You, B.; Guo, W. Scalable molten salt synthesis of platinum alloys planted in metal–nitrogen–graphene for efficient oxygen reduction. *Angew. Chem. Int. Ed.* **2022**, *61*, e202115835. [CrossRef] [PubMed]
74. Xia, Y.-F.; Guo, P.; Li, J.-Z.; Zhao, L.; Sui, X.-L.; Wang, Y.; Wang, Z.-B. How to appropriately assess the oxygen reduction reaction activity of platinum group metal catalysts with rotating disk electrode. *IScience* **2021**, *24*, 103024. [CrossRef] [PubMed]
75. Pozio, A.; De Francesco, M.; Cemmi, A.; Cardellini, F.; Giorgi, L. Comparison of high surface Pt/C catalysts by cyclic voltammetry. *J. Power Sources* **2002**, *105*, 13–19. [CrossRef]
76. Garsany, Y.; Baturina, O.A.; Swider-Lyons, K.E.; Kocha, S.S. *Experimental Methods for Quantifying the Activity of Platinum Electrocatalysts for the Oxygen Reduction Reaction*; ACS Publications: Washington, DC, USA, 2010.
77. Jiang, L.; Sun, G.; Sun, S.; Liu, J.; Tang, S.; Li, H.; Zhou, B.; Xin, Q. Structure and chemical composition of supported Pt–Sn electrocatalysts for ethanol oxidation. *Electrochim. Acta* **2005**, *50*, 5384–5389. [CrossRef]
78. Antolini, E.; Salgado, J.; Giz, M.; Gonzalez, E. Effects of geometric and electronic factors on ORR activity of carbon supported Pt–Co electrocatalysts in PEM fuel cells. *Int. J. Hydrog. Energy* **2005**, *30*, 1213–1220. [CrossRef]
79. Cheng, N.; Liu, J.; Banis, M.N.; Geng, D.; Li, R.; Ye, S.; Knights, S.; Sun, X. High stability and activity of Pt electrocatalyst on atomic layer deposited metal oxide/nitrogen-doped graphene hybrid support. *Int. J. Hydrogen Energy* **2014**, *39*, 15967–15974. [CrossRef]
80. Wu, H.; Wexler, D.; Wang, G. PtxNi alloy nanoparticles as cathode catalyst for PEM fuel cells with enhanced catalytic activity. *J. Alloys Compd.* **2009**, *488*, 195–198. [CrossRef]
81. Justin, P.; Charan, P.H.K.; Rao, G.R. High performance Pt–Nb$_2$O$_5$/C electrocatalysts for methanol electrooxidation in acidic media. *Appl. Catal. B* **2010**, *100*, 510–515. [CrossRef]
82. Saha, M.S.; Zhang, Y.; Cai, M.; Sun, X. Carbon-coated tungsten oxide nanowires supported Pt nanoparticles for oxygen reduction. *Int. J. Hydrogen Energy* **2012**, *37*, 4633–4638. [CrossRef]
83. Song, Y.; Duan, D.; Shi, W.; Wang, H.; Yang, S.; Sun, Z. Promotion effects of ZrO$_2$ on mesoporous Pd prepared by a one-step dealloying method for methanol oxidation in an alkaline electrolyte. *J. Electrochem. Soc.* **2017**, *164*, F1495. [CrossRef]
84. Aguilar-Vallejo, A.; Álvarez-Contreras, L.; Guerra-Balcázar, M.; Ledesma-García, J.; Gerardo Arriaga, L.; Arjona, N.; Rivas, S. Electrocatalytic evaluation of highly stable Pt/ZrO$_2$ electrocatalysts for the methanol oxidation reaction synthesized without the assistance of any carbon support. *ChemElectroChem* **2019**, *6*, 2107–2118. [CrossRef]
85. Gwebu, S.S.; Maxakato, N.W. The influence of ZrO$_2$ promoter in Pd/fCNDs-ZrO$_2$ catalyst towards alcohol fuel electrooxidation in alkaline media. *Mater. Res. Express* **2020**, *7*, 015607. [CrossRef]
86. Wang, X.; Li, W.; Chen, Z.; Waje, M.; Yan, Y. Durability investigation of carbon nanotube as catalyst support for proton exchange membrane fuel cell. *J. Power Sources* **2006**, *158*, 154–159. [CrossRef]
87. Higuchi, E.; Uchida, H.; Watanabe, M. Effect of loading level in platinum-dispersed carbon black electrocatalysts on oxygen reduction activity evaluated by rotating disk electrode. *J. Electroanal. Chem.* **2005**, *583*, 69–76. [CrossRef]
88. Paulus, U.; Schmidt, T.; Gasteiger, H.; Behm, R. Oxygen reduction on a high-surface area Pt/Vulcan carbon catalyst: A thin-film rotating ring-disk electrode study. *J. Electroanal. Chem.* **2001**, *495*, 134–145. [CrossRef]
89. He, Q.; Mukerjee, S. Electrocatalysis of oxygen reduction on carbon-supported PtCo catalysts prepared by water-in-oil microemulsion. *Electrochim. Acta* **2010**, *55*, 1709–1719. [CrossRef]
90. Van Brussel, M.; Kokkinidis, G.; Vandendael, I.; Buess-Herman, C. High performance gold-supported platinum electrocatalyst for oxygen reduction. *Electrochem. Commun.* **2002**, *4*, 808–813. [CrossRef]
91. Van Brussel, M.; Kokkinidis, G.; Hubin, A.; Buess-Herman, C. Oxygen reduction at platinum modified gold electrodes. *Electrochim. Acta* **2003**, *48*, 3909–3919. [CrossRef]
92. Qiao, J.; Lin, R.; Li, B.; Ma, J.; Liu, J. Kinetics and electrocatalytic activity of nanostructured Ir–V/C for oxygen reduction reaction. *Electrochim. Acta* **2010**, *55*, 8490–8497. [CrossRef]
93. Zaman, S.; Tian, X.; Su, Y.-Q.; Cai, W.; Yan, Y.; Qi, R.; Douka, A.I.; Chen, S.; You, B.; Liu, H. Direct integration of ultralow-platinum alloy into nanocarbon architectures for efficient oxygen reduction in fuel cells. *Sci. Bull.* **2021**, *66*, 2207–2216. [CrossRef]
94. Horwood, E. *Instrumental Methods in Electrochemistry*; Series in Physical Chemistry; Horwood, E., Ed.; Southampton Electrochemistry Group, University of Southampton: Southampton, UK, 1985.
95. Durst, J.; Siebel, A.; Simon, C.; Hasché, F.; Herranz, J.; Gasteiger, H. New insights into the electrochemical hydrogen oxidation and evolution reaction mechanism. *Energy Environ. Sci.* **2014**, *7*, 2255–2260. [CrossRef]

96. Zhang, J. *PEM Fuel Cell Electrocatalysts and Catalyst Layers: Fundamentals and Applications*; Springer Science & Business Media: Berlin/Heidelberg, Germany, 2008. [CrossRef]
97. Markovic, N.; Gasteiger, H.; Ross, P.N. Kinetics of oxygen reduction on Pt (hkl) electrodes: Implications for the crystallite size effect with supported Pt electrocatalysts. *J. Electrochem. Soc.* **1997**, *144*, 1591. [CrossRef]
98. Markovic, N.M.; Gasteiger, H.A.; Ross Jr, P.N. Oxygen reduction on platinum low-index single-crystal surfaces in sulfuric acid solution: Rotating ring-Pt (hkl) disk studies. *J. Phys. Chem.* **1995**, *99*, 3411–3415. [CrossRef]
99. Perez, J.; Villullas, H.M.; Gonzalez, E.R. Structure sensitivity of oxygen reduction on platinum single crystal electrodes in acid solutions. *J. Electroanal. Chem.* **1997**, *435*, 179–187. [CrossRef]
100. Marković, N.; Gasteiger, H.; Grgur, B.; Ross, P. Oxygen reduction reaction on Pt (111): Effects of bromide. *J. Electroanal. Chem.* **1999**, *467*, 157–163. [CrossRef]
101. Bett, J.; Lundquist, J.; Washington, E.; Stonehart, P. Platinum crystallite size considerations for electrocatalytic oxygen reduction—I. *Electrochim. Acta* **1973**, *18*, 343–348. [CrossRef]
102. Damjanovic, A.; Sepa, D. An analysis of the pH dependence of enthalpies and Gibbs energies of activation for O_2 reduction at Pt electrodes in acid solutions. *Electrochim. Acta* **1990**, *35*, 1157–1162. [CrossRef]
103. Sarapuu, A.; Kasikov, A.; Laaksonen, T.; Kontturi, K.; Tammeveski, K. Electrochemical reduction of oxygen on thin-film Pt electrodes in acid solutions. *Electrochim. Acta* **2008**, *53*, 5873–5880. [CrossRef]
104. Zignani, S.C.; Antolini, E.; Gonzalez, E.R. Evaluation of the stability and durability of Pt and Pt–Co/C catalysts for polymer electrolyte membrane fuel cells. *J. Power Sources* **2008**, *182*, 83–90. [CrossRef]
105. Salgado, J.R.C.; Antolini, E.; Gonzalez, E.R. Carbon supported Pt–Co alloys as methanol-resistant oxygen-reduction electrocatalysts for direct methanol fuel cells. *Appl. Catal. B Environ.* **2005**, *57*, 283–290. [CrossRef]
106. Mustain, W.E.; Kepler, K.; Prakash, J. $CoPd_x$ oxygen reduction electrocatalysts for polymer electrolyte membrane and direct methanol fuel cells. *Electrochim. Acta* **2007**, *52*, 2102–2108. [CrossRef]
107. Zaman, S.; Huang, L.; Douka, A.I.; Yang, H.; You, B.; Xia, B.Y. Oxygen reduction electrocatalysts toward practical fuel cells: Progress and perspectives. *Angew. Chem.* **2021**, *133*, 17976–17996. [CrossRef]
108. Wang, G.; Shen, X.; Wang, B.; Yao, J.; Park, J. Synthesis and characterisation of hydrophilic and organophilic graphene nanosheets. *Carbon* **2009**, *47*, 1359–1364. [CrossRef]
109. Woo, S.; Kim, I.; Lee, J.K.; Bong, S.; Lee, J.; Kim, H. Preparation of cost-effective Pt–Co electrodes by pulse electrodeposition for PEMFC electrocatalysts. *Electrochim. Acta* **2011**, *56*, 3036–3041. [CrossRef]
110. Antolini, E.; Giorgi, L.; Pozio, A.; Passalacqua, E. Influence of Nafion loading in the catalyst layer of gas-diffusion electrodes for PEFC. *J. Power Sources* **1999**, *77*, 136–142. [CrossRef]

Disclaimer/Publisher's Note: The statements, opinions and data contained in all publications are solely those of the individual author(s) and contributor(s) and not of MDPI and/or the editor(s). MDPI and/or the editor(s) disclaim responsibility for any injury to people or property resulting from any ideas, methods, instructions or products referred to in the content.

Enhanced Tribodegradation of a Tetracycline Antibiotic by Rare-Earth-Modified Zinc Oxide

Dobrina Ivanova [1], Hristo Kolev [2], Bozhidar I. Stefanov [3] and Nina Kaneva [1,*]

[1] Laboratory of Nanoparticle Science and Technology, Department of General and Inorganic Chemistry, Faculty of Chemistry and Pharmacy, University of Sofia, 1164 Sofia, Bulgaria; dobrina.k.ivanova@gmail.com
[2] Institute of Catalysis, Bulgarian Academy of Sciences, Acad. G. Bonchev St., bl. 11, 1113 Sofia, Bulgaria; hgkolev@ic.bas.bg
[3] Department of Chemistry, Faculty of Electronic Engineering and Technologies, Technical University of Sofia, 8 Kliment Ohridski Blvd, 1756 Sofia, Bulgaria; b.stefanov@tu-sofia.bg
* Correspondence: nina_k@abv.bg

Abstract: Tribocatalysis is an emerging advanced oxidation process that utilizes the triboelectric effect, based on friction between dissimilar materials to produce charges that can initiate various catalytic reactions. In this study, pure and rare-earth-modified ZnO powders (La_2O_3, Eu_2O_3, 2 mol %) were demonstrated as efficient tribocatalysts for the removal of the tetracycline antibiotic doxycycline (DC). While the pure ZnO samples achieved 49% DC removal within 24 h at a stirring rate of 100 rpm, the addition of Eu_2O_3 increased the removal efficiency to 67%, and La_2O_3-modified ZnO powder exhibited the highest removal efficiency, reaching 80% at the same stirring rate. Additionally, increasing the stirring rate to 300 and 500 rpm led to 100% DC removal in the ZnO/La case within 18 h, with the pronounced effect of the stirring rate confirming the tribocatalytic effect. All tribocatalysts exhibited excellent recycling properties, with less than a 3% loss of activity over three cycles. Furthermore, a scavenger assay confirmed the importance of superoxide radical generation for the overall reaction rate. The results of this investigation indicate that the rare-earth-modified ZnO tribocatalysts can effectively utilize mechanical energy to decompose pollutants in contaminated water.

Keywords: tribocatalysis; zinc oxide powder; rare earths; doxycycline; water remediation

1. Introduction

The major environmental hazard worldwide and a significant contributor to the decline in human health is wastewater, due to its complex composition and slow degradation rate [1–3]. Degrading organic substances in sewage using widely available environmental energy sources, such as solar, thermal, and mechanical energy, is a promising strategy for environmental remediation [4–7]. Among these, photocatalysts can absorb solar energy, one of the cleanest and most renewable energy sources, to initiate various photocatalytic reactions, including the reduction of carbon dioxide and water splitting for hydrogen production [8,9]. Photocatalysis can also be employed to remediate polluted wastewater under UV or visible light irradiation for the degradation of organic pollutants. However, photocatalysis technology has limited practical applications due to the high recombination rate of photo-excited carriers and challenges in transmitting light through optically opaque environments.

A new avenue to energy-catalyzed organic pollutant removal is pyroelectric catalysis, where heat energy is used for charge generation. However, most pyroelectric catalysts require a high rate of temperature change to efficiently convert thermal energy into chemical energy, and meeting the excitation conditions of thermal catalysis in aqueous environments is challenging [10]. Therefore, developing a novel catalytic approach to address current environmental pollution issues is imperative.

This leads to the possibility of utilizing mechanical energy for catalytic conversion. In this context, the tribocatalytic effect has attracted significant interest due to its repeatability and environmentally friendly nature [11–13]. Tribocatalysis is based on the triboelectric effect, where triboelectric charges are generated when two materials come into contact through friction [14–16]. A promising method to reduce environmental pollution is to use the positive and negative triboelectric charges produced during friction to react with oxygen and hydroxyl ions in water, respectively, for reactive species generation which may in turn be employed to decompose hazardous dyes in effluents from the textile industry [17,18]. For the first time, Li et al. reported that tribocatalytic $Ba_{0.75}Sr_{0.25}TiO_3$ nanoparticles could effectively break down organic dyes in a liquid environment [19]. These nanoparticles absorb frictional energy between glass and PTFE surfaces, breaking down dye molecules through electron–hole pair activation and subsequent redox chemical reactions. Zhao et al. [20] employed ZnO nanorods and PTFE magnetic bar stirring to degrade Rhodamine B (RB) dye; at a rotation speed of 1000 rpm, the RB dye decomposition efficiency reached 99.8% within 60 h. Thus, utilizing the frictional effect of nanomaterials to degrade dyes has emerged as a novel concept; however, the degradation of drugs by ZnO-based tribocatalysts has not been widely studied. Only two publications, by Sun et al. [21] and Li et al. [22], have demonstrated the tribodegradation of tetracycline using alternative materials—FeOOH nanorods and pyrite-based tribocatalysts, respectively.

In this study, ZnO was chosen due to its remarkable semiconductor properties, high chemical stability, environmental friendliness, and piezoelectric properties, which are key for the fabrication of efficient tribocatalysts. Pure and La- or Eu-modified ZnO powders were prepared to examine their tribocatalytic efficiency in decomposing doxycycline (DC), motivated by the expectation that rare-earth elements, known to enhance piezoelectric properties, would improve the triboelectric efficiency of the ZnO catalyst. We demonstrate that doping with rare-earth ions and increasing the magnetic stirring speed enhances degradation activity, confirming that the degradation of DC is driven by mechanical energy. This finding aligns with our previous work, where we showed that rare-earth doping improves the photocatalytic efficiency of ZnO by suppressing the recombination of photogenerated electrons and holes [23].

2. Results and Discussion

2.1. Structural and Morphological Characterization of the Pure and Rare-Earth-Modified ZnO Tribocatalysts

The microstructure and morphology of the ZnO samples modified with Eu and La rare-earth (RE) ions were examined in detail using scanning electron microscopy (SEM). The SEM images (Figure 1) reveal that the ZnO nanocomposites consist of particles with different shapes and sizes. An analysis of the SEM images shows an average particle diameter of 0.7 ± 0.1 μm for pristine ZnO, which remains consistent in the ZnO/Eu sample (0.7 ± 0.2 μm) and increases slightly to 0.8 ± 0.2 μm in the ZnO/La case. Despite the incorporation of Eu_2O_3 and La_2O_3 in the ZnO/Eu and ZnO/La case, respectively, the low treatment temperature (100 °C) preserved the morphology of the samples, with no significant effect observed based on the type of rare-earth element.

The Brunauer–Emmett–Teller (BET) surface area analysis revealed that all the RE-modified samples exhibited a higher surface area than pure ZnO (10.3 ± 1.6 m^2/g). Notably, the ZnO/La powder displayed the largest surface area (32.3 ± 1.8 m^2/g), suggesting the potential for higher tribocatalytic activity compared to ZnO/Eu (30.3 ± 1.7 m^2/g).

Energy-dispersive X-ray spectroscopy (EDS) confirmed the presence of Zn, O, and rare-earth elements in the modified ZnO powders (Figure 2). Peaks corresponding to zinc, oxygen, and the respective rare-earth elements were observed. The RE elements' weight percentage was approximately 3 wt. % in all cases (Table 1). The absence of impurity peaks in the EDS spectrum indicates the high purity of the starting ZnO material. Europium and lanthanum were homogeneously distributed across the ZnO surface, as evidenced by the mapping data in Figure 2c.

Figure 1. SEM micrographs of (**a**) pure ZnO; (**b**) Eu-modified ZnO; and (**c**) La-modified ZnO powder. Insets show the particle size distribution obtained from the respective micrographs.

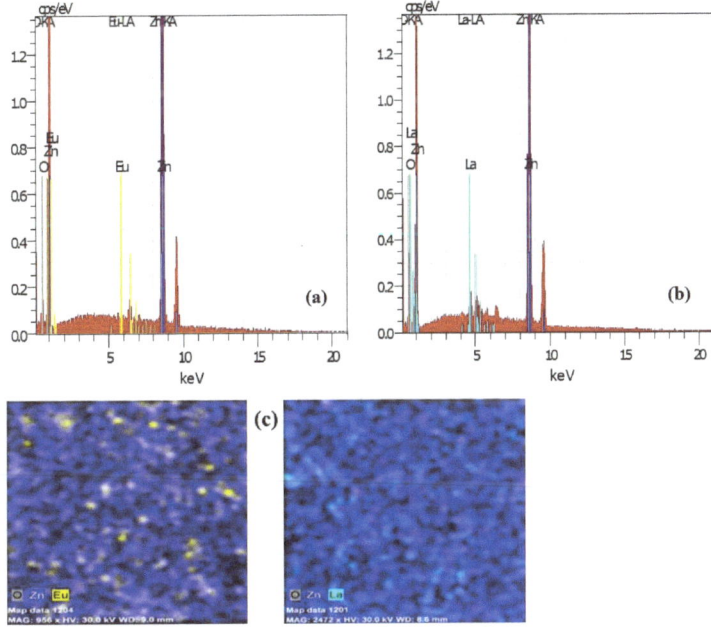

Figure 2. EDS spectra of ZnO powders modified with different rare-earth ions (2 mol %): (**a**) Eu; (**b**) La; (**c**) mapping surface data for rare-earth-modified samples.

Table 1. EDS values of ZnO/RE powders.

Sample Powders	C Norm. [wt. %]		C Atom. [at. %]		C Error, [%]	
ZnO/Eu	O	41.66	O	64.90	O	4.9
	Zn	45.84	Zn	34.26	Zn	1.6
	Eu	3.25	Eu	0.84	Eu	0.1
ZnO/La	O	42.45	O	61.81	O	4.4
	Zn	54.28	Zn	36.45	Zn	1.5
	La	3.27	La	1.74	La	0.2

Transmission electron microscopy (TEM) further revealed the morphology and positioning of the RE-oxide phase on the ZnO surface, with the resulting micrographs presented in Figure 3.

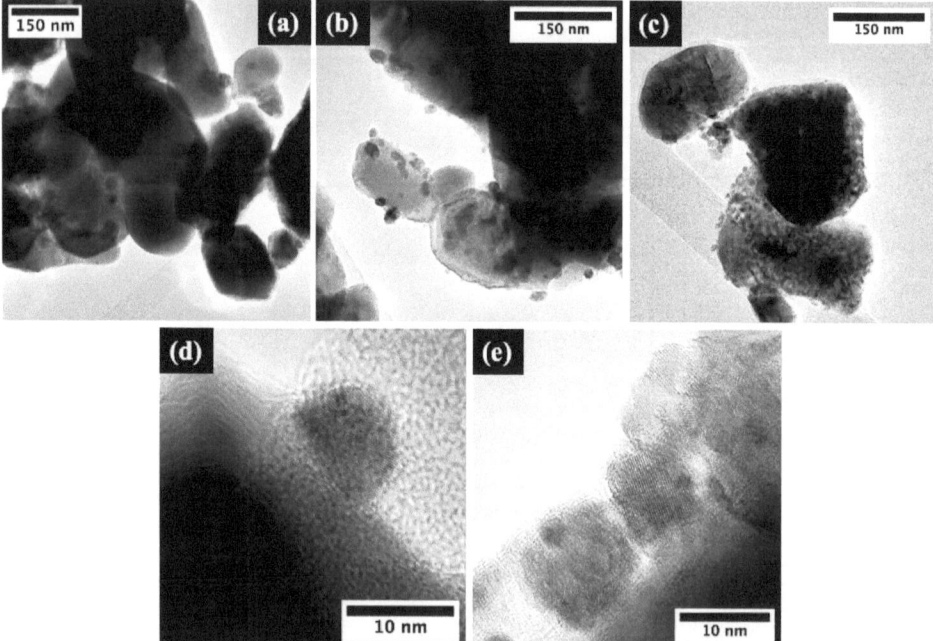

Figure 3. TEM micrographs of (a) pure ZnO, (b,d) ZnO/Eu, and (c,e) ZnO/La powders.

The TEM images confirmed that the tribocatalysts consist of polycrystalline agglomerates several hundred nanometers in diameter, consistent with SEM observations. The ZnO phase is decorated by the respective Eu_2O_3 and La_2O_3 co-catalyst RE-oxide phases, which appear as surface-bound particles approximately 10 nm in diameter in higher magnification images (Figure 3c,d).

The chemical state of the RE-modified ZnO tribocatalysts was investigated via X-ray photoelectron spectroscopy (XPS). Figure 4a shows the Zn 2p region of the XPS spectrum, where a peak with a binding energy (BE) of 1021.7 eV and another band at 1045 eV correspond to Zn $2p_{1/2}$ and Zn $2p_{3/2}$ in ZnO.

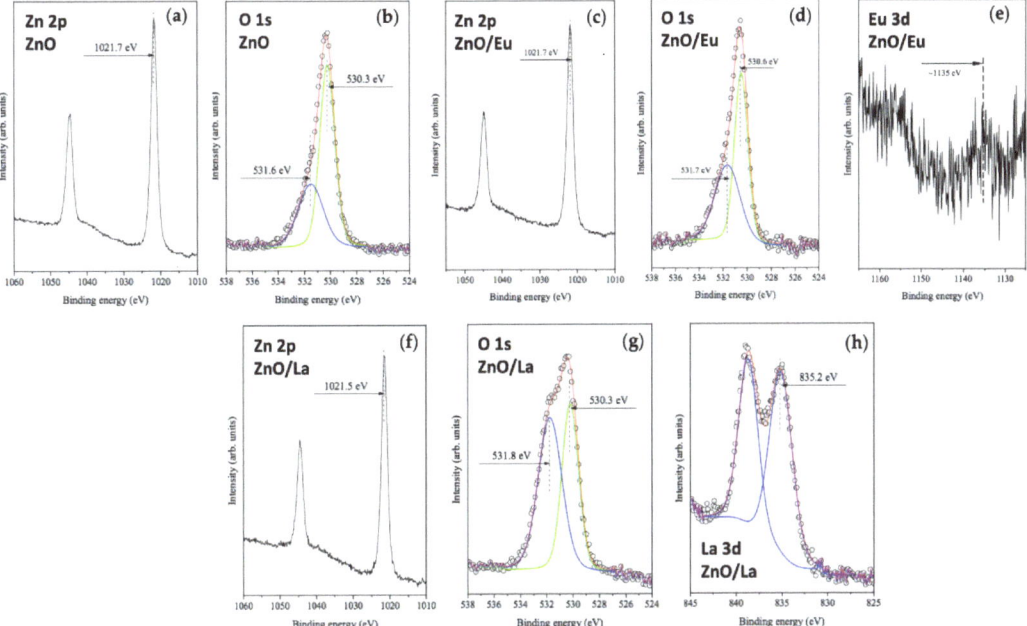

Figure 4. XPS spectra of ZnO and RE-modified ZnO powders showing (**a**) the Zn 2p region in pure ZnO; (**b**) the O 1s region in pure ZnO; (**c**) the Zn 2p region in ZnO/Eu; (**d**) the O 1s region in ZnO/Eu; (**e**) the Eu 3d region in ZnO/Eu^{3+}; (**f**) the Zn 2p region in ZnO/La; (**g**) the O 1s region in ZnO/La; and (**h**) the La 3d region in ZnO/La.

The O 1s region (Figure 4b) shows a doublet at 530.3 eV and 531.6 eV, corresponding to extraneous and lattice oxygen in wurtzite ZnO [24]. No significant changes were observed in the Zn 2p and O 1s XPS spectra for ZnO/Eu (Figure 4c,d) and ZnO/La (Figure 4f,g). In the ZnO/Eu case (Figure 4e), a peak around 1135 eV BE corresponds to the Eu 3d$_{5/2}$ level, indicating oxygen-coordinated Eu^{3+} [25]. For the ZnO/La sample, the La 3d region (Figure 4h) shows a doublet at 835.2 eV with a satellite at ~838 eV BE, consistent with the presence of La^{3+} [26].

XPS analysis revealed an RE concentration of 5.52 at. % for ZnO/La, suggesting uniform coverage of the ZnO particles with the dopant. However, for the ZnO/Eu case, only 0.24 at. % was detected via XPS, which is inconsistent with the EDX results. This discrepancy suggests that ZnO particles accumulate around the Eu$_2$O$_3$ oxide, masking it in the surface-sensitive XPS analysis.

Powder X-ray diffraction (XRD) patterns of the pure and RE-modified ZnO powders after annealing at 100 °C are shown in Figure 5. The crystalline phase of pure ZnO is hexagonal wurtzite, evidenced by intense diffraction peaks at 2θ = 31.94°, 34.67°, 36.51°, 48.23°, 56.84°, 63.22°, 67.53°, and 68.18°. These peaks correspond to the lattice planes (100), (002), (101), (102), (110), (103), (112), and (201), respectively [27]. The peak positions align with JCPDS Card No. 36-1451. No impurities or phase modifications were observed in the crystalline structure, and the XRD patterns of both pure and RE-modified ZnO samples show strong, sharp peaks, indicating a high degree of crystallinity [28]. The RE^{3+} phase is uniformly distributed as tiny oxide clusters among ZnO nanoparticles, and the low concentration of lanthanide ions (2 mol %) in the modified ZnO composite may explain the absence of a distinct phase. Notably, the crystallite size of RE-modified ZnO is larger than that of pure ZnO, as determined by the Scherrer equation using the main peak (101). The crystallite size of pure ZnO is 37 nm, while the modified ZnO catalysts have crystallite sizes of approximately 42 nm. The increase in the crystallite size of ZnO/RE could be attributed to bond formation between the oxides on the surface of the composite samples, which affects the crystallite size [29,30].

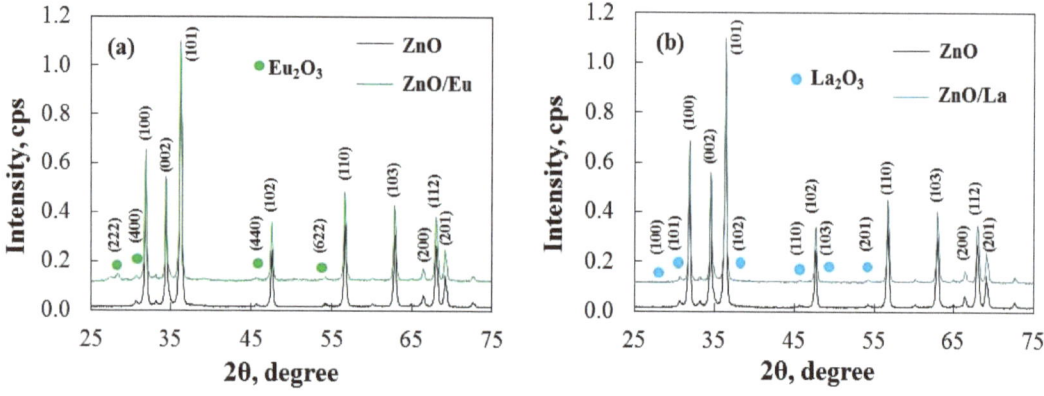

Figure 5. XRD patterns of (**a**) ZnO/Eu and (**b**) ZnO/La powders, with main reflections denoted, according to JCPDS 36-1451 (wurtzite ZnO), JCPDS 43-1008 (Eu_2O_3), and JCPDS 83-1355 (La_2O_3).

The XRD data show no discernible change in crystal size with the addition of lanthanide ions (2 mol %). Except for a minor increase in the average crystallite size after modification, the crystalline lattice parameters remain largely unchanged (Table 2). The calculated lattice parameters closely resemble those of ZnO, indicating that powders modified with rare-earth elements maintain their hexagonal wurtzite structure. A positive value for tensile strain is observed when the microstrain of the samples is calculated using the c-axis lattice parameter. The tensile strain is slightly reduced in modified powders compared to ZnO.

Table 2. Crystallite size and lattice parameters of pure and RE-modified ZnO powders.

Sample Powders	Crystallite Size, nm	Parameters of the Crystalline Lattice, Å	Microstrains, a.u.
ZnO	37	a, b: 3.2531 c: 5.2057	6×10^{-4}
ZnO/Eu	41	a, b: 3.2516 c: 5.1535	4×10^{-4}
ZnO/La	42	a, b: 3.2504 c: 5.1524	4×10^{-4}

Raman spectroscopy further confirmed the phase composition of the ZnO materials, with the spectra depicted in Figure 6. In all cases, the most intensive Raman bands were observed at 331, 439, and 1154 cm^{-1}, corresponding to ZnO's E_2(high −)–E_2(low) mode, the E_2(high) mode, and the $2A_1$(low) + $2E_2$(low) broad band, respectively [31]. These data are consistent with wurtzite ZnO and do not suggest any major modifications of the main tribocatalyst component due to functionalization.

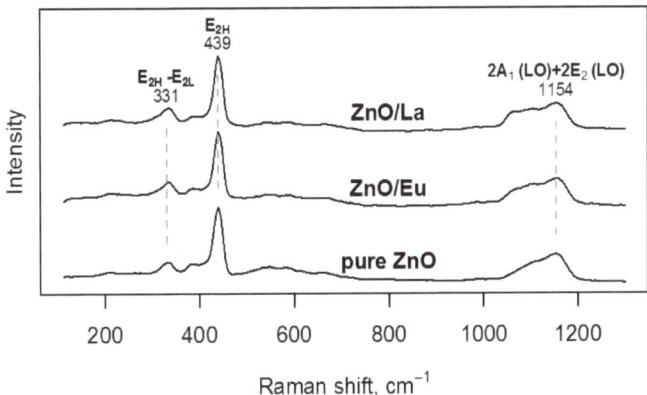

Figure 6. Raman spectra of pristine ZnO, ZnO/Eu, and ZnO/La powders.

2.2. Tribocatalysis for Decomposition of Doxycycline—Effect of Rare-Earth Elements

The tribocatalytic efficiency of pure and rare-earth-modified ZnO powders was evaluated for the degradation of a tetracycline antibiotic, doxycycline (DC), under dark conditions. Magnetic stirring facilitated the tribocatalytic process, and the drug concentration was consistently maintained at 15 mg/L in all experiments. UV/Vis spectroscopy was employed to monitor the degradation of DC by tracking the absorption maxima at 275 nm. To determine how the different rare-earth elements affect ZnO's activity during the tribocatalytic process, spectral changes in the degradation of DC were examined. Figure 7 presents the UV/Vis spectra for DC degradation using pure ZnO, ZnO/Eu, and ZnO/La, respectively.

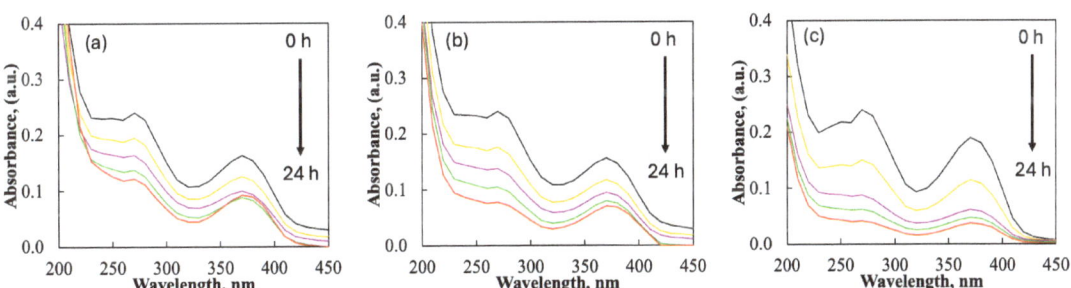

Figure 7. Spectral data for the tribodegradation of DC for (**a**) pristine ZnO; (**b**) Eu-modified ZnO; and (**c**) La-modified ZnO.

Figure 8a displays the tribodegradation results (at 300 rpm), highlighting that ZnO/La powder exhibits the fastest degradation rate, achieving 92.5% degradation after 24 h of friction. In contrast, a control experiment without a tribocatalyst showed negligible degradation (~4%), underscoring the importance of the catalyst in the friction process. The catalytic efficiencies follow the order ZnO/La > ZnO/Eu > ZnO. The higher efficiency of ZnO/La is likely due to its increased specific surface area, which promotes better separation of tribo-generated electron–hole pairs and enhances carrier participation in redox reactions, along with providing an increased number of active sites for DC adsorption [32]. The reaction rate constants, as shown in Figure 8b, were determined by $\ln(C_t/C_0) = -kt$ pseudo-first order kinetics, typically used to describe photo- and tribocatalytic removal [20] and to further confirm this trend, with ZnO/La exhibiting the highest rate constant ($k = 0.1015$ h^{-1}).

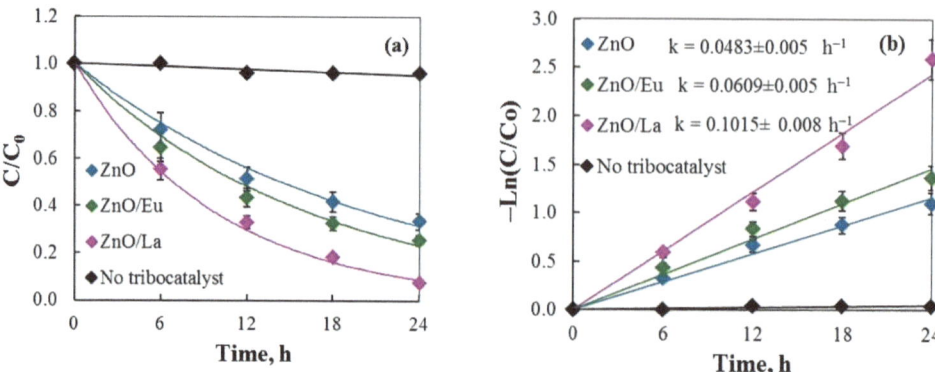

Figure 8. (a) Stirring degradation of DC solution using ZnO and ZnO modified with different rare-earth elements (2 mol%) under magnetic stirring conditions (300 rpm); (b) kinetic fitting.

2.3. Tribocatalysis for Decomposition of Doxycycline—Plausible Mechanism

The mechanism of tribocatalysis is still being explored, but two primary pathways have been proposed [33]: (i) electron transfer from the tribocatalyst to the PTFE bar; and (ii) excitation of electron–hole pairs due to ZnO deformation, similar to photocatalysis [34]. Figure 9 illustrates a schematic representation of tribocatalysis by ZnO and ZnO/RE composites.

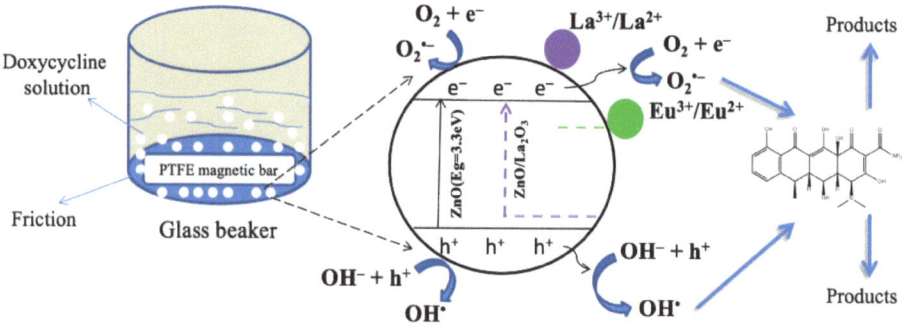

Figure 9. Plausible mechanism of the tribocatalysis degradation of DC by ZnO and ZnO/RE powders by analogy to [33,34].

During rotational friction stirring, the interaction between the PTFE bar and tribocatalyst powders generates positive and negative charges through electron extraction onto the PTFE surface. PTFE absorbs electrons from the ZnO and ZnO/RE surfaces and, additionally, electron–holes are generated in the semiconductor chemical reactions caused with the DC molecule. Electrons represent excited e^-, and holes represent the formed h^+ that results from ZnO absorbing mechanical energy during friction; similarly, heterogeneous photocatalysis in which organic pollutants are broken down by photoexcitation of ZnO electron–hole pairs is comparable to frictional contact-induced catalysis [35,36]. During the decomposition of the drug, the oxygen molecules react with the tribogenerated electrons on the PTFE surface, and superoxide radicals are formed, while the holes remaining on the ZnO surface may interact with OH^- and be transformed into $OH^\bullet$, with both tribogenerated radicals effectively attacking the DC molecule.

Two fundamental questions must be addressed to comprehend the energy transfer in RE-modified ZnO: (i) How does the rare-earth ion energy level relate to the host tribocatalyst's valence and conduction bands? (ii) How do the locations of rare-earth ion energy levels impact the processes of charge migration and trapping?

Duffy's oxide model [37] is thus used to calculate the energy band gaps of ZnO and RE ions as a function of their optical electronegativity: $E_g = 3.71 \times \Delta\chi$, where $\Delta\chi$ is the optical electronegativity of the binary oxide, which is 3.15 for ZnO, 2.54 for Eu_2O_3 and 2.5 for La_2O_3. The calculated band gap values are discovered to increase in the following order: $E_{g[ZnO]} = 3.3$ eV $< E_{g[ZnO/Eu]} = 4.3$ eV $< E_{g[ZnO/La]} = 5.5$ eV. To confirm this expectation, UV/Vis diffuse reflectance spectra were obtained for the pure ZnO and ZnO/RE composites, and then converted using the Kubelka–Munk approach: $F(R) = (1-R)^{-2}/2R$, where $F(R)$ is the Kubelka–Munk (K-M) function, which can be approximated functionally to an absorption coefficient, i.e., $F(R) \propto \alpha$, and used directly to obtain the optical bandgap (E_g) of the powders via Tauc analysis. As ZnO is a direct bandgap semiconductor, E_g can be obtained as the cross-section of the functional dependence $(F(R)h\upsilon)^2$ vs. $h\upsilon$, where $h\upsilon$ is the energy corresponding to the wavelength at which the reflectance value of the obtained $F(R)$ was measured. Figure 10 depicts the resulting Tauc plots for ZnO and the three ZnO/RE composites.

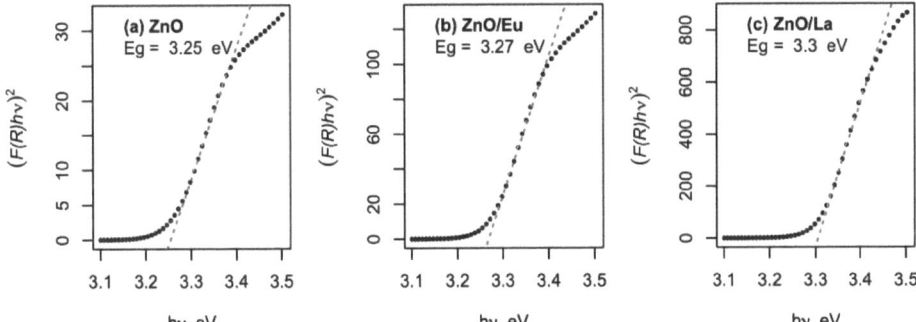

Figure 10. Tauc plots for (**a**) ZnO, (**b**) ZnO/Eu, and (**c**) ZnO/La tribocatalyst powders. The optical bandgap value is presented as an insert in each plot.

As shown in Figure 10, the experimentally observed optical bandgaps closely follow the expected arrangement, predicted by Duffy's model: $E_{g[ZnO]} < E_{g[ZnO/Eu]} < E_{g[ZnO/La]}$; however, only a modest difference of 0.05 eV is observed across the ZnO and ZnO/La case. It should be noted, however, that in the model case, a binary oxide is assumed, while as seen by the TEM evidence in Figure 3, a heterostructure between ZnO and RE oxide is formed; hence, the optical absorption will be governed mainly by the ZnO semiconductor. As the most active tribocatalyst, namely ZnO/La, also exhibits the highest bandgap, it could be suggested that the main contribution to its enhanced activity is expected to be the improved triboelectric charge separation between the ZnO and the RE oxide phase and the formation of a heterojunction between the two dissimilar semiconductors, which has been demonstrated as an effective strategy for improved tribocatalytic activity in the literature [38]. Additionally, the RE ion's 4f-shell can take electrons from or transfer them to the energy bands of a compound through the RE^{3+}/RE^{2+} or RE^{4+}/RE^{3+} valence change, in which, as illustrated in Figure 9, the La and Eu ions in ZnO composite catalysts could contribute to electron trapping and transfer.

Apart from the enhanced activity of the La and Eu modification potentially attributed to electron–hole separation, the generation of $O_2^{\bullet -}$ and $OH^{\bullet}$ radicals is among the main mechanisms in tribocatalysis [18,38]. The highest efficiency is seen in the La-modified ZnO sample, which can be guessed as possibly being due to the higher number of oxygen vacancies in this instance (caused by the differing charge and electronegativity of lanthanum and zinc ions) and the stronger hydroxyl ion adsorption onto the ZnO surface [39]. The reaction between the hole and OH^- promotes the formation of $OH^{\bullet}$. Degradation of the organic pollutant at the surface of La-modified ZnO can thus be linked to the tribogeneration of the extremely potent non-selective oxidants OH• [40]. Since the added RE energy levels in the case of Eu-modified ZnO are near but below the energy of the ZnO conduction band, the reaction of the tribogenerated electron and O_2 molecules favors the formation of $O_2^{•-}$ radicals.

Introducing the RE oxide phase into ZnO creates distinct energy levels and potentially suppresses tribogenerated charge recombination, further boosting catalytic efficiency. The RE phase helps to trap electrons, prevent electron–hole recombination, and produce more superoxide and hydroxyl radicals, all of which contribute to pollutant degradation.

To confirm the involvement of hydroxyl and superoxide radicals, a radical scavenger assay was performed using ascorbic acid (AA) and isopropyl alcohol (IPA) as scavengers for superoxide ($O_2^{•-}$) and hydroxyl (OH•), respectively [41,42].

Figure 11 shows that three tribocatalyst systems responded similarly to the addition of AA and IPA scavengers. The results show that superoxide radicals have a more significant impact on the DC tribodegradation rate in all three tribocatalyst systems, as evidenced by the pronounced inhibition with AA.

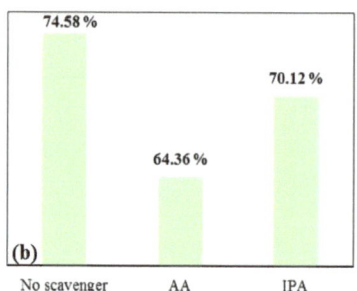
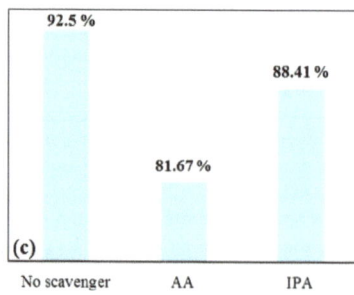

Figure 11. Effect of scavengers on DC degradation in tribocatalysis process using (a) pure, (b) europium, and (c) lanthanum-modified ZnO powder.

2.4. Tribocatalysis for Decomposition of Doxycycline—Effect of Magnetic Stirring and Catalyst Recycling

The impact of varying rotational speeds on the tribocatalytic breakdown of doxycycline (DC) using pure and RE-modified ZnO powders was studied at different speeds: 100, 300, and 500 rpm, and as illustrated in Figure 12 there is a concomitant increase in DC removal with stirring speed. The pure ZnO sample exhibited degradation efficiencies of 49.2%, 66.7%, and 80.4% at 100, 300, and 500 rpm respectively.

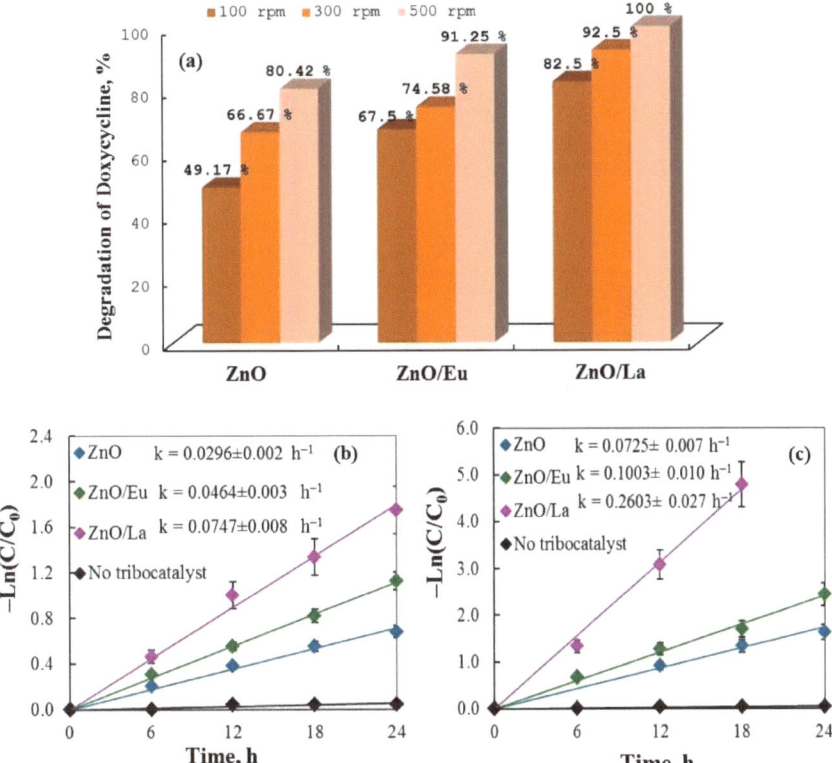

Figure 12. (a) Degradation of DC at different rotation speeds using pure and RE-modified ZnO composites; (b,c) ln(C/Co) vs. plot showing the rate of drug decomposition in tribocatalysis experiments by semiconductors stirred at 100 and 500 rpm.

The ZnO/La sample demonstrated the highest catalytic performance, achieving 100% drug degradation in less than 20 h at a rotation speed of 500 rpm, outperforming the other modified samples (Figure 12a). In all cases, the tribocatalytic activity followed the order of ZnO/La > ZnO/Eu > ZnO and the increased rotation speed enhanced the rate of drug decomposition (Figure 12b,c).

Table 3 presents the rate constants and percentages of DC decomposition after the first tribocatalytic cycle. The data support the conclusion that rare-earth elements enhance the tribocatalytic process, allowing for effective drug degradation even in the absence of light.

Table 3. The values of rate constants and percent of DC decomposition using the tribocatalytic process after the first cycle.

Sample Powders	100 rpm		300 rpm		500 rpm	
	k, h^{-1}	D, %	k, h^{-1}	D, %	k, h^{-1}	D, %
ZnO	0.0296	49	0.0483	67	0.0725	80
ZnO/Eu	0.0464	68	0.0609	75	0.1003	91
ZnO/La	0.0747	83	0.1015	93	0.2603	100

Figure 13 shows the results of a study on the recyclability of ZnO, Eu/ZnO, and La/ZnO powders over three consecutive cycles. The catalytic properties of the powders declined slightly with each cycle, with the tribocatalytic degradation of DC decreasing by approximately 3% for each type of catalyst after three cycles in distilled water. Despite this decrease, the hydrothermal powders demonstrated good cycling stability for DC decomposition. Notably, the ZnO/La nanostructures maintained the highest degree of tribocatalytic activity across all cycles, confirming their potential for repeated use in DC degradation. These findings indicate that while the tribocatalytic efficiency of the powders decreases slightly with repeated use, the ZnO/La composite remains the most effective and stable catalyst over multiple cycles.

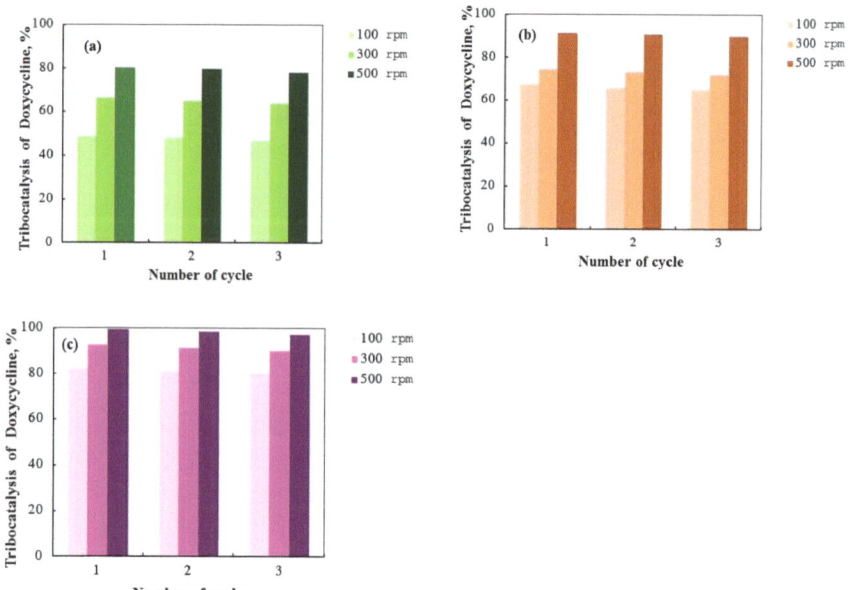

Figure 13. Tribocatalytic degradation rate of DC for three consecutive cycles using ZnO (**a**), Eu/ZnO (**b**), and La/ZnO (**c**) powders at different rotation speeds.

3. Materials and Methods

3.1. Reagents and Preparation of RE-Modified ZnO Powders

Zinc oxide commercial powder (>99.0%), La_2O_3 (>99.0%), Eu_2O_3 (>99.0%), and absolute C_2H_5OH were obtained from Fluka, Burlington, MA, USA).

Doxycycline ($C_{22}H_{24}N_2O_8$, λ_{max} = 275 nm, Teva, Sofia, Bulgaria) was selected for the tribocatalytic experiments as the modal pollutant because of its widespread use in real-world settings.

A straightforward and environmentally friendly hydrothermal process was used to create three series of ZnO/RE composite powders. In a glass vessel, the appropriate amounts of commercial ZnO powder and La_2O_3 (2 mol %) were combined, and ethanol was added as a mixing medium to create La-modified tribocatalysts. The materials were combined for ten minutes, sonicated for thirty more minutes, and then dried for one hour at 100 °C to produce the ZnO/La powders needed for tribocatalytic testing. The remaining catalyst (Eu) was prepared under the same ideal conditions, with a concentration of 2 mol% of RE ions.

3.2. Instrumental Methods

The surface morphology of pure ZnO and ZnO modified by RE was examined using SEM (JSM-5510, Krefeld, Germany) operating at 10 kV of acceleration voltage. For elemental analysis or chemical characterization of the samples, energy-dispersive X-ray spectroscopy (EDXdetector: Quantax 200, Bruker Resolution 126 eV, Berlin, Germany) was employed. Transmission electron microscopy was performed on a JEOL JEM-2100 (JEOL Ltd., Tokyo, Japan), operating at 200 kV. Based on Brunauer–Emmett–Teller (BET) N_2 absorption (Quantachrome Instruments NOVA 1200e, Boyton Beac, FL, USA), the surface area of pure and RE/ZnO composite powders was estimated. The samples were degassed at 150 °C for four hours before N_2 adsorption for the BET analysis. XRD (Siemens D500 with Cu Kα radiation, Karlsruhe, Germany) was used to analyze the crystallinity and phase composition of the catalysts. Scherrer's equation was used to estimate the average crystallite sizes. X-ray photoelectron measurements were performed using the ESCAAB MkII electron spectrometer (VG Scientific, now Thermo Scientific, Manchester, UK) equipped with a twin anode MgKα/AlKα non-monochromated X-ray source that used excitation energies of 1253.6 and 1486.6 eV, respectively. The base pressure in the analysis chamber was 5×10^{-10} mbar. The only non-monochromated X-ray source used for the measurements was AlKα. There was roughly 1 eV in the instrumental solution. SpecsLabl2 CasaXPS software (2.3.25PR1) was used to analyze the data. Shirley-type background and X-ray satellite subtraction were used in the processing of the measured spectrum. Using a symmetric Gaussian–Lorentzian curve fitting, the peak positions and areas were assessed. Raman spectrometry was carried on a ThunderOptics Eddu TO-ERS-532 spectrometer, Montpellier, France), equipped with a 532 nm laser source and a 20× microscope objective lens. UV/Vis spectra were obtained on an Evolution 300 Thermo Scientific spectrophotometer (Madison, WI, USA), equipped with a DRA-EV-300 Diffuse Reflectance Accessory.

3.3. Tribocatalytic Experiments and Radical Assay

The tribocatalytic experiments were carried out with 50 mL DC solution prepared with distilled water in a 100 mL glass beaker, equipped with a magnetic stirrer. The tribocatalytic reaction was conducted at constant room temperature (23 ± 2 °C) in the dark. The initial concentration of DC was 15 ppm. In total, 50 mg catalyst (pure or RE-modified ZnO) was added to a glass reactor containing DC solution and the suspension was magnetically stirred using a PTFE-coated magnetic bar (ø 8 mm, L = 35 mm). To attain the adsorption equilibrium between the doxycycline solution and tribocatalysts, the resultant mixture was soaked for 30 min without any magnetic stirring. After that, the reactor was turned on, initially rotating at a constant speed of 300 rpm. At regular intervals, aliquot samples of 2 mL of the reaction solution were taken. The tribocatalyst was then centrifuged at 6000 rpm. UV–Vis spectra of aliquots from the reaction media were recorded in the range of 200–450 nm. The peak at maximal drug absorption was at 275 nm (absorbance decreased as a function of stirring time for each catalyst). At this wavelength, not only was the degradation of doxycycline observed but also its degradation products (phenolic compounds) [43,44]. This method was similar to all other decomposition performance tests, except for differences in the type of catalyst (pure and ZnO/RE powders) and magnetic stirring conditions (100 and 500 rpm).

The following formula was used to estimate the drug tribodegradation degree (D%):

$$Decomposition\% = \frac{C}{C_0} \times 100\% \qquad (1)$$

where C_0 is the initial concentration of doxycycline and C is the concentration (absorbance) of drug at time = t (min) [18,22].

Additionally, blank experiments without catalysts were carried out—there was no sign of any removal of the tetracycline antibiotic under PTFE stirring without a tribocatalyst.

The reactive species causing the degradation of the DC were investigated using a scavenger test. Isopropyl alcohol (IPA) and ascorbic acid (AA) were used as scavengers to absorb superoxide and hydroxyl radicals, respectively. To identify the specific reactive species that underwent tribocatalysis-induced degradation of the organic dye (50 mL), 6 mM of each scavenger was used separately.

4. Conclusions

In this study, the tetracycline antibiotic doxycycline (DC) was successfully degraded using magnetic stirring in the dark, facilitated by three types of ZnO powders modified with rare-earth elements (Eu^{3+} and La^{3+}). Among these, the ZnO/La composite, which exhibited the highest specific surface area, demonstrated the most effective degradation. The degradation rate significantly increased with higher stirring speeds, emphasizing the role of mechanical energy in the process. The results reveal that mechanical energy absorbed during friction effectively excites the electrons and holes in ZnO and ZnO/RE composites, leading to efficient drug breakdown. This tribocatalytic effect represents a promising, eco-friendly pathway for harnessing mechanical energy from the environment to address pollution. By enabling drug degradation through tribocatalysis, this method opens new possibilities for controlling environmental contamination.

Author Contributions: Conceptualization, N.K.; methodology, N.K.; formal analysis, N.K.; investigation, D.I., H.K. and B.I.S.; data curation, N.K.; writing—original draft preparation, N.K.; writing—review and editing, N.K. and B.I.S.; visualization, D.I., N.K., H.K. and B.I.S.; supervision, N.K. All authors have read and agreed to the published version of the manuscript.

Funding: This research received no external funding.

Institutional Review Board Statement: Not applicable.

Informed Consent Statement: Not applicable.

Data Availability Statement: Data are contained within the article.

Acknowledgments: Research equipment of the distributed research infrastructure INFRAMAT, supported by the Bulgarian Ministry of Education and Science under contract D01-284/17.12.2019, was used.

Conflicts of Interest: The authors declare no conflicts of interest.

References

1. Amaechi, I.; Youssef, A.; Dorfler, A.; Gonzalez, Y.; Katoch, R.; Ruediger, A. Catalytic Applications of Non-Centrosymmetric Oxide Nanomaterials. *Angew. Chem. Int. Ed. Engl.* **2022**, *61*, e202207975. [CrossRef] [PubMed]
2. Das, R.; Vecitis, C.; Schulze, A.; Cao, B.; Ismail, A.; Lu, X.; Chen, J.; Ramakrishna, S. Recent advances in nanomaterials for water protection and monitoring. *Chem. Soc. Rev.* **2017**, *46*, 6946–7020. [CrossRef]
3. Jeon, I.; Ryberg, E.; Alvarez, P.; Kim, J. Technology assessment of solar disinfection for drinking water treatment. *Nat. Sustain.* **2022**, *5*, 801–808. [CrossRef]
4. Alsharyani, A.; Muruganandam, L. Fabrication of zinc oxide nanorods for photocatalytic degradation of docosane, a petroleum pollutant, under solar light simulator. *RSC Adv.* **2024**, *14*, 9038–9049. [CrossRef] [PubMed]
5. Huang, H.; Wang, H.; Jiang, W. Solar-driven $Bi_6O_5(OH)_3(NO_3)_5(H_2O)_3/Bi_2WO_6$ heterojunction for efficient degradation of organic pollutants: Insights into adsorption mechanism, charge transfer and degradation pathway. *Sep. Purif. Technol.* **2024**, *349*, 127747. [CrossRef]
6. Olasupo, A.; Corbin, D.; Shiflett, M. Trends in low temperature and non-thermal technologies for the degradation of persistent organic pollutants. *J. Hazard. Mater.* **2024**, *468*, 133830. [CrossRef] [PubMed]
7. Chen, X.; Wang, J.; Wang, Z.; Xu, H.; Liu, C.; Huo, B.; Meng, F.; Wang, Y.; Sun, C. Low-frequency mechanical energy in the environment for energy production and piezocatalytic degradation of organic pollutants in water: A review. *J. Water Process Eng.* **2023**, *56*, 104312. [CrossRef]
8. Verma, A.; Fu, Y. The prospect of CuxO-based catalysts in photocatalysis: From pollutant degradation, CO_2 reduction, and H_2 production to N_2 fixation. *Environ. Res.* **2024**, *241*, 117656. [CrossRef]
9. Kumar, A.; Rana, S.; Sharma, G.; Dhiman, P.; Shekh, M.; Stadler, F. Recent advances in zeolitic imidazole frameworks based photocatalysts for organic pollutant degradation and clean energy production. *J. Environ. Chem. Eng.* **2023**, *11*, 110770. [CrossRef]
10. Xu, H.; Yang, H.; Zhou, J.; Yin, Y. A Route Choice Model with Context-Dependent Value of Time. *Transp. Sci.* **2017**, *51*, 536–548. [CrossRef]
11. Li, P.; Tang, C.; Xiao, X.; Jia, Y.; Chen, W. Flammable gases produced by TiO_2 nanoparticles under magnetic stirring in water. *Friction* **2021**, *10*, 1127–1133. [CrossRef]
12. Cui, X.; Li, P.; Lei, H.; Tu, C.; Wang, D.; Wang, Z.; Chen, W. Greatly enhanced tribocatalytic degradation of organic pollutants by TiO_2 nanoparticles through efficiently harvesting mechanical energy. *Sep. Purif. Technol.* **2022**, *289*, 120814. [CrossRef]

13. Hu, J.; Ma, W.; Pan, Y.; Chen, Z.; Zhang, Z.; Wan, C. Resolving the Tribocatalytic reaction mechanism for biochar regulated Zinc Oxide and its application in protein transformation. *J. Colloid Interface Sci.* **2022**, *607*, 1908–1918. [CrossRef] [PubMed]
14. Fan, F.; Xie, S.; Wang, G.; Tian, Z. Tribocatalysis: Challenges and perspectives. *Sci. China Chem.* **2021**, *64*, 1609–1613. [CrossRef]
15. Geng, L.; Qian, Y.; Song, W.; Bao, L. Enhanced tribocatalytic pollutant degradation through tuning oxygen vacancy in BaTiO$_3$ nanoparticles. *Appl. Surf. Sci.* **2023**, *637*, 157960. [CrossRef]
16. Wu, J.; Qin, N.; Bao, D. Effective enhancement of piezocatalytic activity of BaTiO$_3$ nanowires under ultrasonic vibration. *Nano Energy* **2018**, *45*, 44–51. [CrossRef]
17. Feng, Y.; Ling, L.; Wang, Y.; Xu, Z.; Cao, F.; Li, H. Engineering spherical lead zirconate titanate to explore the essence of piezo-catalysis. *Nano Energy* **2017**, *40*, 481–486. [CrossRef]
18. Yang, B.; Chen, H.; Guo, X.; Wang, L.; Xu, T.; Bian, J. Enhanced tribocatalytic degradation using piezoelectric CdS nanowires for efficient water remediation. *J. Mater. Chem. C* **2020**, *8*, 14845–14854. [CrossRef]
19. Li, P.; Wu, J.; Wu, Z.; Jia, Y.; Ma, J.; Chen, W.; Zhang, L.; Yang, J.; Liu, Y. Strong tribocatalytic dye decomposition through utilizing triboelectric energy of barium strontium titanate nanoparticles. *Nano Energy* **2019**, *63*, 103832. [CrossRef]
20. Zhao, J.; Chen, L.; Luo, W.; Li, H.; Wu, Z.; Xu, Z.; Zhang, Y.; Zhang, H.; Yuan, G.; Gao, J.; et al. Strong Tribo-Catalysis of Zinc Oxide Nanorods Via Triboelectrically-Harvesting Friction Energy. *Ceram. Int.* **2020**, *46*, 25293–25298. [CrossRef]
21. Sun, S.; Sui, X.; Yu, H.; Zheng, Y.; Zhu, X.; Wu, X.; Li, Y.; Lin, Q.; Zhang, Y.; Ye, W.; et al. High Tribocatalytic Performance of FeOOH Nanorods for Degrading Organic Dyes and Antibiotics. *Small Methods* **2024**, 2301784. [CrossRef] [PubMed]
22. Li, X.; Tong, W.; Song, W.; Shi, J.; Zhang, Y. Performance of tribocatalysis and tribo-photocatalysis of pyrite under agitation. *J. Clean. Prod.* **2023**, *414*, 137566. [CrossRef]
23. Kaneva, N.; Bojinova, A.; Papazova, K.; Dimitrov, D. Photocatalytic purification of dye contaminated sea water by lanthanide (La^{3+}, Ce^{3+}, Eu^{3+}) modified ZnO. *Catal. Today* **2015**, *252*, 113–119. [CrossRef]
24. Qu, G.; Fan, G.; Zhou, M.; Rong, X.; Li, T.; Zhang, R.; Sun, J.; Chen, D. Graphene-Modified ZnO Nanostructures for Low-Temperature NO$_2$ Sensing. *ACS Omega* **2019**, *4*, 4221–4232. [CrossRef]
25. Zhao, S.; Shen, Y.; Li, A.; Chen, Y.; Gao, S.; Liu, W.; Wei, D. Effects of rare earth elements doping on gas sensing properties of ZnO nanowires. *Ceram. Int.* **2021**, *47*, 24218–24226. [CrossRef]
26. Jia, T.; Wang, W.; Long, F.; Fu, Z.; Wang, H.; Zhang, Q. Fabrication, characterization and photocatalytic activity of La-doped ZnO nanowires. *J. Alloys Compd.* **2009**, *484*, 410–415. [CrossRef]
27. Ada, K.; Gökgöz, M.; Önal, M.; Sarıkaya, Y. Preparation and characterization of a ZnO powder with the hexagonal plate particles. *Powder Technol.* **2008**, *181*, 285–291. [CrossRef]
28. Kumar, S.; Kavitha, R. Lanthanide ions doped ZnO based photocatalysts. *Sep. Purif. Technol.* **2021**, *274*, 118853. [CrossRef]
29. Tsuji, T.; Terai, Y.; Kamarudin, M.; Kawabata, M.; Fujiwara, Y. Photoluminescence properties of Sm-doped ZnO grown by sputtering-assisted metalorganic chemical vapor deposition. *J. Non-Cryst. Solids* **2012**, *358*, 2443–2445. [CrossRef]
30. Khatamian, M.; Khandar, A.; Divband, B.; Haghighi, M.; Ebrahimiasl, S. Heterogeneous photocatalytic degradation of 4-nitrophenol in aqueous suspension by Ln (La^{3+}, Nd^{3+} or Sm^{3+}) doped ZnO nanoparticles. *J. Mol. Catal. A Chem.* **2012**, *365*, 120–127. [CrossRef]
31. Peleš, A.; Pavlović, V.P.; Filipović, S.; Obradović, N.; Mančić, L.; Krstić, J.; Mitrić, M.; Vlahović, B.; Rašić, G.; Kosanović, D.; et al. Structural investigation of mechanically activated ZnO powder. *J. Alloys Compd.* **2015**, *648*, 971–979. [CrossRef]
32. Xu, Y.; Yin, R.; Zhang, Y.; Zhou, B.; Sun, P.; Dong, X. Unveiling the Mechanism of Frictional Catalysis in Water by Bi$_{12}$TiO$_{20}$: A Charge Transfer and Contaminant Decomposition Path Study. *Langmuir* **2022**, *38*, 14153–14161. [CrossRef] [PubMed]
33. Che, J.; Gao, Y.; Wu, Z.; Ma, J.; Wang, Z.; Liu, C.; Jia, Y.; Wang, X. Review on tribocatalysis through harvesting friction energy for mechanically-driven dye decomposition. *J. Alloys Compd.* **2024**, *1002*, 175413. [CrossRef]
34. Li, X.; Tong, W.; Shi, J.; Chen, Y.; Zhang, Y.; An, Q. Tribocatalysis mechanisms: Electron transfer and transition. *J. Mater. Chem. A* **2023**, *11*, 4458–4472. [CrossRef]
35. Soares, A.; Araujo, F.; Osajima, J.; Guerra, Y.; Viana, B.; Peña-Garcia, R. Nanotubes/nanorods-like structures of La-doped ZnO for degradation of Methylene Blue and Ciprofloxacin. *J. Photochem. Photobiol. A Chem.* **2024**, *447*, 115235. [CrossRef]
36. Campos, S.; Calzadilla, W.; Salazar-González, R.; Venegas-Yazigi, D.; León, J.; Fuentes, S. Photocatalytic activity of barium titanate composites with zinc oxide doped with lanthanide ions for sulfamethoxazole degradation. *J. Environ. Chem. Eng.* **2024**, *12*, 112896. [CrossRef]
37. Duffy, J. Trends in energy gaps of binary compounds: An approach based upon electron transfer parameters from optical spectroscopy. *J. Phys. C Solid State Phys.* **1980**, *13*, 2979–2989. [CrossRef]
38. Wang, Y.; Shen, S.; Liu, M.; He, G.; Li, X. Enhanced tribocatalytic degradation performance of organic pollutants by Cu$_{1.8}$S/CuCo$_2$S$_4$ p-n junction. *J. Colloid Interface Sci.* **2024**, *655*, 187–198. [CrossRef] [PubMed]
39. Anandan, S.; Vinu, A.; Lovely, K.; Gokulakrishnan, N.; Srinivasu, P.; Mori, T.; Murugesan, V.; Sivamurugan, V.; Ariga, K. Photocatalytic activity of La-doped ZnO for the degradation of monocrotophos in aqueous suspension. *J. Mol. Catal. A Chem.* **2007**, *266*, 149–157. [CrossRef]
40. Korake, P.; Dhabbe, R.; Kadam, A.; Gaikwad, Y.; Garadkar, K. Highly active lanthanum doped ZnO nanorods for photodegradation of metasystox. *J. Photochem. Photobiol. B Biol.* **2014**, *130*, 11–19. [CrossRef]
41. Zhao, J.; Dang, Z.; Muddassir, M.; Raza, S.; Zhong, A.; Wang, X.; Jin, J. A New Cd(II)-Based Coordination Polymer for Efficient Photocatalytic Removal of Organic Dyes. *Molecules* **2023**, *28*, 6848. [CrossRef] [PubMed]

42. Xiang, R.; Zhou, C.; Liu, Y.; Qin, T.; Li, D.; Dong, X.; Muddassir, M.; Zhong, A. A new type Co(II)-based photocatalyst for the nitrofurantoin antibiotic degradation. *J. Mol. Struct.* **2024**, *1312*, 138501. [CrossRef]
43. Chopra, I.; Roberts, M. Tetracycline antibiotics: Mode of action, applications, molecular biology, and epidemiology of bacterial resistance. *Microbiol. Mol. Biol. Rev.* **2001**, *65*, 232–260. [CrossRef]
44. Thomas, M.; Nałęcz-Jawecki, G.; Giebułtowicz, J.; Drzewicz, P. Degradation of oxytetracycline by ferrate (VI): Treatment optimization, UHPLC-MS/MS and toxicological studies of the degradation products, and impact of urea and creatinine on the removal. *Chem. Eng. J.* **2024**, *485*, 149802. [CrossRef]

Disclaimer/Publisher's Note: The statements, opinions and data contained in all publications are solely those of the individual author(s) and contributor(s) and not of MDPI and/or the editor(s). MDPI and/or the editor(s) disclaim responsibility for any injury to people or property resulting from any ideas, methods, instructions or products referred to in the content.

Article

Transformation of Graphite Recovered from Batteries into Functionalized Graphene-Based Sorbents and Application to Gas Desulfurization

Rodolfo Fernández-Martínez [1,*], Isabel Ortiz [2], M. Belén Gómez-Mancebo [1], Lorena Alcaraz [3], Manuel Fernández [1], Félix A. López [3], Isabel Rucandio [1] and José María Sánchez-Hervás [2]

1 Chemistry Division, Technology Department, Research Centre for Energy, Environment and Technology (CIEMAT), 28040 Madrid, Spain; mariabelen.gomez@ciemat.es (M.B.G.-M.); m.fernandez@ciemat.es (M.F.); isabel.rucandio@ciemat.es (I.R.)
2 Sustainable Thermochemical Valorization Unit, Energy Department, Research Centre for Energy, Environment and Technology (CIEMAT), 28040 Madrid, Spain; isabel.ortiz@ciemat.es (I.O.); josemaria.sanchez@ciemat.es (J.M.S.-H.)
3 National Centre for Metallurgical Research (CENIM), Spanish National Research Council (CSIC), Avda. Gregorio del Amo 8, 28040 Madrid, Spain; alcaraz@cenim.csic.es (L.A.); f.lopez@csic.es (F.A.L.)
* Correspondence: rodolfo.fernandez@ciemat.es

Abstract: The recycling and recovery of value-added secondary raw materials such as spent Zn/C batteries is crucial to reduce the environmental impact of wastes and to achieve cost-effective and sustainable processing technologies. The aim of this work is to fabricate reduced graphene oxide (rGO)-based sorbents with a desulfurization capability using recycled graphite from spent Zn/C batteries as raw material. Recycled graphite was obtained from a black mass recovered from the dismantling of spent batteries by a hydrometallurgical process. Graphene oxide (GO) obtained by the Tour's method was comparable to that obtained from pure graphite. rGO-based sorbents were prepared by doping obtained GO with NiO and ZnO precursors by a hydrothermal route with a final annealing step. Recycled graphite along with the obtained GO, intermediate (rGO-NiO-ZnO) and final composites (rGO-NiO-ZnO-400) were characterized by Wavelength Dispersive X-ray Fluorescence (WDXRF) and X-ray diffraction (XRD) that corroborated the removal of metal impurities from the starting material as well as the presence of NiO- and ZnO-doped reduced graphene oxide. The performance of the prepared composites was evaluated by sulfidation tests under different conditions. The results revealed that the proposed rGO-NiO-ZnO composite present a desulfurization capability similar to that of commercial sorbents which constitutes a competitive alternative to syngas cleaning.

Keywords: Zn/C battery waste; reduced graphene oxide; rGO-based composites; desulfurization; syngas cleaning; sustainability

Citation: Fernández-Martínez, R.; Ortiz, I.; Gómez-Mancebo, M.B.; Alcaraz, L.; Fernández, M.; López, F.A.; Rucandio, I.; Sánchez-Hervás, J.M. Transformation of Graphite Recovered from Batteries into Functionalized Graphene-Based Sorbents and Application to Gas Desulfurization. *Molecules* **2024**, *29*, 3577. https://doi.org/10.3390/molecules29153577

Academic Editors: Sake Wang, Nguyen Tuan Hung and Minglei Sun

Received: 6 June 2024
Revised: 15 July 2024
Accepted: 19 July 2024
Published: 29 July 2024

Copyright: © 2024 by the authors. Licensee MDPI, Basel, Switzerland. This article is an open access article distributed under the terms and conditions of the Creative Commons Attribution (CC BY) license (https://creativecommons.org/licenses/by/4.0/).

1. Introduction

The global production and demand for graphite has increased in recent years, largely because of the use of graphite for producing batteries of electric vehicles. In 2022, the global consumption of graphite reached 3.8 million tons, compared to 3.6 million tons in 2021 [1]. Actually, mining continues to be the main source of graphite with 1.6 million tons in 2023 [2]. It is essential to preserve carbonaceous natural resources by the application of sustainable approaches for graphite production whilst reducing the environmental impact. According to the principles of a circular economy, recycling graphite from devices where it is present may contribute to the key challenge for achieving the objective of closed-loop system materials and sustainability [3].

Cathodes from Zn/C batteries are made of graphite carbon. The recycling process is based on a hydrometallurgical process in a preliminary step of the dismantling of batteries followed by acidic leaching [4]. This yields a black mass consisting of a mixture of graphite and metallic oxides. In spite of the presence of metallic impurities, recovered graphite can be used as a precursor to prepare high-added-value products such as graphene-related materials and graphene-based composites [5–7].

Graphene is a 2D material that has recently received an enormous amount of attention due to their exceptional physicochemical properties. It is constituted of a single-atom-thick layer of conjugated sp2 carbon atoms arranged in a honeycomb structure. The top-down approach based on chemical exfoliation constitutes the most standardized method to synthesize graphene-related materials, such as graphene oxide (GO) or reduced graphene oxide (rGO) [8,9]. Among its strengths is its versatility, scalability and tunability in terms of the surface area, size and functionality while its weaknesses lie in the presence of defects and remaining functional groups due to the incomplete reduction of oxygen groups [10]. To date, chemical exfoliation represents the most suitable route to produce large amounts of graphene at a reasonable cost from graphite oxide [11]. Graphene oxide (GO) is synthesized from graphite under strong oxidative conditions and subsequent exfoliation. Afterwards, GO may be converted to reduced graphene oxide (rGO) by a reduction process where the oxygen-containing groups are partially removed.

rGO presents characteristics that make it an ideal support to prepare hybrid sorbents. Among them, the most crucial in order to prepare rGO-based composites are the high surface-to-volume ratio, owing to its 2D nature, its porous structure as well as its remaining oxygenated groups, which act as anchoring sites to active groups, and the presence of vacancies that serve as a trap for incoming atoms and nanoclusters [12]. This allows the possibility of incorporating a wide variety of groups that determine the functionality of the resulting composites. Potential applications of rGO-based sorbents include the retention and removal of pollutants and energy storage [13–17]. In a recent study, the suitability of rGO-NiO-ZnO-based sorbents for hydrogen sulfide removal from syngas has been demonstrated [18].

Gasification is a thermochemical process that under sub-stoichiometric oxygen converts biomass or waste into a gaseous chemical energy carrier, known as synthesis gas, which can be used to produce bioenergy or further transformed into a variety of gaseous and liquids fuels such as hydrogen, synthetic natural gas (SNG), jet fuels, or chemicals, e.g., methanol or dimethyl ether (DME) [19].

One issue of concern is the presence of sulfur species in the raw syngas since all the above transformation processes rely on the use of catalysts, which are extremely sensitive to deactivation by sulfur compounds. Zinc oxide-based sorbents have been demonstrated to achieve gas desulfurization targets. In addition to that, sulfur removal with such sorbents can be nicely coupled with thermochemical biofuel catalytic production given that they work in similar temperature windows. Their main drawback is their cost. There is a clear need for the development of cheaper materials and their synthesis from recovered materials is a good way [20].

In the recent literature, there have been several works based on the use of recycled graphite from spent Zn/C batteries to synthesize graphene-related materials [21]. Composite materials have been used as a support to energy applications such as supercapacitors [22] or to fuel hydrogen production [6]. However, to the best of our knowledge, no prior studies have considered the preparation of rGO-based sorbent for the removal of sulfur pollutants from renewable fuels produced by catalytic processes.

In this work, the synthesis of cost-effective and efficient rGO-based sorbents with a gas desulfurization capability is proposed. The black mass recovered from exhausted Zn/C battery rods is leveraged as a low-cost and sustainable graphite source to be used as a precursor of graphene oxides. Previous studies from Zn/C batteries are typically based on the use of strong leaching conditions that generate waste with difficult management [5]. In the proposed work, a previously assessed process to recovery the metals from Zn/C batteries, allowing the production of Zn/Mn oxides with varying stoichiometry for various applications [23], has been performed. This method results in a carbonaceous insoluble residue, which is suitable for its use as a raw material to obtain carbon-based materials. Furthermore, the use of this solid residue allows the achievement of a fully sustainable zero-waste process. Further functionalization with Ni and Zn reagents allows the obtainment of rGO-NiO-ZnO composites. The developed composites have been evaluated as sorbents to remove hydrogen sulfide and compared to other commercially available sorbents. rGO-based sorbents from recycled graphite may contribute to the circular economy by the valorization of wastes that are typically scrapped.

2. Results

Starting from graphite previously expanded by treatment with sulfuric acid and hydrogen peroxide, functionalized reduced graphene oxides have been synthesized, as described in Section 3.3. The synthesized materials as well as the starting materials have been chemically and structurally characterized to confirm the presence, in each case, of the desired material with both the catalytic activity (ZnO and NiO) and support capacity (rGO).

The synthesized materials have been named as set out in Table 1 below:

Table 1. Denomination of the materials involved in this work.

Denomination	Type of Material	Treatment
GRERV21-MX-SA	Recycled graphite	Recycled graphite from batteries treated with sulfuric acid
GO-17	Graphene oxide	Graphene oxide obtained using the Tour method and ultrasonic exfoliation
rGO-NiO-ZnO	Reduced graphene oxide functionalized with NiO and ZnO	Hydrothermally reduced graphene oxide, pre-functionalized with NiO and ZnO
rGO-NiO-ZnO-400	NiO and ZnO functionalized reduced graphene oxide with calcination at 400 °C.	Graphene oxide reduced by hydrothermal reduction, pre-functionalized with NiO and ZnO and subsequently subjected to 400 °C

2.1. Chemical Characterization of the Materials

All the materials were analyzed using the X-ray fluorescence (XRF) technique, which allows the direct analysis of the solid. The analysis was carried out using helium gas as a medium and introducing the samples directly into a specific sample holder for powder samples.

The elemental characterization performed by the Wavelength Dispersive X-ray Fluorescence (WDXRF) technique is able to determine the elements present in the sample between oxygen and uranium and in concentrations ranging from a percentage (%) to parts per million (mg/kg). The semi-quantitative method used in the analysis, developed by Malvern Panalytical, allows the rapid analysis of the elements in the sample. The results of the analyses carried out on the materials involved in this work are shown in Table 2, expressed as a %.

All samples show moderate-to-high C concentrations (Table 2). The GRERV21-MX-SA sample, recycled graphite, shows the highest percentage (63%). As the oxygenated groups have been introduced between the graphite sheets upon oxidation, to obtain the graphite oxide, GO-17 sample, as expected, the concentration of carbon in the GO samples decreases to around 40%. These C concentrations are significantly lower than those observed in GO samples obtained by the Tour method from high-purity natural graphite [24,25]. This means that the oxidation process takes place in a higher extent resulting in a more oxidized product with a greater abundance of oxygen groups. In the functionalized reduced-graphene oxide material, the rGO-NiO-ZnO sample, the C concentration has been considerably reduced, as expected [26]. This result corroborates the tendency of rGO-NiO-ZnO to recover the original graphite structure. On the other hand, high concentrations of the elements Ni and Zn are observed, corresponding to 5 and 37%, respectively, demonstrating the successful introduction of Ni and Zn into the structure of the rGO material. These two metal oxides are the active species for hydrogen sulfide removal at an intermediate temperature. In the thermal reductions of graphene oxides, a violent process of CO and CO_2 emission occurs [27] which causes part of the oxygen along with part of the C to be removed as a gas. This effect along with the introduction of a large proportion of Zn and Ni phases produces an important reduction of the C concentration in the rGO-NiO-ZnO-400 sample. The removal of carbon and oxygen in the form of CO and CO_2 results in a pre-concentration of the elements Ni and Zn, whose concentrations correspond to 6.2 and 47%, respectively, which are comparable to those observed for composites synthesized from natural graphite in the same experimental conditions [18]. However, the C percentage in the composite (rGO-NiO-ZnO-400) is significantly lower than that obtained from natural graphite (3.9% vs. 11.0%). The reason might be the presence of remaining unreduced oxygen groups probably because of the great abundance of oxygen groups in the GO which results in a lesser degree of reduction.

Table 2. Results of the chemical characterization (%) carried out by WDXRF on the materials involved in this work. (-) means the absence of the element.

Element (%)	GRERV21-MX-SA	GO-17	rGO-NiO-ZnO	rGO-NiO-ZnO-400
C	63	39	8.5	3.9
Zn	0.080	-	37	47
Ni	-	-	5.0	6.2
Mn	9.6	0.31	-	-
Ba	1.3	1.2	0.16	0.22
Al	0.14	0.080	0.029	0.039
Br	0.037	0.020	-	-
Ca	0.0095	0.0041	-	-
Ce	-	-	0.07	0.12
Cl	0.94	0.31	-	0.022
Fe	0.46	0.020	0.011	0.016
K	0.15	0.15	-	-
La	0.054	0.032	-	-
P	0.017	0.055	0.0087	0.0089
Pb	0.041	0.043	-	-
S	0.66	1.8	0.18	0.26
Si	0.53	0.28	0.091	0.11
Ti	0.20	0.090	0.055	0.062

The results presented in Table 2 show that in the GRERV21-MX-SA sample, the most abundant element after C is Mn (9.6%), an element coming from the cathode of Zn/C batteries. A high concentration of Ba is also observed, corresponding to 1.3%. However, in the GO-17 sample, all the elements, with the exception of C, are in a low concentration, below 2%. The most abundant element after C is found to be S, which may be partly due to the oxidation process. Ba, with a concentration of 1.2% in the GO-17 sample, presents a concentration equivalent to that in the starting graphite (GRERV21-MX-SA). The rest of the elements, except C, Ni and Zn, appearing in the rGO samples are in a very low concentration, including Ba. This means that the oxidation process applied to the sample is able to remove the major elements present in the original sample with the exception of barium. The concentration of this element strongly decreases in the rGO samples as a consequence of thermal annealing.

2.2. Structural and Textural Characterization of the Materials

To determine the crystalline structures present in the materials, X-ray diffraction (XRD) was performed. Table 3 shows the crystalline phases found in the synthesized materials.

Table 3. The crystalline phases found in the synthesized materials.

Material	Compounds
GRERV21-MX-SA	C, $BaSO_4$, SiO_2
GO-17	BaK_xSO_4
rGO-NiO-ZnO	ZnO, $Zn_4(CO_3)(OH)_6H_2O$, $Zn_5(OH)_6(CO_3)_2$, $ZnSO_4$, $3Zn(OH)_2$
rGO-NiO-ZnO-400	ZnO, $Ni_{0.7}Zn_{0.3}$

As can be seen in Table 3, the starting sample (GRERV21-MX-SA) is mainly composed of graphite (Figure 1), with a main peak around 26° (2θ). $BaSO_4$, a rather insoluble compound, and SiO_2 phases are also identified. Figure 1 shows that in the GO-17 sample, a main peak can be seen at approximately 10° (2θ) corresponding to graphene oxide [28]. This result corroborates the complete transformation of graphite to graphene oxide, with the disappearance of the peak around 26° (2θ). This peak is quite broad, with a width at mid-height corresponding to 1.71° (2θ) and a low intensity, which indicates that adequate oxidation and exfoliation of the graphite has occurred [28], corroborating the data obtained by the WDXRF technique. In addition, a BaK_xSO_4 phase similar to the one found in the GRERV21-MX-SA sample appears, since it has not been possible to remove the Ba with the treatment employed, as it corresponds to a rather insoluble compound, which may come from the starting graphite.

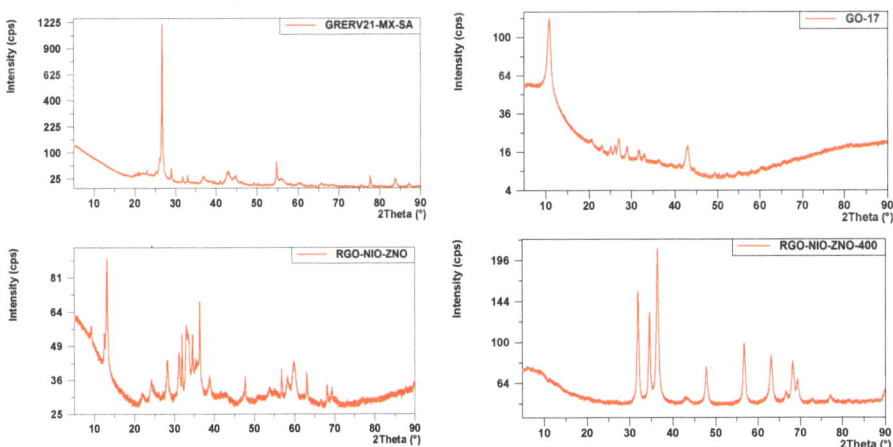

Figure 1. XRD patterns for the synthesized materials.

The functionalization and reduction process has resulted in the sample labeled rGO-NiO-ZnO. This sample also shows a poorly crystalline profile with low intensity peaks (Figure 1). Furthermore, a very broad band between 20 and 30° (2θ) appears, corresponding to the reduced graphene oxide (rGO) [27].

The crystalline phases found in rGO-NiO-ZnO correspond, in their totality, to compounds with Zn (Table 3); however, from Table 2 it can be deduced that also 5% Ni has been introduced in the sample during the hydrothermal treatment. Ni when is introduced into graphene structures gives rise to amorphous structures (probably nickel oxyhydroxides), which could be anchored via covalent bonds to the graphene sheets [26]. The presence of these partially oxidized phases is typically from composites derived from hydrothermal routes and it has been previously observed in similar structures [18,29]. The crystalline Zn compounds that have been identified in this sample (Table 3) correspond to zinc oxide as well as other compounds such as oxyhydroxides formed during the hydrothermal process [30] that are the ultimate precursors of zinc oxide.

The analysis of the rGO-NiO-ZnO-400 material is shown in Figure 1. The profile of this material is more crystalline with more defined peaks. The phases found (Table 3) correspond to Zn (ZnO) and Ni ($Ni_{0.7}Zn_{0.3}O$) oxides, although the latter is doped with a small proportion of Zn. This corroborates the results obtained in the WDXRF analysis and demonstrates the effectiveness of the annealing treatment at 400 °C to effectively complete the oxidation of oxyhydroxide's intermediate phases into ZnO and NiO.

Textural characterization was carried out by assessing the specific surface area (SSA), applying the physical adsorption of nitrogen (N2) at 196 °C, and the results were calculated according to the Brunauer–Emmett–Teller (BET) method. Also, the pore volume was calculated by applying t-plot methodology.

This textural characterization has been applied on the rGO-NiO-ZnO-400 sample as this is the sample used in the proposed application. This sample has a specific surface area (SSA) of 46.48 $m^2 \cdot g^{-1}$ and t-Plot micropore volume of 0.003385 $cm^3 \cdot g^{-1}$, which is similar to that obtained with the same treatment on pure graphite [18]. In a previous work, J. M. Sánchez-Hervas et al. synthesized several materials similar to ZnO-NiO-400 from pure graphite. In particular, the so-called rGO (5) (Zn-Ni-rGO) had the same synthesis as rGO-ZnO-NiO-400. In this case, the SSA corresponded to 46.42 $m^2 \cdot g^{-1}$ and a t-Plot micropore volume of 0.0028 $cm^3 \cdot g^{-1}$.

The BET and t-Plot measurements show the rGO specific surface area and micropore volume data, which are comparable to those measured in rGO obtained from non-recycled graphite precursors.

2.3. Sulfidation Tests

The reduced graphene rGO-NiO-ZnO-400 sample has been studied as a desulfurization sorbent.

The H_2S removal ability of the same type of sorbents synthetized from pure graphite was demonstrated in a previous study [18]. The reactive adsorption of H_2S on rGO/metal hydroxides occurs via acid–base reactions. Through this mechanism, H_2S is effectively retained on the surface of the adsorbent by the direct replacement of OH groups and the acid–base reaction with the metal (hydr)oxides, resulting in the formation of sulfites and sulfates. In this work, the performance of the rGO-NiO-ZnO-400 sorbent in three different atmospheres is presented. Firstly, a simplified atmosphere with 0.9% (v/v) in nitrogen is used to compare the performance of this sorbent with the ones studied in our previous research. Then, two different syngas compositions representative of biomass gasification processes were employed.

To compare the performance of the new sorbent with those previously evaluated, the same conditions were studied, namely: 400 °C of temperature, 10 bar of pressure and a gas space velocity of 3500 h^{-1} with a gas stream containing 9000 ppmv of H_2S/N_2.

The gas' hourly space velocity, GHSV, is the ratio of the volumetric gas-flow-rate in normal conditions to the bulk sorbent volume loaded into the reactor. The selected values for the desulfurization operating conditions were set in accordance with previous studies published by the authors [31].

The performance of the sorbents has been evaluated by the S loading capacity and actual breakthrough times compared to the theoretical values. To determine these theoretical values, two sulfidation reactions were considered:

$$ZnO + H_2S \rightarrow ZnS + H_2O \qquad (1)$$

$$NiO + H_2S \rightarrow NiS + H_2O \tag{2}$$

The theoretical S load capacity (S_0) is, therefore, calculated following the equation:

$$S_0(\%) = \left(\frac{\%ZnO}{MW_{ZnO}} + \frac{\%NiO}{MW_{NiO}}\right) \cdot MW_S \tag{3}$$

And the theoretical sorption time, t_0, when complete sulfidation is achieved, was calculated as the ratio between the theoretical amounts of S that each material can adsorb (g) based on its composition and the S mass flow rate used in each experiment (g/min). This procedure assumes that S is totally retained by the sorbent and no S escapes in the gas outlet.

$$t_0(min) = \left(\frac{S_0(\%) \cdot M_{sorb}(g)}{MW_S \cdot \frac{P \cdot Q_{H2S}}{R \cdot T}}\right) \tag{4}$$

where M_{sorb} is the mass of sorbent used for desulfurization, MW is the molecular weight of S, P is the absolute pressure in sulfidation conditions, Q_{H2S} is the volumetric gas flow rate of H_2S in the process conditions, R is the universal gas constant and T is the absolute temperature in sulfidation conditions. The sulfidation breakthrough point was set at 0.01% (v/v).

Dimensionless breakthrough curves and the utilization yield for two sample sorbents obtained from pure graphite (rGO(5)) and from recycled graphite as well as commercial sorbents (Z-Sorb III™) are depicted in Figures 2 and 3. As can be observed, the sorbent from recycled graphite shows a lower desulfurization capacity than the sorbents obtained from pure graphite at the same experimental conditions. This is expected since the lesser reduction observed in rGO from recycled graphite with respect to the same rGO obtained from pure graphite should result in a lower degree of recovery of the characteristic π-conjugated structure and, therefore, a lower electron mobility [32,33]. However, the desulfurization capacity remains at the levels of the commercial sorbent, Z-Sorb III™, indicating the suitability of the proposed sorbents for desulfurization applications.

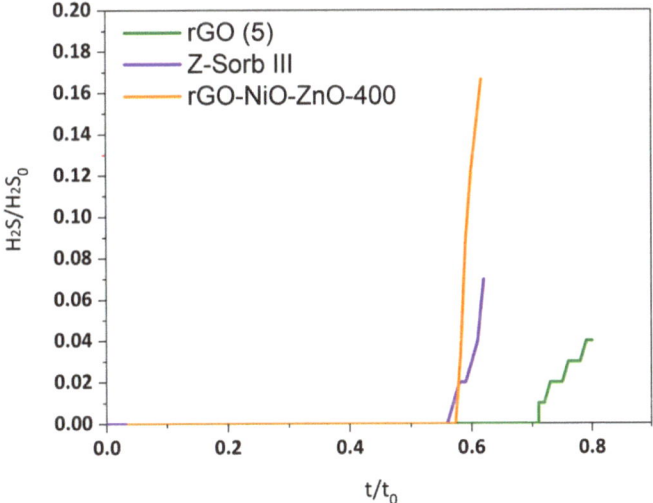

Figure 2. Dimensionless sulfidation breakthrough curves.

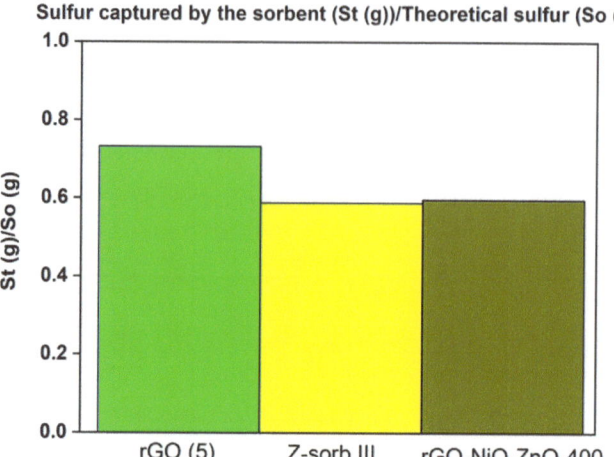

Figure 3. Comparison of the sorbents in terms of utilization yield.

As many gasifiers operate at atmospheric pressure and because of the difficulties of compressing a dirty syngas, the performance of the rGO-NiO-ZnO-400 sorbent was evaluated at 1 bar in a simplified atmosphere with 0.9% (v/v) in nitrogen.

The breakthrough curves for both operating pressures are shown in Figure 4. A closer value to the theoretical time is achieved at atmospheric pressure. Unlike the commercial sorbent which exhibits a better performance at high pressure values (2MPa) [34], the desulfurization capacity of the rGO-NiO-ZnO sorbent increases from 17.8% to 24.1% when the pressure decreases from 10 bar to 1 bar. This result is clearly advantageous since it means that in the case of the commercial application of this technology, there would be no need for gas compression upstream to the desulfurization reactor.

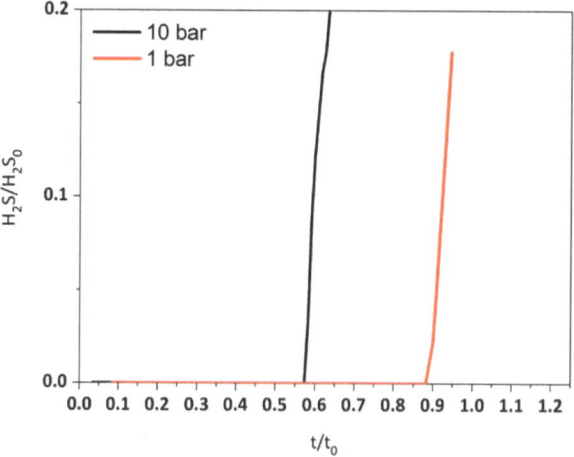

Figure 4. Dimensionless sulfidation breakthrough curves at different pressures.

Since better results were obtained when a lower pressure was applied, it was decided to study the sorbent performance with synthetic syngas mixtures at 1 bar. The operating conditions, including the atmosphere composition, of the three different types of experiment carried out to determine the desulfurization capacity of the sample are summarized in Table 4.

Table 4. Experimental conditions.

N°	1	2	3
Reactor temperature (°C)	400	400	400
Pressure (bar)	1	1	1
Gas hourly space velocity GHSV (h^{-1})	3500	3500	3500
Sufidation gas composition (% v/v)			
H_2S	0.9	0.3	0.3
N_2	99.1	39.8	33.0
H_2	-	26.1	22.7
CO	-	13.6	24.9
CO_2	-	7.4	8.5
CH_4	-	2.8	0.6
H_2O	-	10	10

Table 5 summarizes the results obtained. The actual S loading capacity and the breakthrough times determined in the experiment (in test 2) were close to the theoretical ones. Therefore, the utilization of the sorbent at a breakthrough time provides a very high value (efficiency).

Table 5. Summary of experimental results and comparison of desulfurization performance.

N°	1	2	3
Theoretical Sulfur load capacity S_0 (%)	26.4	26.4	26.4
Real Sulfur load capacity S_t (%)	24.1	27.1	22.8
Theoretical sorption time, t_0 (min)	466	1389	1200
Breakthrough time (min)	414	1385	1008
Efficiency (S_t/S_0) (%)	91	102	86

Regarding the syngas composition, some components of the gas can interfere with sorbent sulfidation. Many examples of CO_2 interference can be found in the literature [35–39] as well as CO, CH_4 and H_2O [35,40–43]. In this work, different composition of CO, CO_2 and CH_4 were used while the water content was kept constant.

By analyzing the effect of the gas atmosphere, no significant differences were observed between the sorbent performances in tests under nitrogen and syngas atmospheres. Under a full syngas atmosphere, the sorbent did not lose its S retention capacity very significantly, which means that there is no strong competitive adsorption of or deactivation by any of the syngas components and, therefore, it does not interfere in the desulfurization process.

The mixture with a low quantity of CO and CO_2 and higher CH_4 (test N°2: 13.6%, 7.4% and 2.8%, respectively) exhibited a good sorbent performance. However, when a syngas composition with a higher content of CO and CO_2 and lower CH_4 (test N°3: 24.9%, 8.5% and 0.6%) was used, a slight decrease in the S loading capacity was observed. The S loading capacity decreased from 27% to 23%. This can be attributed to the reducing power of the gas for the second mixture which was a little bit higher. The reducing power is expressed as the ratio of reducing compounds in the syngas (sum of H_2 + CO + CH_4) to oxidized compounds (CO_2 + H_2O). For the syngas mixture, denoted as number 2, the reducing power is 2.44, whereas for the syngas mixture, number 3, it is 2.6. Moreover, in a previous study [34], the authors also observed that for rich CO syngas, the Boudouard reaction, CO disproportionation, occurred under a specific gas velocity, leading to a poor desulfurization performance. Coking due to methane cracking would also decrease the S removal capacity due to hindering access to zinc oxide and nickel oxide desulfurization sites.

Figure 5 shows the sorbent's dimensionless breakthrough curves of rGO-NiO-ZnO-400 for the three gas mixtures. As can be seen, there is almost no H_2S in the exit stream prior to the breakthrough point, which is then followed by a sharp increase in the hydrogen sulfide concentration in the reactor outlet.

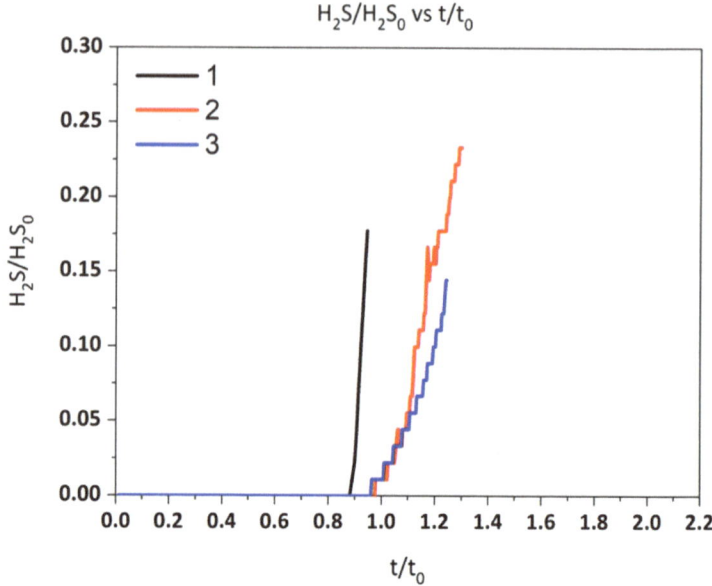

Figure 5. Breakthrough curves of sorbent at different conditions.

2.4. Regeneration and Ciclability of Sorbents

The capability of the sorbent was tested under the three atmospheres evaluated in order to have a preliminary insight on its performance over repeated sulfidation and regeneration cycles. To that aim, first, in a nitrogen atmosphere, three sulfidation and regeneration cycles were performed. Regeneration was carried out using a mixture of nitrogen with 2% of oxygen at 550 °C. These conditions were maintained until no SO_2 was detected in the outlet stream and then another subsequent sulfidation test was conducted. Then, the same number of cycles was undertaken for the syngas mixture atmospheres.

In Figure 6a–c, breakthrough curves are presented. A slow decrease in the sulfur retention capacity can be observed when nitrogen and the mixture with a low quantity of CO and CO_2 and higher CH_4 (13.6%, 7.4% and 2.8%, respectively) are used. On the other hand, when the mixture with a higher CO and CO_2 and lower CH_4 content (24.9%, 8.5% and 0.6%) was used, no significant difference was observed in the sorbent's performance. Despite this negligible reduction in the sulfur retention capacity for the first syngas mixture, the efficiency of the sorbent remained high enough (see Table 6) to allow its use in several cycles.

Table 6. Summary of cyclability experimental results.

N° Test Type		1			2			3	
Cycle n°	1	2	3	1	2	3	1	2	3
Theoretical Sulfur load capacity S_0 (%)		26.4			26.4			26.4	
Real Sulfur load capacity S_t (%)	24.1	23.7	22.8	27.1	24	24	22.8	23.1	23.7
Theoretical sorption time, t_0 (min)		466			1389			1200	
Breakthrough time (min)	414	406	392	1385	1232	1230	1008	1021	1048
Efficiency (S_t/S_0) (%)	91	90	86	102	91	91	86	87	90

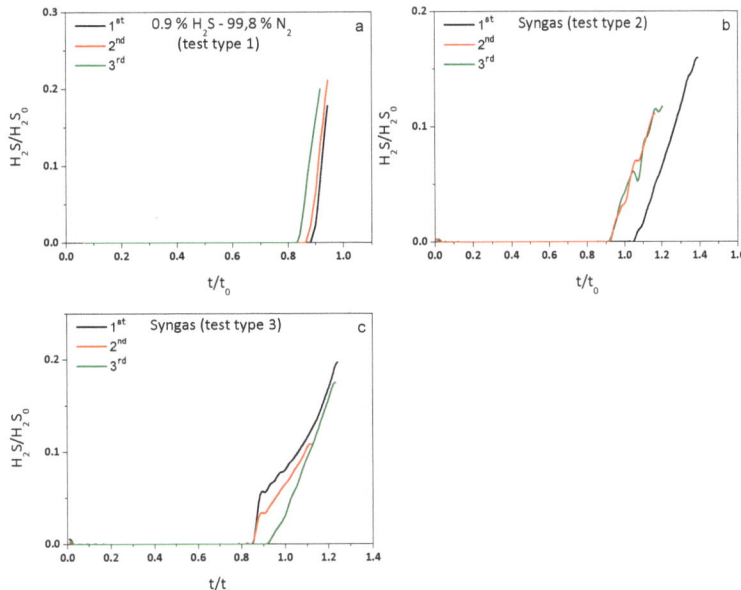

Figure 6. Cyclability of sorbent, (**a**–**c**) present breakthrough curves.

3. Materials and Methods

3.1. Chemicals

All reagents used for graphene oxide and rGO-NiO-ZnO synthesis were of analytical reagent grade, mostly supplied by Fisher Scientific (Hampton, NH, USA). Ultrapure water (resistivity $\geq$ 18.2 MΩ·cm) from a Milli-Q system (Millipore Bedford, Bedford, MA, USA) was used throughout.

3.2. Apparatus and Instruments

Equipment used for the synthesis of graphene oxide and rGO-NiO-ZnO composites includes a thermostatic bath with heating control (Huber KISS225B, Offenburg, Germany), a rod stirrer (Selecta SE-100, Barcelona, Spain), a vacuum freeze-dryer (LaboGene Coolsafe Touch 1110-4, Allerød, Denmark), a high-capacity floor centrifuge (Gyrozen 1736R, Daejeon, Republic of Korea), an ultrasonic probe (Fisherbrand model 505, Pittsburgh, PA, USA), an automatic mill (Bosch TSM6A011W, Stuttgart, Germany), a magnetic stirrer (Selecta Multimatic 9N, Barcelona Spain), 250 mL PTFE lined autoclave reactors, a vacuum drying oven (WITEG WOV 70, Wertheim, Germany) and a gradient tube furnace (CARBOLITE TZF 1200 °C, Sheffield, United Kingdom).

X-ray diffraction (XRD) measurements were carried out using a X'Pert Pro (Malvern-Panalytical, Almelo, Netherlands) using Cu Kα radiation (λ = 1.54056 Å). The instrument was configured with Bragg–Brentano geometry and with the operating parameters of 40 kV and 45 mA. Diffractograms were acquired in the range 5–120° 2θ, with scanning steps of 0.02° 2θ.

Phase composition and structural analyses on the materials were obtained using HighScore Plus 4.8 software (Malvern-Panalytical). Compound identification was carried out using information available from the Crystallography Open Database (COD) and the International Center for Diffraction Data (ICDD).

Chemical characterization (% and mg/kg concentrations) of all samples involved in this study was performed employing wavelength dispersive X-ray fluorescence (XRF) instrumentation by using an automated AXIOS Malvern-PANalytical spectrometer with a Rh tube.

To assess the C content of the samples, an Elemental analyzer, LECO TruSpec CHN elemental analyzer (St. Joseph, MI, USA, was employed, which was then heated up to at least 900 °C in the presence of oxygen gas. Mineral and organic compounds were oxidized and/or volatilized to carbon dioxide, which was measured by an infrared detection method.

An ASAP 2020 analyzer (Micromeritics, Norcross, GA, USA) was employed to evaluate the specific surface area and pore volume of graphene material.

3.3. Recovery and Treatment of Graphite from Spent Zn/C Batteries

The recycled graphite used as starting material was obtained following a procedure described below. The recovered black mass from spent Zn/C batteries was subjected to acidic leaching using a mixture of 250 mL of milliQ water, 250 mL of H_2O_2 and 500 mL of HCl at room temperature for 1 h [44]. After that, the final mixture was filtered using a pressure filter obtaining the carbonaceous material which will be used as a precursor. In order to remove the possible small quantities of metals that may be present, the non-soluble residue was subjected to acidic leaching using a sulfuric acid 2 M concentration solution at 80 °C. Finally, the mixture was filtered and the obtained solid was dried to achieve the corresponding recycled graphite material precursor.

3.4. Synthesis Methods

3.4.1. Synthesis of Graphene Oxide

Graphene oxide was synthesized following the guidelines provided in the Marcano–Tour method [45] with slight modifications in order to manage the highly reactive graphite. Briefly, 3 g of recycled graphite were weighed and a 9:1 mixture of concentrated H_2SO_4 and H_3PO_4 (360:40 mL) was added. In order to avoid an explosive reaction because of the presence of remaining trace metal from the original black mass, the mixture was cooled onto crushed ice and then 18 g of $KMnO_4$ was slowly added in small portions, usually four portions. The mixture was then heated to 50 °C using a temperature-controlled water bath and stirred for 18 hours, turning out into a dark purple paste. Afterwards, the content was cooled down to room temperature by adding 400 mL of ultrapure ice-water to the mixture to stop the oxidation process. Finally, 10 mL of 30 wt.% H2O2 was added in order to reduce residual $KMnO_4$ to soluble $MnSO_4$ in an acidic medium, as described in the following reaction:

$$2\ KMnO_4 + 5\ H_2O_2 + 3\ H_2SO_4 \rightarrow 2\ MnSO_4 + K_2SO_4 + H_2O + 5\ O_2 \qquad (5)$$

After H_2O_2 addition, bubbling occurred and suspension turned dark yellow, which is indicative of a high oxidation level. The obtained yellow–brown suspension was then cooled down to room temperature overnight. After its transfer to two 400 mL centrifuge tubes, the suspension was centrifuged at 8000 rpm for 1h and the supernatants removed. The obtained graphite oxide was washed repeatedly with 250 mL 1M HCl and ultrapure water until the pH of the supernatant achieved 3.5–4. The final sample was placed in a Petri dish to be deep-frozen at −80 °C for 48 h and subsequently freeze-dried under high vacuum. Finally, the obtained graphite oxide was ground using an automatic mill.

3.4.2. Synthesis of rGO-NiO-ZnO Composites

rGO-NiO-ZnO composites were prepared according to experiment 5 from Sanchez-Hervas et al. [18] with several modifications in order to scale up the production. In spite of not presenting the highest adsorption capacity, the synthesis conditions were selected as a compromising solution between a high surface area, chemical stability and easiness to scalability. In a typical synthesis, 1 L of 5 mg/mL homogeneous aqueous dispersion of graphite oxide was prepared from four batches of 250 mL. Then, each 250 mL dispersion was successively sonicated in a low-power sonication bath for 1 h and with probe sonication for 3 h, producing clear graphene oxide (GO) dispersions with no visible particulate matter. After sonication process was complete, the four dispersions were joined and equally distributed in five beakers and subjected to magnetic stirring. Corresponding quantities of $Zn(NO_3)_2$, $Ni(NO_3)_2$ and urea were successively added in small portions across 3 h to the GO dispersions while stirring. The mass ratio of $Zn(NO_3)_2$:GO is 12.4:1 while the mass ratio of Ni(NO3)2:GO is 2:1. The mixture was stirred for 2 more hours. The solutions of GO and metal oxide precursors were then transferred to 250 mL PTFE-lined stainless steel autoclaves and subjected to hydrothermal treatment at 120 °C for 48 h. The obtained greyish precipitates were washed successively two times with ethanol and two times with ultrapure water and dried at 60 °C under vacuum. Finally, a thermal annealing treatment at 400 °C under Ar atmosphere was applied to obtain the NiO-ZnO-decorated rGO using a gradient tube furnace.

3.5. Desulfurization Test

3.5.1. Test Rig

Desulfurization tests were carried out in a Microactivity Pro Unit. A full description of the system can be found elsewhere [46]. The unit can work at up to 700 °C and 20 bar with a maximum operating gas flow rate of 4.5 NL/min. Three mass flow controllers and a HPLC pump (Gilson 307) produce the desired gas mixture composition. Dry gas and water are preheated separately up to 200 °C in two independent loops, in which water was vaporized and then mixed before entering the reactor. The sulfur-resistant tubular reactor of 9.2 mm OD and 30 cm long was placed in one single-zone SS304 oven to heat up the full gas stream to the operating temperature and controlled by a 1.5 mm thermocouple.

For the tests presented in this work, three different atmospheres were employed: one simplified atmosphere with H_2S and nitrogen and two resembling different gasification gases. The simplified atmosphere consisted of a mixture of hydrogen sulfide (9000 ppmv) in nitrogen and no water was fed to the system during this experiment. The gasification gas composition selected was (1) 5% CH_4, 24% CO, 13% CO_2, 46% H_2, 12% N_2, (2) 1% CH_4, 44% CO, 15% CO_2 and 40% H_2. For those experiments, the dry stream mimicking the gasification gas was mixed with a stream of H_2S in nitrogen to produce a final concentration of H_2S of 3000 ppmv. Gas humidity was set at 10% that which was achieved by the liquid feeding system which feeds water, vaporizes it and mixes it with the dry gas stream before reaching the reactor.

Gas stream compositions were measured after water removal by gas chromatography using a CP4900 Varian gas microchromatograph equipped with two columns, a Porapack HP-PLOT Q and a Molecular Sieve HP-PLOT and with two thermal conductivity detectors.

3.5.2. Experimental Methodology

rGO-NiO-ZnO-400 composites prepared as powder were pelletized, weighed and sieved to 0.5–1 mm fraction to avoid excessive pressure drop in the reactor. Z-sorb III is commercially available in pellets with 0.3 cm diameter and 0.6–0.9 cm long, with bulk density of 0.88 g/cc, but in order to test it in the same conditions, prior to the desulfurization experiments, Z-sorb III pellets were milled and sieved to obtain the 0.5–1 mm fraction. Graphene sorbents had a bulk density of approx. 1.1 g/cc. During the tests, no mechanical degradation was observed during the adsorption or regeneration tests. Briefly, 2.5 g of this fraction was placed in the reactor and the gas and water feeding systems were set to the desired values so that gas hourly space velocity was maintained at 3500 h^{-1}. The reaction system was heated to the desulfurization temperature with a continuous flow of nitrogen through the reactor. When the desired temperature was reached, the nitrogen flow was stopped, and switched to the sulfidation atmosphere to start the experiment. Inlet and outlet gas composition were continuously determined by micro-GC and sulfidation progressed until a sharp increase in H_2S concentration at the reactor outlet was noticed, which meant that the breakthrough point was achieved. At the end of each run, the used sorbent was discharged from the reactor, weighed and characterized.

3.5.3. Regeneration Test

Oxidative regeneration of spent rGO-NiO-ZnO-400 was applied to bring the sorbent back to its original oxidation state. During regeneration, the following reactions took place (Equations (6) and (7)):

$$NiS + 3/2\, O_2 \rightarrow NiO + SO_2 \qquad (6)$$

$$ZnS + 3/2\, O_2 \rightarrow ZnO + SO_2 \qquad (7)$$

Regeneration conditions were selected in accordance with previous studies [33] and are shown in Table 7.

Table 7. Regeneration conditions.

Regeneration Conditions	
Reactor temperature (°C)	550
Pressure (bar)	1
Gas hourly space velocity GHSV (h^{-1})	2000
Regeneration gas composition (% v/v) N_2 O_2	98 2

Sorbent regeneration is maintained until SO_2 concentration in the regeneration stream gas is not detected by microGC. At this point, the regeneration gas is switched to nitrogen to cool down the system. When the reactor reaches 400 °C, a new sulfidation cycle of the material begins again. The first set of experiments consisted of 3 sulfidations in nitrogen during 10 cycles while sulfidation in syngas streams were performed during 3 cycles.

4. Conclusions

We have prepared a NiO- and ZnO-doped rGO composite from graphite recovered by recycling spent Zn/C batteries.

The oxidation and reduction processes of recycled graphite are able to remove impurities from the recycled graphite, resulting in high-quality and low-cost graphene-related materials fairly close to those synthesized from pure graphite.

The desulfurization properties of the sorbent were investigated by exposing them to various operating conditions and atmospheres. Although there was a decrease in the desulfurization capacity compared to the sorbent prepared from pure graphite because a weaker reduction occurred, the proposed composite exhibited a capacity similar to the commercially available sorbents with fairly good response times.

In conclusion, this work demonstrates that rGO-NiO-ZnO composites from recycled graphite have a great potential as a suitable and sustainable alternative to commercial desulfurization sorbents that encourages the development of more environmentally sustainable technologies for the industrial scale-up process. The proposed sorbent exhibited a superior performance at a low pressure which is clearly favorable, allowing its use without the need of gas compression upstream to the desulfurization reactor. Furthermore, the cycling tests showed that the rGO-NiO-ZnO sorbent shows acceptable stability with no drastic decay in the available capacity. Further studies should focus on the improvement of the desulfurization capacity and stability of the sorbents by exploring different experimental conditions for the reduction of GO during the hydrothermal and annealing treatments.

Author Contributions: Conceptualization, R.F.-M., I.O. and J.M.S.-H.; methodology and characterization, R.F.-M., I.O., M.B.G.-M., M.F., L.A. and F.A.L.; resources, J.M.S.-H., F.A.L. and I.R.; data curation, R.F.-M., I.O., M.B.G.-M. and L.A.; writing—original draft preparation, R.F.-M., I.O. and M.B.G.-M.;

writing—review and editing, R.F.-M., I.O., M.B.G.-M., L.A., F.A.L., I.R. and J.M.S.-H. All authors have read and agreed to the published version of the manuscript.

Funding: This research was funded by the Spanish Ministry of Science and Innovation under the projects GONDOLA grant number [PDC2021-120799-I00] and BIOENH2, grant number, [PLEC2023-010216].

Institutional Review Board Statement: Not applicable.

Informed Consent Statement: Not applicable.

Data Availability Statement: The original contributions presented in the study are included in the article, further inquiries can be directed to the corresponding author.

Acknowledgments: The authors gratefully acknowledge Elena Pérez Garrido from the Division of Materials of Energy Interest and Veronica Marti Jimenez from the Sustainable Thermochemical Valorisation Unit for their assistance in the experimental work.

Conflicts of Interest: The authors declare no conflicts of interest.

References

1. Government of Canada. Graphite Facts. Available online: https://natural-resources.canada.ca/our-natural-resources/minerals-mining/mining-data-statistics-and-analysis/minerals-metals-facts/graphite-facts/24027 (accessed on 1 May 2024).
2. Statista, Mine Production of Graphite Worldwide from 2010 to 2023 (in 1000 Metric Tons). In Statista. Available online: https://www.statista.com/statistics/1005851/global-graphite-production/ (accessed on 1 May 2024).
3. Kara, S.; Hauschild, M.; Sutherland, J.; McAloone, T. Closed-loop systems to circular economy: A pathway to environmental sustainability? *CIRP Ann.* **2022**, *71*, 505–528. [CrossRef]
4. Chen, W.-S.; Liao, C.-T.; Lin, K.-Y. Recovery Zinc and Manganese from Spent Battery Powder by Hydrometallurgical Route. *Energy Procedia* **2017**, *107*, 167–174. [CrossRef]
5. Loudiki, A.; Mustapha, M.; Azriouil, M.; Laghrib, F.-E.; Farahi, A.; Bakasse, M.; Lahrich, S.; Mhammedi, M. Graphene oxide synthesized from zinc-carbon battery waste using a new oxidation process assisted sonication: Electrochemical properties. *Mater. Chem. Phys.* **2021**, *275*, 125308. [CrossRef]
6. Sperandio, G.; Machado Junior, I.; Bernardo, E.; Lopes, R. Graphene Oxide from Graphite of Spent Batteries as Support of Nanocatalysts for Fuel Hydrogen Production. *Processes* **2023**, *11*, 3250. [CrossRef]
7. Vadivel, S.; Tejangkura, W.; Sawangphruk, M. Graphite/Graphene Composites from the Recovered Spent Zn/Carbon Primary Cell for the High-Performance Anode of Lithium-Ion Batteries. *ACS Omega* **2020**, *5*, 15240–15246. [CrossRef]
8. Chen, I.W.; Chen, Y.-S.; Kao, N.-J.; Wu, C.-W.; Zhang, Y.-W.; Li, H.-T. Scalable and high-yield production of exfoliated graphene sheets in water and its application to an all-solid-state supercapacitor. *Carbon* **2015**, *90*, 16–24. [CrossRef]
9. Kumar, N.; Salehiyan, R.; Chauke, V.; Joseph Botlhoko, O.; Setshedi, K.; Scriba, M.; Masukume, M.; Sinha Ray, S. Top-down synthesis of graphene: A comprehensive review. *FlatChem* **2021**, *27*, 100224. [CrossRef]
10. Tian, S.; Sun, J.; Yang, S.; He, P.; Wang, G.; Di, Z.; Ding, G.; Xie, X.; Jiang, M. Controllable Edge Oxidation and Bubbling Exfoliation Enable the Fabrication of High Quality Water Dispersible Graphene. *Sci. Rep.* **2016**, *6*, 34127. [CrossRef] [PubMed]
11. Mbayachi, V.B.; Ndayiragije, E.; Sammani, T.; Taj, S.; Mbuta, E.R.; Khan, A. Graphene synthesis, characterization and its applications: A review. *Results Chem.* **2021**, *3*, 100163. [CrossRef]
12. Yam, K.M.; Guo, N.; Jiang, Z.; Li, S.; Zhang, C. Graphene-Based Heterogeneous Catalysis: Role of Graphene. *Catalysts* **2020**, *10*, 53. [CrossRef]
13. Fadlalla, M.; Ganesh Babu, S. Role of graphene in photocatalytic water splitting for hydrogen production. In *Graphene-Based Nanotechnologies for Energy and Environmental Applications*; Elsevier: Amsterdam, The Netherlands, 2019; pp. 81–108.
14. Hasani, A.; Teklagne, M.; Do, H.; Hong, S.; Le, Q.; Ahn, S.H.; Kim, S.Y. Graphene-based catalysts for electrochemical carbon dioxide reduction. *Carbon Energy* **2020**, *2*, 158–175. [CrossRef]
15. Karimi, S.; Tavasoli, A.; Mortazavi, Y.; Karimi, A. Enhancement of cobalt catalyst stability in Fischer–Tropsch synthesis using graphene nanosheets as catalyst support. *Chem. Eng. Res. Des.* **2015**, *104*, 713–722. [CrossRef]
16. Wang, S.; Sun, H.; Ang, H.M.; Tadé, M.O. Adsorptive remediation of environmental pollutants using novel graphene-based nanomaterials. *Chem. Eng. J.* **2013**, *226*, 336–347. [CrossRef]
17. Yaengthip, P.; Siyasukh, A.; Payattikul, L.; Kiatsiriroat, T.; Punyawudho, K. The ORR activity of nitrogen doped-reduced graphene oxide below decomposition temperature cooperated with cobalt prepared by strong electrostatic adsorption technique. *J. Electroanal. Chem.* **2022**, *915*, 116366. [CrossRef]
18. Sánchez-Hervás, J.M.; Maroño, M.; Fernández-Martínez, R.; Ortiz, I.; Ortiz, R.; Gómez-Mancebo, M.B. Novel ZnO-NiO-graphene-based sorbents for removal of hydrogen sulfide at intermediate temperature. *Fuel* **2022**, *314*, 122724. [CrossRef]
19. I.E.A. Bioenergy. *Task 33 Workshop: Biomass and Waste Gasification for the Production of Fuels*. Available online: https://task33.ieabioenergy.com (accessed on 1 May 2024).
20. Sánchez-Hervás, J.; Ortiz, I.; Martí, V.; Andray, A. Removal of Organic Sulfur Pollutants from Gasification Gases at Intermediate Temperature by Means of a Zinc–Nickel-Oxide Sorbent for Integration in Biofuel Production. *Catalysts* **2023**, *13*, 1089. [CrossRef]

21. Le, P.A.; Nguyen, N.T.; Nguyen, P.L.; Phung, T.V.B. Minireview on Cathodic and Anodic Exfoliation for Recycling Spent Zinc-Carbon Batteries To Prepare Graphene Material: Advances and Outlook of Interesting Strategies. *Energy Fuels* **2023**, *37*, 7062–7070. [CrossRef]
22. Thirumal, V.; Sreekanth, T.V.M.; Yoo, K.; Kim, J. Facile Preparations of Electrochemically Exfoliated N-Doped Graphene Nanosheets from Spent Zn-Carbon Primary Batteries Recycled for Supercapacitors Using Natural Sea Water Electrolytes. *Energies* **2022**, *15*, 8650. [CrossRef]
23. Alcaraz, L.; Jiménez-Relinque, E.; Plaza, L.; García-Díaz, I.; Castellote, M.; López, F.A. Photocatalytic Activity of $Zn_xMn_{3-x}O_4$ Oxides and ZnO Prepared From Spent Alkaline Batteries. *Front. Chem.* **2020**, *8*, 661. [CrossRef]
24. Al-Gaashani, R.; Najjar, A.; Zakaria, Y.; Mansour, S.; Atieh, M.A. XPS and structural studies of high quality graphene oxide and reduced graphene oxide prepared by different chemical oxidation methods. *Ceram. Int.* **2019**, *45*, 14439–14448. [CrossRef]
25. Aliyev, E.M.; Khan, M.M.; Nabiyev, A.M.; Alosmanov, R.M.; Bunyad-zadeh, I.A.; Shishatskiy, S.; Filiz, V. Covalently Modified Graphene Oxide and Polymer of Intrinsic Microporosity (PIM-1) in Mixed Matrix Thin-Film Composite Membranes. *Nanoscale Res. Lett.* **2018**, *13*, 359. [CrossRef] [PubMed]
26. Bayoumy, A.M.; Gomaa, I.; Elhaes, H.; Sleim, M.; Ibrahim, M.A. Application of Graphene/Nickel Oxide Composite as a Humidity Sensor. *Egypt. J. Chem.* **2021**, *64*, 85–91.
27. Gómez-Mancebo, M.B.; Fernández-Martínez, R.; Ruiz-Perona, A.; Rubio, V.; Bastante, P.; García-Pérez, F.; Borlaf, F.; Sánchez, M.; Hamada, A.; Velasco, A.; et al. Comparison of Thermal and Laser-Reduced Graphene Oxide Production for Energy Storage Applications. *Nanomaterials* **2023**, *13*, 1391. [CrossRef] [PubMed]
28. Bukovska, H.; García-Perez, F.; Brea Núñez, N.; Bonales, L.J.; Velasco, A.; Clavero, M.Á.; Martínez, J.; Quejido, A.J.; Rucandio, I.; Gómez-Mancebo, M.B. Evaluation and Optimization of Tour Method for Synthesis of Graphite Oxide with High Specific Surface Area. *C* **2023**, *9*, 65. [CrossRef]
29. Kottegoda, I.R.M.; Idris, N.H.; Lu, L.; Wang, J.-Z.; Liu, H.-K. Synthesis and characterization of graphene–nickel oxide nanostructures for fast charge–discharge application. *Electrochim. Acta* **2011**, *56*, 5815–5822. [CrossRef]
30. Rabchinskii, M.K.; Sysoev, V.V.; Brzhezinskaya, M.; Solomatin, M.A.; Gabrelian, V.S.; Kirilenko, D.A.; Stolyarova, D.Y.; Saveliev, S.D.; Shvidchenko, A.V.; Cherviakova, P.D.; et al. Rationalizing Graphene-ZnO Composites for Gas Sensing via Functionalization with Amines. *Nanomaterials* **2024**, *14*, 735. [CrossRef]
31. Sánchez, J.M.; Ruiz, E.; Otero, J. Selective Removal of Hydrogen Sulfide from Gaseous Streams Using a Zinc-Based Sorbent. *Ind. Eng. Chem. Res.* **2005**, *44*, 241–249. [CrossRef]
32. Das, D.; Das, M.; Sil, S.; Sahu, P.; Ray, P.P. Effect of Higher Carrier Mobility of the Reduced Graphene Oxide–Zinc Telluride Nanocomposite on Efficient Charge Transfer Facility and the Photodecomposition of Rhodamine B. *ACS Omega* **2022**, *7*, 26483–26494. [CrossRef] [PubMed]
33. Ozer, L.Y.; Garlisi, C.; Oladipo, H.; Pagliaro, M.; Sharief, S.A.; Yusuf, A.; Almheiri, S.; Palmisano, G. Inorganic semiconductors-graphene composites in photo(electro)catalysis: Synthetic strategies, interaction mechanisms and applications. *J. Photochem. Photobiol. C Photochem. Rev.* **2017**, *33*, 132–164. [CrossRef]
34. Sánchez-Hervás, J.M.; Otero, J.; Ruiz, E. A study on sulphidation and regeneration of Z-Sorb III sorbent for H_2S removal from simulated ELCOGAS IGCC syngas. *Chem. Eng. Sci.* **2005**, *60*, 2977–2989. [CrossRef]
35. Frilund, C.; Simell, P.; Kaisalo, N.; Kurkela, E.; Koskinen-Soivi, M.-L. Desulfurization of Biomass Syngas Using ZnO-Based Adsorbents: Long-Term Hydrogen Sulfide Breakthrough Experiments. *Energy Fuels* **2020**, *34*, 3316–3325. [CrossRef]
36. Kawase, M.; Otaka, M. Removal of H_2S using molten carbonate at high temperature. *Waste Manag.* **2013**, *33*, 2706–2712. [CrossRef]
37. Rahim, D.A.; Fang, W.; Wibowo, H.; Hantoko, D.; Susanto, H.; Yoshikawa, K.; Zhong, Y.; Yan, M. Review of high temperature H_2S removal from syngas: Perspectives on downstream process integration. *Chem. Eng. Process.-Process Intensif.* **2023**, *183*, 109258. [CrossRef]
38. Sasaoka, E.; Hirano, S.; Kasaoka, S.; Sakata, Y. Characterization of reaction between zinc oxide and hydrogen sulfide. *Energy Fuels* **1994**, *8*, 1100–1105. [CrossRef]
39. Selim, H.; Gupta, A.K.; Al Shoaibi, A. Effect of CO_2 and N_2 concentration in acid gas stream on H_2S combustion. *Appl. Energy* **2012**, *98*, 53–58. [CrossRef]
40. Dhage, P.; Samokhvalov, A.; McKee, M.L.; Duin, E.C.; Tatarchuk, B.J. Reactive adsorption of hydrogen sulfide by promoted sorbents Cu-ZnO/SiO2: Active sites by experiment and simulation. *Surf. Interface Anal.* **2013**, *45*, 865–872. [CrossRef]
41. Gil-Lalaguna, N.; Sánchez, J.L.; Murillo, M.B.; Gea, G. Use of sewage sludge combustion ash and gasification ash for high-temperature desulphurization of different gas streams. *Fuel* **2015**, *141*, 99–108. [CrossRef]
42. Lee, J.; Feng, B. A thermodynamic study of the removal of HCl and H_2S from syngas. *Front. Chem. Sci. Eng.* **2012**, *6*, 67–83. [CrossRef]
43. Thao Ngo, T.N.L.; Chiang, K.-Y. Hydrogen sulfide removal from simulated synthesis gas using a hot gas cleaning system. *J. Environ. Chem. Eng.* **2023**, *11*, 109592. [CrossRef]
44. Romo, L.A.; López-Fernández, A.; García-Díaz, I.; Fernández, P.; Urbieta, A.; López, F.A. From spent alkaline batteries to $Zn_xMn_{3-x}O_4$ by a hydrometallurgical route: Synthesis and characterization. *RSC Adv.* **2018**, *8*, 33496–33505. [CrossRef]

45. Marcano, D.C.; Kosynkin, D.V.; Berlin, J.M.; Sinitskii, A.; Sun, Z.Z.; Slesarev, A.; Alemany, L.B.; Lu, W.; Tour, J.M. Improved Synthesis of Graphene Oxide. *ACS Nano* **2010**, *4*, 4806–4814. [CrossRef] [PubMed]
46. Maroño, M.; Sánchez, J.M.; Ruiz, E.; Cabanillas, A. Study of the Suitability of a Pt-Based Catalyst for the Upgrading of a Biomass Gasification Syngas Stream via the WGS Reaction. *Catal. Lett.* **2008**, *126*, 396–406. [CrossRef]

Disclaimer/Publisher's Note: The statements, opinions and data contained in all publications are solely those of the individual author(s) and contributor(s) and not of MDPI and/or the editor(s). MDPI and/or the editor(s) disclaim responsibility for any injury to people or property resulting from any ideas, methods, instructions or products referred to in the content.

Article

Ethylene Elimination Using Activated Carbons Obtained from Baru (*Dipteryx alata* vog.) Waste and Impregnated with Copper Oxide

Ana Carolina de Jesus Oliveira [1,*], Camilla Alves Pereira Rodrigues [2], Maria Carolina de Almeida [3], Eliane Teixeira Mársico [4], Paulo Sérgio Scalize [5], Tatianne Ferreira de Oliveira [1], Victor Andrés Solar [6] and Héctor Valdés [6,*]

1. School of Agronomy, Federal University of Goiás, Goiania 74690-900, Brazil
2. Faculty of Nutrition, Federal University of Goiás, Goiania 74605-080, Brazil
3. Federal Institute of Education, Science and Technology of Goiás, Inhumas 75402-556, Brazil
4. Faculty of Veterinary, Fluminense Federal University, Niteroi 24230-231, Brazil
5. School of Civil and Environmental Engineering, Federal University of Goiás, Goiania 74605-220, Brazil
6. Clean Technologies Laboratory, Engineering Faculty, Universidad Católica de la Santísima Concepción, Alonso de Ribera 2850, Concepcion 4030000, Chile
* Correspondence: carolina_jesus@discente.ufg.br (A.C.d.J.O.); hvaldes@ucsc.cl (H.V.)

Citation: Oliveira, A.C.d.J.; Rodrigues, C.A.P.; de Almeida, M.C.; Mársico, E.T.; Scalize, P.S.; de Oliveira, T.F.; Solar, V.A.; Valdés, H. Ethylene Elimination Using Activated Carbons Obtained from Baru (*Dipteryx alata* vog.) Waste and Impregnated with Copper Oxide. *Molecules* **2024**, *29*, 2717. https://doi.org/10.3390/molecules29122717

Academic Editors: Sake Wang, Minglei Sun and Nguyen Tuan Hung

Received: 20 April 2024
Revised: 31 May 2024
Accepted: 2 June 2024
Published: 7 June 2024

Copyright: © 2024 by the authors. Licensee MDPI, Basel, Switzerland. This article is an open access article distributed under the terms and conditions of the Creative Commons Attribution (CC BY) license (https://creativecommons.org/licenses/by/4.0/).

Abstract: Ethylene is a plant hormone regulator that stimulates chlorophyll loss and promotes softening and aging, resulting in a deterioration and reduction in the post-harvest life of fruit. Commercial activated carbons have been used as ethylene scavengers during the storage and transportation of a great variety of agricultural commodities. In this work, the effect of the incorporation of copper oxide over activated carbons obtained from baru waste was assessed. Samples were characterized by X-ray diffraction (XRD), N_2 adsorption-desorption at $-196\ °C$, field-emission scanning electron microscopy (FESEM) coupled with energy-dispersive X-ray spectroscopy (EDS), and infrared (IR) spectroscopy. The results showed that the amount of ethylene removed using activated carbon obtained from baru waste and impregnated with copper oxide (1667 µg g^{-1}) was significantly increased in comparison to the raw activated carbon (1111 µg g^{-1}). In addition, carbon impregnated with copper oxide exhibited better adsorption performance at a low ethylene concentration. Activated carbons produced from baru waste are promising candidates to be used as adsorbents in the elimination of ethylene during the storage and transportation of agricultural commodities at a lower cost.

Keywords: activated carbon; adsorption; baru waste; ethylene; porous materials

1. Introduction

Ethylene is a natural hormone that regulates a wide variety of developmental processes in plants and accumulates during the growth of fruit and vegetables [1]. In addition to accelerated ripening, ethylene synthesis can promote fruit softening, senescence, and rot, thereby reducing the post-harvest shelf-life of fruit with consequent economic losses [2,3]. Ethylene is physiologically active, even at low concentrations, measured in the range of parts per million (ppm) to parts per billion (ppb), which can cause the rapid deterioration of fresh agricultural commodities during transportation and storage [4].

According to the Institute for Applied Economic Research (IPEA), there is a global challenge foreseen among the twelve Sustainable Development Goals (SDGs) to halve the post-harvest losses and waste of fruit and vegetables by 2030. Thus, better management of ethylene during the storage and distribution chain becomes essential. Recent studies have reported various technologies and methods to inhibit ethylene production at the plant level or in closed fruit storage environments. Adsorption appears as one of the most promising processes to reduce ethylene content due to its low energy consumption, cost-effectiveness, and flexibility of operation [5,6]. In particular, the efficiency of the adsorption process mainly depends on the proper selection of the adsorbent [7].

To date, the main adsorbents applied for the removal of ethylene are activated carbons, zeolites, organic polymers, and organometallic structures, with an emphasis on activated carbons, which are the most used adsorption material in the elimination of Volatile Organic Compounds (VOCs) [8–10]. The advantages of using activated carbons (ACs) include their well-developed micropore structure, large surface area, high mechanical strength, as well as being easily regenerable materials compared to other adsorbents [10]. However, these characteristics alone may not be sufficient to ensure better efficiency in the adsorption process. One way to increase the adsorption selectivity of ACs toward target molecules is to modify their chemical surface properties accordingly [11].

The chemical modification of ACs by doping with high-valent metal species has been applied to increase adsorption selectivity toward some VOCs. First, the porous structure of ACs physically adsorbs the selected metal, and the high valence is reduced by the chemical surface groups of ACs [10]. Therefore, ACs loaded with magnesium, zinc, copper, and zirconium oxides demonstrated strong adsorption of VOCs, such as acetone and toluene, which can be explained by the acid–base interactions of VOC-metal oxide, surface functional groups, and the polarity of the adsorbate [12]. In addition, it was reported that on the surface of AC impregnated with copper oxide, some oxygen-containing functional groups were covered, which led to the generation of more surface sites, where toluene adsorption took place [13]. Another adsorption experiment demonstrated that the introduction of CuO on activated carbon can significantly improve the adsorption performance of siloxanes through hydrogen bonding [14].

So far, no work has sought to evaluate and explain the mechanism of ethylene adsorption on activated carbon doped with CuO, a common, simple, and low-cost transition metal oxide [15]. Since ACs can be prepared from different carbonaceous materials, such as agro-industrial wastes, a species of Brazilian fruit known as baru (*Dipteryx alata* vog.) is presented as a favorable source for the preparation of adsorbents. Baru is a fruit composed of a thin and rough shell (epicarp), a fleshy and fibrous pulp (mesocarp), and a single kernel that is surrounded by a woody structure (endocarp). The kernel is intended for consumption in natural form or oil extraction, while the shell and pulp, which represent around 95% of the fresh weight of the fruit, are generally discarded as waste, causing environmental impacts due to their excessive release into the environment [16,17].

In this context, this study aimed to bring a new use to wasted baru biomass, making sustainable use of baru (*Dipteryx alata* vog.) shells in the production of activated carbons impregnated with CuO species for the adsorptive removal of ethylene during the transportation and storage of agricultural commodities. The ACs were characterized physico-chemically using different techniques. Moreover, the adsorption capacities of the developed CA materials were determined through dynamic adsorption studies, and the influence of the presence of moisture during the adsorption process was evaluated. Finally, using *operando* transmission IR spectroscopy assays, an adsorption mechanism was proposed to describe the surface chemical interactions that take place between ethylene molecules and the tested ACs.

2. Results and Discussion

2.1. Characterization of the Precursor Material and the Prepared Activated Carbons

Figure 1A shows the thermogravimetric (TGA) curve and the first derivate of the TGA curve (DTG curve) of dried baru waste. The TGA curve displays mass loss of 73% from room temperature to 800 °C, which corresponds to a heat treatment yield of 27%.

The DTG curve in Figure 1A shows two main thermal degradation events; the first, which took place in a temperature range between 200 °C and 250 °C, can be attributed to the volatilization and/or decomposition of unsaturated triacylglycerides (oleic and linoleic acid) present in the dry baru waste. The second event, which occurred between 350 °C and 450 °C, may be related to the beginning of the decomposition of biomass, whose main components are hemicellulose, cellulose, and lignin [18,19]. The lignocellulosic fraction found in the dry baru waste was characterized by having 24% cellulose, 18% lignin, and 12% hemicellulose. Previous studies have demonstrated that precursors with high lignin content lead to the generation of ACs with higher preparation yields, whereas cellulose-rich precursors result in materials with higher surface areas [20]. In general, baru waste could be used as a precursor material for the production of porous materials due to its significant lignocellulosic content (>54%) and low ash content.

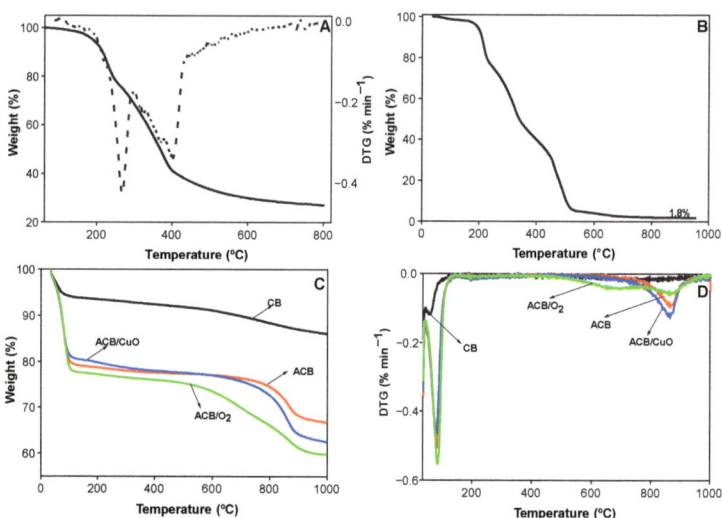

Figure 1. Thermogravimetric results: (**A**) solid line is the TGA curve and dashed line represents the DTG curve of baru waste, (**B**) TGA curves in air of baru waste, (**C**) TGA and (**D**) DTG curves of prepared materials: non-activated baru carbon (CB); baru activated carbon (ACB); baru activated carbon impregnated with copper oxide (ACB/CuO); and activated carbon in oxygen atmosphere (ACB/O_2).

The TGA analysis in the presence of air confirms the formation of a residue (inorganic content) of around 1.8 wt.% (Figure 1B). TGA curves of prepared materials from baru waste shown in Figure 1C indicate two main stages of thermal decomposition. The first, between 40 °C and 100 °C, resulted from the desorption of water from the samples that was absorbed during the storage period. All materials, except the CB sample (non-activated baru carbon), exhibited a similar rate of thermal decomposition in the first stage. The CB sample revealed the greatest thermal stability up to the final carbonization temperature (1000 °C), with a mass loss of 13.9%, while the ACB, ACB/CuO, and ACB/O_2 samples presented losses of 33.3%, 37.5%, and 40.2%, respectively. The thermal stability of the prepared activated carbons compared to non-activated baru carbon (CB) was significantly reduced. The main mass loss is due to the evolution of water that occurs during the condensation stage of phosphoric acid and to the reactions between phosphoric acid and the lignocellulosic fraction (hemicellulose and lignin) of the biomass, which begin around 50 °C. Since the CB material did not undergo any activation step, it was more stable to degradation, condensation, and dehydration reactions [21]. In the second stage of thermal decomposition, the ACB and ACB/CuO samples revealed similar behavior as the temperature increased from 600 °C to 900 °C. Both samples were activated with H_3PO_4 in an inert atmosphere; therefore, mass losses in this temperature range can be explained by carbon combustion and phosphoric acid volatilization [22]. Phosphoric acid interacts with biomass to form phosphate and polyphosphate bonds that bind and cross-link polymer fragments, decreasing the losses of volatile material during pyrolysis [23,24]. The DTG curves presented in Figure 1D confirm that the initial phase of thermal decomposition of the samples is governed by the loss of moisture, in addition to the existence of another important thermal event in the range of 750 °C to 900 °C for ACB and ACB/CuO. The maximun intensity increases in the ACB/CuO sample, indicating that the surface functional groups formed during chemical modification are less stable in this temperature range [25].

Figure 2a displays the X-ray diffraction patterns of porous materials. All AC samples show an amorphous halo at 2θ = 25°, which can be attributed to the carbon (002) plane coming from the decomposition of complex compounds present in the baru husk, such as carbohydrates, lipids, and proteins, indicating an amorphous structure [26,27].

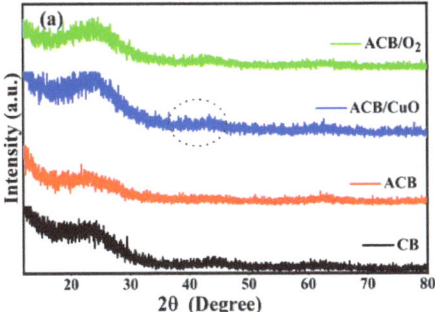 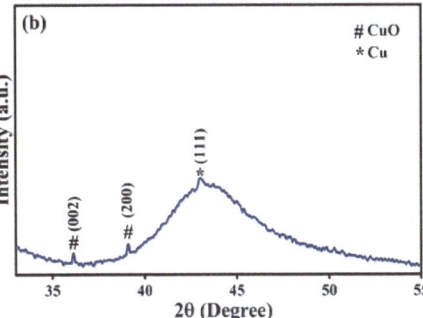

Figure 2. (a) X-ray diffraction patterns of non-activated baru carbon (CB); baru activated carbon (ACB); baru activated carbon impregnated with copper oxide (ACB/CuO); activated carbon in oxygen atmosphere (ACB/O$_2$); (b) maximized image of the dashed circle illustrated in the X-ray diffraction pattern of baru activated carbon impregnated with copper oxide (ACB/CuO) from (a).

In the diffractograms of ACB and CB, a second halo is observed at 2θ = 12.3°, indicating that the applied carbonization process at high temperature is responsible for the decomposition of some organic compounds present in the precursor matrix, consequently contributing to the formation of a semicrystalline structure. Additionally, a weak peak appeared between 35° and 45°, which corresponds to the carbon (101) plane. Despite the fact that weak copper oxide and metallic copper planes were observed in the ACB/CuO sample, due to the overlap with the carbon planes, in the 2θ range of 33°–55° (see Figure 2b), peaks related to the presence of CuO (at 36° and 39°), as well as metallic copper nanoparticles (at 42.7°), were registered. For CuO, all diffraction peaks can be indexed to the crystalline monoclinic structure of CuO nanoparticles (JCPDS card No.: 80-1916), whereas, in the case of Cu nanoparticles, they can be indexed to the face-centered cubic Cu (JCPDF Card No.: 85-1326) [28]. The carbon matrix could be responsible for hiding the intensity of such peaks [29]. Further confirmation of Cu and O elements is presented in the EDS spectra.

Textural features of carbonaceous materials were analyzed by nitrogen adsorption and desorption at −196 °C, and the results are compiled in Figure 3 and Table 1. The CB material presented a type II isotherm, and the other materials (ACB, ACB/CuO and ACB/O$_2$) presented type I isotherms, according to the IUPAC classification. The type I isotherm is common in adsorption measurements and occur mainly in microporous materials; the type II isotherm is characteristic of macroporous or low-pore-volume materials.

All adsorbents record BET surface (S$_{BET}$) values between 291 and 886 m^2 g^{-1} and a total pore volume between 0.16 and 0.46 cm^3 g^{-1} (see Table 1). As expected, the CB sample, which did not undergo any activation step, presented a low surface area and a reduced pore volume, denoting that the chemical activation process is the main factor that contributes to the development of the microporosity and mesoporosity of the other carbons that were activated in the same proportions of precursor material and acid agent (1:2). Moreover, other factors, such as the synthesis temperature, the carbonization atmosphere (nitrogen or air), and the metal oxide used in the impregnation step, were also determining factors in the generation of the observed pore structure. The largest BET surface area (S_{BET} = 886 m^2 g^{-1}) and the greatest development of porosity and micropore volume (0.37 cm^3 g^{-1}) were obtained after chemical activation with H_3PO_4 (1:2) and carbonization in an N_2 atmosphere (sample ACB). In the case of sample ACB/O_2, which differs from the ACB sample only in terms of the carbonization atmosphere, it presented a greater volume of mesopores, which indicates that carbonization in an air atmosphere was decisive in the generation of the mesoporous structure of this material. Chatir et al. [30] indicated that atmospheric air contributes to the formation of water-soluble phosphorus compounds on the carbon surface, which allows for obtaining a porous and adjustable structure, depending on the temperature used. Loading the activated carbon surface with CuO has a significant effect on the micropore volume (0.30 cm^3 g^{-1}). Thus, compared with the activated carbon (ACB), the total pore volume, surface area, and average width of micropores decreased. It seems that the loaded CuO particles blocked part of the micropore structure, having less of an effect on the mesopores. This resulted in a decrease in the observed values of specific surface area and pore volume. The results found in the present work are comparable and even superior to other activated carbons generated from fruit wastes. For example, activated carbons produced from palm kernel/shell and activated in a nitrogen and air atmosphere showed an S_{BET} value of 457 m^2 g^{-1} [31], while activated carbon based on lemon peel produced an S_{BET} value of around 500 m^2 g^{-1} [32] and activated carbon derived from mangosteen peel generated S_{BET} values between 460 and 1039 m^2 g^{-1} [33]. Regarding the applied copper impregnation process, we also observed that the textural characteristics of the ACB/CuO sample were similar and, in some cases, superior to other porous materials impregnated with copper, such as natural zeolite [5] and porous boron nitride [34], as shown in Table 2.

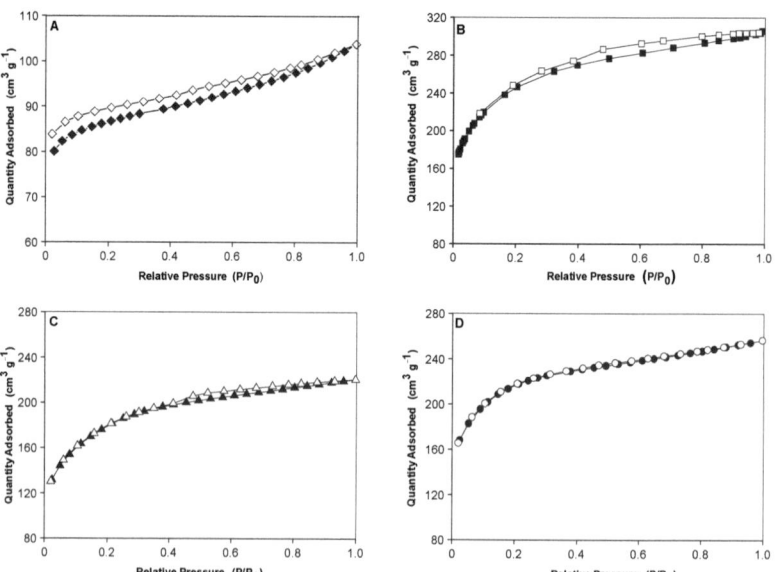

Figure 3. Nitrogen adsorption–desorption isotherms at −196 °C: (**A**) non-activated baru carbon (CB), (**B**) baru activated carbon (ACB), (**C**) baru activated carbon impregnated with copper oxide (ACB/CuO) and (**D**) activated carbon in oxygen atmosphere (ACB/O_2). ♦ CB, ■ ACB, ▲ ACB/CuO, ● ACB/O_2 (filled symbols: adsorption; empty symbols: desorption).

Table 1. Textural characteristics of adsorbent materials.

Samples	S_{BET} [m^2 g^{-1}]	V_T [cm^3 g^{-1}]	V_{meso} [cm^3 g^{-1}]	V_{micro} [cm^3 g^{-1}]	L_0 [nm]	pH$_{PZC}$
CB	291	0.16	0.05	0.11	4.35	6.98
ACB	886	0.46	0.09	0.37	1.72	3.81
ACB/CuO	628	0.34	0.04	0.30	1.37	3.65
ACB/O$_2$	747	0.39	0.21	0.18	2.71	3.53

S_{BET} = BET surface area; V_T = total pore volume; V_{micro} = micropore volume; V_{meso} = mesopore volume; L_0 = average micropore width.

Table 2. Comparison of S_{BET} surface areas of different copper-modified porous materials reported in the literature.

Precursor	Adsorbents	S_{BET} [m^2 g^{-1}]	References
Baru activated carbon	ACB/CuO	628	This work
Natural zeolite	NH$_4$Z$_2$-Cu	351	[5]
Commercial activated carbon	10-CuO/AC-800	667	[14]
Commercial activated carbon	CuO/AC08	947	[13]
Porous boron nitride	BN-Cu	626	[34]

The surface charge distribution of carbonaceous materials was evaluated by determining the pH of the point of zero charge (pH$_{PZC}$). The pH$_{PZC}$ consists of the pH value at which the net charge densities on the surface of the material become zero [22]. The results for this parameter indicate that pH$_{PZC}$ values (see Table 1) decrease with increasing precursor/acid ratios. As the activated carbon samples ACB, ACB/CuO, and ACB/O$_2$ underwent the same degree of activation (1:2), very close values were found for the pH$_{PZC}$. On the contrary, non-activated baru carbon (CB) (without any activation step) resulted in a pH$_{PZC}$ value close to 7, indicating that at this pH value, the surface charges are neutral. Lower pH$_{PZC}$ values denote that porous materials have a higher concentration of acidic groups, such as carboxylic and phenolic groups, resulting from the dissociation of surface oxygen complexes [35].

The porous nature and elemental characteristics of the produced carbon-based materials can be clearly observed in the field-emission scanning electron microscopy (FESEM) images obtained coupled to energy-dispersive spectroscopy (EDS) (see Figure 4), which show the presence of microporous and mesoporous structures. EDS spectra indicate the presence of elements, such as carbon, oxygen, calcium, sodium, and silicon. These elements are associated with the nature of the raw material and could also be related to the applied carbonization, activation, and washing processes. Moreover, EDS analysis demonstrates the incorporation of copper on the activated carbon surface after the applied modification procedure (Figure 4i).

In Figure 4a,b, microstructural analysis of non-activated baru carbon (CB) reveals its unevenly agglomerated morphology, and the absence of porosity is confirmed. Figure 4c shows the EDS result, which confirms the presence of elemental constituents in the CB sample.

Figure 4d,e depict FESEM images of baru activated carbon (ACB), illustrating structural and porosity changes compared to the non-activated CB sample. They further confirm the successful carbon activation with good pore structures. Figure 4f displays the EDS analysis, which confirms the presence of activated carbon elements in the structures and the associated changes in elemental composition resulting from the activation process.

Figure 4g shows FESEM images of baru activated carbon impregnated with copper oxide (ACB/CuO), which exhibits uneven morphological structure and agglomerated CuO nanoparticles on the surface. A magnification image in Figure 4h further confirms the deposition of copper oxide (CuO) nanosheets on the ACB surface. An inserted image reconfirms the CuO morphology. The marked circle confirms the porosity and morphology of CuO nanosheets, showcasing the distribution of copper oxide nanoparticles over the activated carbon matrix. Figure 4i presents the EDS analysis, which reveals the elemental composition of carbon and copper within the produced material, providing information on the impregnation process and the resulting composite structure.

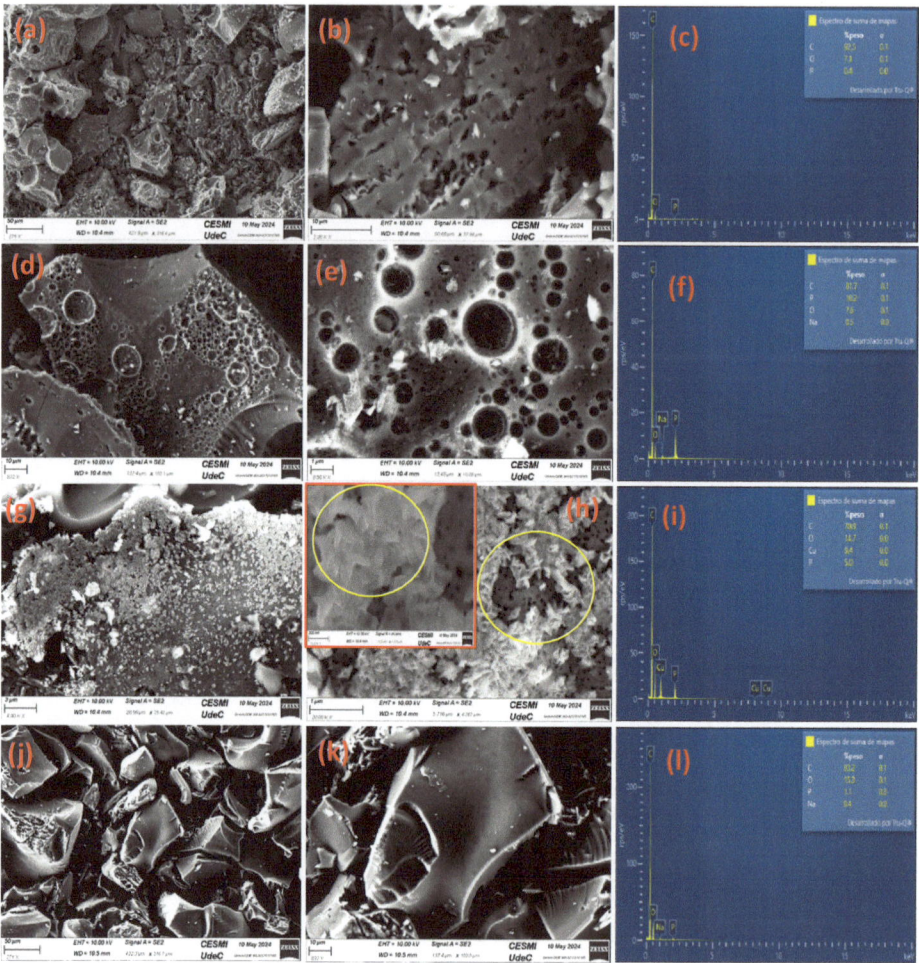

Figure 4. Field-emission scanning electron microscopy coupled to energy-dispersive spectroscopy of (**a–c**) non-activated baru carbon (CB), (**d–f**) baru activated carbon (ACB), (**g–i**) baru activated carbon impregnated with copper oxide (ACB/CuO), (**j–l**) activated carbon in oxygen atmosphere (ACB/O$_2$).

In Figure 4j,k, FESEM images of activated carbon in an oxygen atmosphere (ACB/O$_2$) reveal the morphological changes induced by the exposure to oxygen. Figure 4l depicts the EDS analysis, which reveals the elemental composition of the activated carbon due to oxidative processes.

In Figure 5, the elemental mapping analysis corresponds to the FESEM image of baru activated carbon impregnated with copper oxide (ACB/CuO). Elemental mapping analysis provides valuable information on the distribution and composition of different elements within this material. Elemental mapping reveals the spatial distribution of key elements, including carbon (C), oxygen (O), and copper (Cu), across the surface of the ACB/CuO sample. This analysis allows us to visualize the dispersion of copper oxide nanoparticles within the activated carbon matrix and evaluate the uniformity of the impregnation process.

Notably, the elemental mapping highlights regions of high copper concentration, indicating the successful impregnation of copper oxide onto the activated carbon substrate.

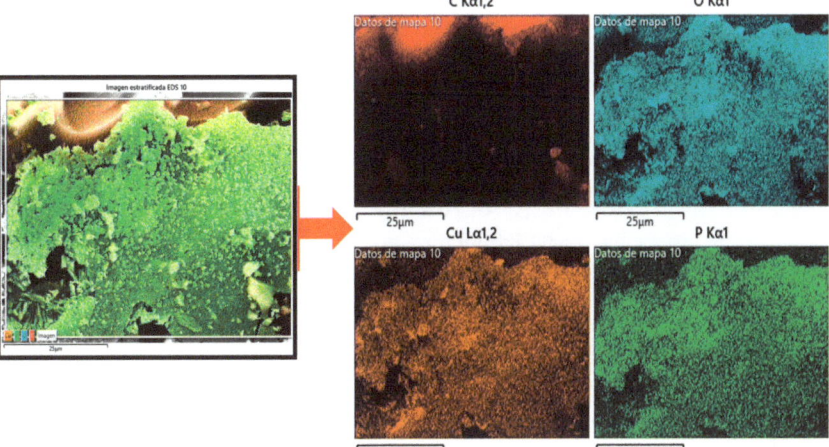

Figure 5. Elemental mapping analysis corresponding to the FESEM image of baru activated carbon impregnated with copper oxide (ACB/CuO).

The surface chemistry of non-activated and activated baru carbon samples was characterized by IR spectroscopy using ATR, and the results are given in Figure 6. The characteristic IR vibrations of lignocellulosic compounds are registered at approximately 3442 cm^{-1}, attributed to O-H of phenolic hydroxyl groups [36]. As can be seen from the spectra, the introduction of CuO into the ACB/CuO sample does not significantly change the surface chemistry compared to the ACB sample and to the other developed materials. The formation and shift of the IR band at 1634 cm^{-1} are due to the bending vibrations of water related to adsorbed water and remain the same for all the AC materials [5]. The most obvious differences in the vibrational spectra between ACB and ACB/CuO were evident in the interval between 400 and 700, where new IR bands attributed to stretching vibrations of the Cu-O bond of copper oxide were formed [37]. The IR bands observed in a range between 1381 cm^{-1} and 615 cm^{-1} resulted from the angular deformation of OH and angular vibration (OH) of the water molecule, respectively [38].

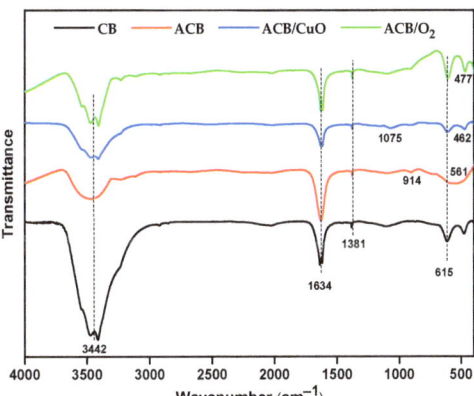

Figure 6. FTIR spectra of non-activated baru carbon (CB); baru activated carbon (ACB); baru activated carbon impregnated with copper oxide (ACB/CuO); and activated carbon in oxygen atmosphere (ACB/O$_2$).

2.2. Chemical Surface Interaction Assessments by Operando IR Spectroscopy

Real-time analysis of the AC samples using IR spectroscopy (*operando* mode) provided useful information on the surface interactions with ethylene molecules through the adsorption process. The evolution of IR spectra is provided in Figure 7, starting at time zero and ending after 200 min of contact time, where ethylene saturation is evident in all samples. IR vibrations corresponding to ethylene gas were observed at IR bands located at 950 cm^{-1}, 1420 cm^{-1}, 1870 cm^{-1}–1914 cm^{-1}, 2344 cm^{-1}, 2970 cm^{-1}, and 3129 cm^{-1}, which are in correspondence with the results reported by other authors [5]. It can also be noticed that the IR band observed at 3442 cm^{-1} ascribed to hydroxyl groups before coming into contact with ethylene (Figure 6) lost its intensity during continuous exposure to ethylene and was almost completed consumed. These results suggest that hydroxyl functional groups may be mainly responsible for the adsorption of ethylene molecules, changing the electrostatic potential distribution on the surface of activated carbon and facilitating important interactions with ethylene [39]. Hence, hydrogen-bonded adducts could be formed between hydroxyl functional groups on the surface AC and ethylene molecules in a similar way, as has been described for the interaction between ethylene molecules and the Brønsted acid sites of zeolitic frameworks [5]. Moreover, the mesoporous and microporous structure of all the samples allows for easy diffusion of small molecules like ethylene with a kinetic diameter of ~3.9 Å and its consequent interaction with the OH groups on the AC surface.

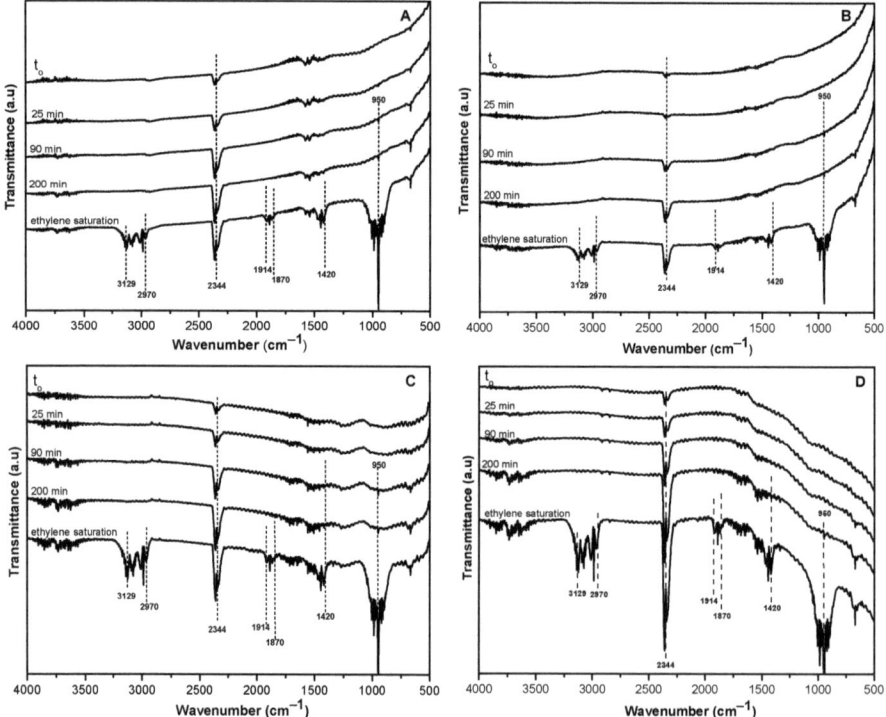

Figure 7. Evolution of IR spectra during ethylene adsorption on AC samples obained by *operando* transmission IR spectroscopy analyses: (**A**) non-activated baru carbon (CB), (**B**) baru activated carbon (ACB), (**C**) baru activated carbon impregnated with copper oxide (ACB/CuO), (**D**) activated carbon in oxygen atmosphere (ACB/O$_2$).

In the case of samples CB and ACB (Figure 7A,B), the IR absorption bands characteristic of ethylene vibrations gradually evolve as the contact time increases. In particular, the ACB/O$_2$ sample (Figure 7D) exhibits more intense vibrations in the first minutes of contact time, reaching a saturation point more quickly, in which the entire surface structure of the adsorbent is covered, and well-developed bands are observed mainly between 2970–3129 cm^{-1} and 950 cm^{-1}. In the case of the sample impregnated with CuO (Figure 7C), the IR band formed between 500 cm^{-1} and 700 cm^{-1} could be related to interactions of ethylene with copper incorporated on the AC surface. Experimental studies have suggested that interactions of ethylene with metal oxides could occur through the π electrons of the C=C bonds of ethylene and the metal orbitals present in the supporting structure. This is a type of interaction known as σ-donation, in which the π molecular orbital of the adsorbed ethylene donates electron density to the empty s-orbital of the metal oxide [5,40].

2.3. Mechanistic Approach of Ethylene Adsorption onto Baru-Based Carbon Samples

Adsorption equilibrium isotherms at 20 °C for the four carbon samples are displayed in Figure 8. Experimental data were fitted to the Langmuir model as global evidence of the surface interactions between ethylene molecules and carbon samples, as follows:

$$q_e = \frac{K_L C_e q_{max}}{1 + K_L C_e} \quad (1)$$

where q_e is the amount of ethylene adsorbed at equilibrium, K_L stands for the Langmuir adsorption constant, C_e represents the ethylene concentration at equilibrium, and q_{max} is the maximum adsorption capacity. The results depicted in Figure 8 are in agreement with those obtained by other authors for the removal of VOCs using activated carbons modified with metallic oxides, showing forms of a classic type I adsorption isotherm [41].

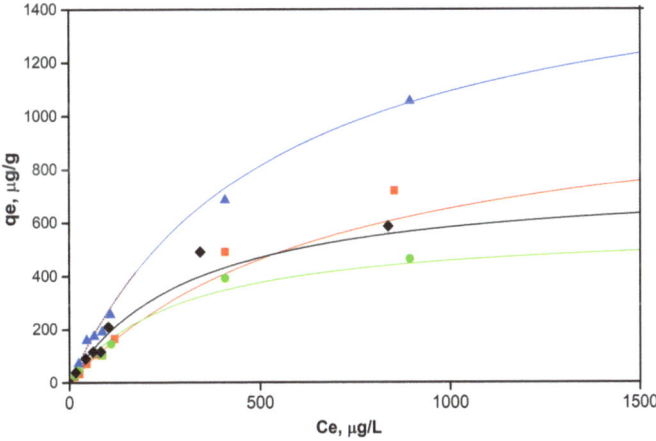

Figure 8. Ethylene adsorption isotherms: (♦) non-activated baru carbon (CB), (■) baru activated carbon (ACB), (▲) baru activated carbon impregnated with copper oxide (ACB/CuO), (●) activated carbon in oxygen atmosphere (ACB/O$_2$). Thin lines are the result of fitting data to Langmuir isotherm model.

At constant temperature, the amount of ethylene adsorbed at equilibrium is low when low concentrations are applied. As the applied concentration increases, there is an increase in the adsorption capacities, reaching maximum values for each carbon sample. As can be seen in Table 3, the Langmuir adsorption model provided a good fit to the ethylene adsorption data with $R^2 > 0.98$. The ACB/CuO and ACB samples showed the best ethylene adsorption capabilities. A slightly higher performance was observed for the CuO-impregnated sample, due to the new surface sites provided on this carbon sample. Both samples presented the largest volumes of micropores, which indicates that the adsorption performance is also related to the textural properties of the activated carbon. In addition, the CuO-impregnated AC sample resulted in an activated carbon with a smaller average micropore diameter (1.37 nm), which is an important parameter that can explain the interactions between the ethylene molecule (kinetic diameter of approximately 3.9 Å) and the chemical structure of carbon. Micropores that are approximately twice the diameter of ethylene are more accessible, avoiding adsorbate blocking and favoring adsorption performance [42].

Table 3. Langmuir parameters for ethylene adsorption at 20 °C on non-activated baru carbon and modified carbon samples.

Samples	q_{max} [µg g^{-1}]	K_L [dm^3 µg^{-1}]	R^2
CB	769	0.00314	0.98
ACB	1111	0.00143	0.98
ACB/CuO	1667	0.00191	0.98
ACB/O$_2$	588	0.00358	0.99

Furthermore, the results for the influence of the presence of moisture on ethylene adsorption are illustrated in Figure 9. As can be noted, the presence of moisture does not affect ethylene adsorption on all activated carbon samples assessed here. However, a great difference is observed in the case non-activated baru carbon, where a reduction in the adsorption capacity was obtained. These results imply that the surface sites of baru activated carbons where ethylene adsorption occurs are not blocked by the presence of water molecules. Humidity is an important component in fruit storage systems, and below the ideal range (usually ~85%), dehydration and wilting increase [8]. Therefore, adsorbents must be able to adsorb ethylene in high-relative-humidity environments. Activated carbons generated from baru waste appear as alternative adsorbent materials to remove ethylene from closed environments with a high content of humidity.

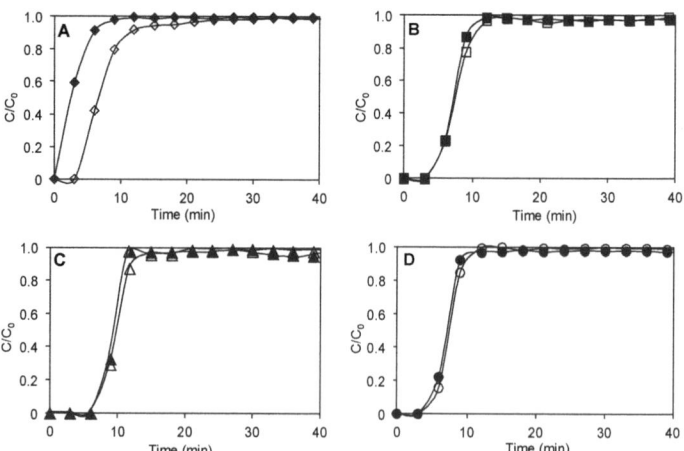

Figure 9. Influence of the presence and absence of moisture on ethylene adsorption: (**A**) non-activated baru carbon, (**B**) baru activated carbon, (**C**) baru activated carbon impregnated with copper oxide, (**D**) activated carbon in oxygen atmosphere. Experimental conditions: 0.1 g of sample, 25 cm^3 min^{-1} ethylene (75 µgL^{-1}) at 20 °C. ♦ CB, ■ ACB, ▲ ACB/CuO, ● ACB/O$_2$ (filled markers represent experiments conducted in the presence of moisture (RH 98%) and open markers represent in the absence of moisture).

The results obtained here are in agreement with those obtained by *operando* IR spectroscopy. The removal of ethylene using baru-based activated carbons seems to take place through a combination of adsorption mechanisms that include interactions of ethylene with hydroxyl groups and with copper incorporated on the AC surface. Thus, hydrogen-bonded adducts could be formed between the hydroxyl functional groups on the AC surface and ethylene molecules. Additionally, the π molecular orbital of the adsorbed ethylene donates electron density to the empty s-orbital of the metal oxide, leading to the observed increase in the adsorption capacity of the CuO-impregnated AC sample.

3. Materials and Methods

3.1. Raw Materials and Reagents

The baru waste was obtained from a local commercial producer located in the state of Minas Gerais, Brazil (17°11′39″ S 44°48′49″ O), in June 2022. Argon (99.9% purity) was supplied by Praxair (Santiago, Chile). Ethylene (C_2H_4; 99.99% purity) was provided by Air Liquide S.A. (Houston, TX, USA). Copper (II) nitrate trihydrate ($Cu(NO_3)_2 \cdot 3\,H_2O$; 99.5%) and phosphoric acid (H_3PO_4; 85%) were obtained from Sigma Aldrich, St. Louis, MO, USA.

3.2. Preparation of Porous Materials

3.2.1. Preparation and Characterization of the Precursor Material

First, the baru waste was washed with distilled water to remove impurities and then dried in a forced convection oven (TE-394/3, Tecnal, Piracicaba, Brazil) for 48 h at 65 °C. The dried waste was then crushed using a shredder and a chipper (TL 1200, Lippel, Agrolândia, Brazil), and the particle size was reduced to 2 mm. The crushed materials were kept in a desiccator as a raw material precursor to produce activated carbons. The baru waste was characterized for ash content, according to standardized methods [43]. The fractions of the raw lignocellulosic components of the biomass (hemicellulose, cellulose, and lignin) were determined using a protocol established for raw materials by the *Unité d'Amelioration Génétique et Physiologie Forestières* (AGPF) using the *INRA GénoBois* technical platform [18,44]. Thermogravimetric analyses were performed using a DTG thermobalance (60/60H-Shimadzu, Kyoto, Japan), according to a methodology proposed by other authors [18]. The procedure included a temperature rise at a rate of 20 °C min^{-1}, from room temperature up to 800 °C (to study a larger temperature domain), followed by a 1 h interval at this final temperature and then cooling to room temperature at 10 °C min^{-1}.

3.2.2. Generation of Activated Carbon Materials

Activated carbons were prepared from the precursor material according to a methodology described by other authors [38]. First, the sample was impregnated with phosphoric acid (H_3PO_4, 85%) at an impregnation ratio of 1:2 (raw material/acid). Then, the mixture was heated up to 80 °C and kept under mechanical stirring at this temperature for 30 min. The resulting material was filtered and dried in an oven at 110 °C for 15 h. After this step, the impregnated material was carbonized in a tubular oven (FT 1200, Sanchis, Porto Alegre, Brazil) under the following conditions: temperature of 800 °C for 40 min; heating rate of 20 °C min^{-1} under nitrogen flow or in air (160 cm^3 min^{-1}). After that, the carbonized sample was first washed with 37% hydrochloric acid and then with distilled water until the pH was close to neutral. Finally, the activated carbon generated from baru waste was oven-dried at 50 °C until the complete evaporation of water was ensured. Samples activated in nitrogen flow were called ACB, and samples activated in air were named ACB/O_2. In addition to these samples, non-activated baru carbon (CB) was also obtained. In case of CB, the process is the same as for the preparation of ACB but in the absence of H_3PO_4 as activating agent.

3.3. Preparation of Baru Activated Carbon Impregnated with Copper Oxide (ACB/CuO)

Surface modification of activated carbon with copper oxide (CuO) was performed using a simple wet impregnation method described elsewhere [14]. In a typical modification procedure, 5.0 g of carbon (ACB) was added to 100 cm^3 of 0.1 M copper nitrate ($Cu(NO_3)_2$)

solution. The mixture was kept under stirring at 25 °C overnight and then filtered and dried at 110 °C for 12 h. Then, the precursor was transferred to a tubular oven and heated from room temperature up to 280 °C at 10 °C min^{-1} under N_2 flow (200 cm^3 min^{-1}) for 2 h. After cooling, the ACB/CuO sample was obtained.

3.4. Characterization of AC Samples

Nitrogen adsorption and desorption isotherms of AC samples were obtained in a Micrometrics ASAP instrument (Gemini V2380, Norcross, GA, USA) at -196 °C. Specific surface area (A_{BET}) was determined from the adsorption curve in the range $0.05 \leq p/p_0 \leq 0.15$, using the Brunauer–Emmett–Teller (BET) theory. Total pore volume (V_T) was recorded at $p/p_0 = 0.95$, whereas micropore volume (V_{micro}) was determined according to Barrett–Joyner–Halenda (BJH) approach [45]. Mesopore volume (V_{meso}) was calculated by the difference between V_T and V_{micro}. Morphology and chemical composition of carbons were determined by field-emission scanning electron microscopy (FESEM) coupled to X-ray energy-dispersive spectroscopy (EDS)—ZEISS Gemini SEM 360 (Jena, Germany). Thermogravimetric analyses (TGA) of the samples were performed on a NET-ZSCH thermobalance ST409PC (Pomerode, Brazil). Samples of 0.025 g of each AC were heated up to 1000 °C (heating rate of 10 °C min^{-1}) under N_2 flow (160 cm^3 min^{-1}), and the change in sample weight in relation to change in temperature was registered (TG curve). A derivative weight loss curve was also obtained as function of temperature (DTG curve). X-ray diffraction (XRD) patterns were obtained on a Bruker D8 Discover diffractometer (Billerica, MA, USA), using a monochromatic radiation from a tube with a copper anode coupled to a Johansson monochromator operating at 40 kV and 40 mA, Bragg Brentano θ–2θ configuration, Lynxeye one-dimensional detector, range of 2θ from 2° to 80°, with a step of 0.01°. The pH of the point of zero charge, defined as the pH at which the carbon surface has a neutral charge, was determined following the procedure described by Kuśmierek et al. [46]. Functional surface groups were directly evaluated by IR spectroscopy (PERKIN ELMER Spectrum 400 FT-IR spectrometer, Waltham, MA, USA) using attenuated total reflectance (ATR) technique in the infrared region of 4000–500 cm^{-1}, with a resolution of 4 cm^{-1}.

3.5. Chemical Surface Interaction Assessments by Operando Transmission IR Spectroscopy

Chemical interactions between the surface groups of ACs and ethylene were monitored as a function of time in a homemade transmittance cell developed by the Catalysis and Spectrochemistry Laboratory (LCS, Caen, France) and set in a Nicolet™ iS™50 spectrometer (Thermo Fisher Scientific Inc., Waltham, MA, USA) equipped with a DTGS detector. Pellets (Ø = 16 mm, $m \approx 20$ mg cm^{-2}) with a mass of approximately 40 mg (5% w/w activated carbon/KBr) were formed. A pellet made of KBr was used as a background. Before the measurements, samples were thermally activated inside the cell at 100 °C under Ar flow (5 cm^3 min^{-1}) for 15 min. Then, a spectrum of the adsorbate free sample was measured. After that, a stream made of 1% ethylene in Ar was passed through the pellet for 200 min. Subsequently, the diluted ethylene stream was changed to pure ethylene, and the analyses continued until reaching saturation. IR transmittance spectra were registered as a function of time. Spectra were obtained with 60 scans at a resolution of 4 cm^{-1} in a range from 4000 to 500 cm^{-1}.

3.6. Determination of Ethylene Adsorption Isotherms

Ethylene adsorption isotherms were conducted through dynamic adsorption tests on a quartz fixed-bed flow adsorber, as per the procedure described by Abreu et al. [5]. The adsorber was loaded with 0.1 g of sample. Before contact with ethylene, the samples were thermally degassed at 150 °C (3 °C min^{-1}) under argon flow (100 cm^3 min^{-1}) for 1 h. The inlet concentration of ethylene was fixed by diluting a pure stream of C_2H_4 with argon using mass flow controllers. A total flow rate of 25 cm^3 min^{-1} of ethylene at the desired concentration was continuously delivered over the AC sample. The ethylene concentration was determined by gas chromatography (PERKIN ELMER CLARUS 500 gas chromatograph, Waltham, MA, USA) equipped with a flame ionization detector (FID). The adsorption capacity of each AC sample toward ethylene was determined by calculating the areas of different breakthrough curves. For experiments in the presence of humidity, the entire stream was bubbled in a humidification chamber maintained with water at 293 K, allowing for a constant wet flow rate at 98% relative humidity (RH). The adsorption capacities of AC samples towards ethylene ($q_{ethylene}$, $\mu mol_{ethylene}$ g carbon^{-1}) were determined by calculating the areas of different breakthrough curves, as follows:

$$q_{ethylene} = \frac{F \, C_{in}}{m} \int_0^{t_s} \left(1 - \frac{C_{out_t}}{C_{in}}\right) dt \qquad (2)$$

where F (cm^3 min^{-1}) is the gas flow rate, m (g) is the mass of AC placed inside the quartz adsorber, C_{in} and C_{out} (μmol dm^{-3}) are the ethylene inlet and outlet concentrations as a function of time, respectively, and t_s (min) is the adsorption time to reach saturation.

4. Conclusions

Baru waste has the potential to be used as a low-cost precursor in the generation of activated carbons, since it has significant levels of lignin, cellulose, and hemicellulose and a low ash content, which leads to a considerable carbonization yield, even in high-temperature conditions. It was possible to produce activated carbons and carry out surface modifications by impregnation with copper species, which were confirmed by characterization analyses (XRD and FESEM/EDS). Surface modification with copper oxide improved the ethylene adsorption capacity compared to other generated porous materials. The efficient adsorption of ethylene on the CuO-impregnated AC sample was favored by the surface acidity of this AC, where hydrogen-bonded adducts were formed due to the interactions of the surface OH groups and ethylene molecules and also by the interactions between the empty s-orbital of the copper oxide and π-electrons of the C=C bonds of the ethylene molecules. The presence of moisture did not affect the adsorption capacity of baru activated carbon samples, which makes this type of activated carbon an excellent adsorbent option to eliminate ethylene from closed containers with a high content of humidity. Future works should focus on evaluating the regeneration performance and operating conditions for the practical application of this new adsorbent during the storage of climacteric fruit.

Author Contributions: Conceptualization, A.C.d.J.O., E.T.M., T.F.d.O. and H.V.; methodology, A.C.d.J.O., M.C.d.A., P.S.S., V.A.S., T.F.d.O. and H.V.; formal analysis, A.C.d.J.O., T.F.d.O. and H.V.; investigation, A.C.d.J.O., C.A.P.R., V.A.S., T.F.d.O. and H.V.; resources, T.F.d.O. and H.V.; data curation, A.C.d.J.O.; writing—original draft preparation, A.C.d.J.O., T.F.d.O. and H.V.; writing—review and editing, A.C.d.J.O., T.F.d.O. and H.V.; visualization, A.C.d.J.O., E.T.M., P.S.S. and T.F.d.O.; supervision, T.F.d.O. and H.V.; project administration, T.F.d.O. and H.V.; funding acquisition, T.F.d.O. and H.V. All authors have read and agreed to the published version of the manuscript.

Funding: This research was funded by the Federal University of Goiás, Project 18/2020 CAPES, by the Goiás State Research Support Foundation (FAPEG), by CNPQ, grant number 441839/2023-1 and by the Chilean *Agencia Nacional de Investigación y Desarrollo* ANID, FONDECYT Regular, grant number 1200858, to whom the authors are indebted.

Institutional Review Board Statement: Not applicable.

Informed Consent Statement: Not applicable.

Data Availability Statement: The data that support the findings of this study are available from the corresponding author upon reasonable request.

Acknowledgments: We gratefully acknowledge the instrumentation support provided by the Regional Center for Technological Development and Innovation (CRTI) and the Multi-User Analysis Center (CAM), Federal University of Goiás, Goiania, Brazil and by the *Centro de Espectroscopía y Microscopía* (CESMI), *Universidad de Concepción*, Concepcion, Chile. The authors also want to thank Padmanaban Annamalai from *Laboratorio de Tecnologías Limpias, Universidad Católica de la Santísima Concepción* for his valuable collaboration.

Conflicts of Interest: The authors declare no conflicts of interest.

References

1. Saltveit, M.E. Effect of Ethylene on Quality of Fresh Fruits and Vegetables. *Postharvest Biol. Technol.* **1999**, *15*, 279–292. [CrossRef]
2. Ku, V.V.V.; Shohet, D.; Wills, R.B.H.; Kim, G.H. Importance of Low Ethylene Levels to Delay Senescence of Non-Climacteric Fruit and Vegetables. *Aust. J. Exp. Agric.* **1999**, *39*, 221–224. [CrossRef]
3. Wei, H.; Seidi, F.; Zhang, T.; Jin, Y.; Xiao, H. Ethylene Scavengers for the Preservation of Fruits and Vegetables: A Review. *Food Chem.* **2021**, *337*, 127750. [CrossRef] [PubMed]
4. Keller, N.; Ducamp, M.-N.; Robert, D.; Keller, V. Ethylene Removal and Fresh Product Storage: A Challenge at the Frontiers of Chemistry. Toward an Approach by Photocatalytic Oxidation. *Chem. Rev.* **2013**, *113*, 5029–5070. [CrossRef] [PubMed]
5. Abreu, N.J.; Valdés, H.; Zaror, C.A.; Azzolina-Jury, F.; Meléndrez, M.F. Ethylene Adsorption onto Natural and Transition Metal Modified Chilean Zeolite: An *Operando* DRIFTS Approach. *Microporous Mesoporous Mater.* **2019**, *274*, 138–148. [CrossRef]
6. Hu, B.; Sun, D.-W.; Pu, H.; Wei, Q. Recent Advances in Detecting and Regulating Ethylene Concentrations for Shelf-Life Extension and Maturity Control of Fruit: A Review. *Trends Food Sci. Technol.* **2019**, *91*, 66–82. [CrossRef]
7. An, Y.; Fu, Q.; Zhang, D.; Wang, Y.; Tang, Z. Performance Evaluation of Activated Carbon with Different Pore Sizes and Functional Groups for VOC Adsorption by Molecular Simulation. *Chemosphere* **2019**, *227*, 9–16. [CrossRef] [PubMed]
8. Shenoy, S.; Pathak, N.; Molins, A.; Toncheva, A.; Schouw, T.; Hemberg, A.; Laoutid, F.; Mahajan, P.V. Impact of Relative Humidity on Ethylene Removal Kinetics of Different Scavenging Materials for Fresh Produce Industry. *Postharvest Biol. Technol.* **2022**, *188*, 111881. [CrossRef]
9. Regadera-Macías, A.M.; Morales-Torres, S.; Pastrana-Martínez, L.M.; Maldonado-Hódar, F.J. Ethylene Removal by Adsorption and Photocatalytic Oxidation Using Biocarbon–TiO_2 Nanocomposites. *Catal. Today* **2023**, *413–415*, 113932. [CrossRef]
10. Zhu, L.; Shen, D.; Luo, K.H. A Critical Review on VOCs Adsorption by Different Porous Materials: Species, Mechanisms and Modification Methods. *J. Hazard. Mater.* **2020**, *389*, 122102. [CrossRef]
11. Zhang, X.; Gao, B.; Creamer, A.E.; Cao, C.; Li, Y. Adsorption of VOCs onto Engineered Carbon Materials: A Review. *J. Hazard. Mater.* **2017**, *338*, 102–123. [CrossRef]
12. Zhou, J.; Luo, A.; Zhao, Y. Preparation and Characterisation of Activated Carbon from Waste Tea by Physical Activation Using Steam. *J. Air Waste Manag. Assoc.* **2018**, *68*, 1269–1277. [CrossRef] [PubMed]
13. Lei, B.; Liu, B.; Zhang, H.; Yan, L.; Xie, H.; Zhou, G. CuO-Modified Activated Carbon for the Improvement of Toluene Removal in Air. *J. Environ. Sci.* **2020**, *88*, 122–132. [CrossRef] [PubMed]
14. Yang, Z.; Chen, Z.; Gong, H.; Wang, X. Copper Oxide Modified Activated Carbon for Enhanced Adsorption Performance of Siloxane: An Experimental and DFT Study. *Appl. Surf. Sci.* **2022**, *601*, 154200. [CrossRef]
15. He, X.; Gui, Y.; Xie, J.; Liu, X.; Wang, Q.; Tang, C. A DFT Study of Dissolved Gas (C_2H_2, H_2, CH_4) Detection in Oil on CuO-Modified BNNT. *Appl. Surf. Sci.* **2020**, *500*, 144030. [CrossRef]
16. de Miranda Monteiro, G.; Carvalho, E.E.N.; Boas, E.V.B.V. Baru (*Dipteryx Alata Vog.*): Fruit or Almond? A Review on Applicability in Food Science and Technology. *Food Chem. Adv.* **2022**, *1*, 100103. [CrossRef]
17. Rambo, M.K.D.; Nemet, Y.K.S.; Júnior, C.C.S.; Pedroza, M.M.; Rambo, M.C.D. Comparative Study of the Products from the Pyrolysis of Raw and Hydrolyzed Baru Wastes. *Biomass Convers. Biorefinery* **2021**, *11*, 1943–1953. [CrossRef]
18. Boundzanga, H.M.; Cagnon, B.; Roulet, M.; de Persis, S.; Vautrin-Ul, C.; Bonnamy, S. Contributions of Hemicellulose, Cellulose, and Lignin to the Mass and the Porous Characteristics of Activated Carbons Produced from Biomass Residues by Phosphoric Acid Activation. *Biomass Convers. Biorefinery* **2022**, *12*, 3081–3096. [CrossRef]
19. Babas, H.; Khachani, M.; Warad, I.; Ajebli, S.; Guessous, A.; Guenbour, A.; Safi, Z.; Berisha, A.; Bellaouchou, A.; Abdelkader, Z.; et al. Sofosbuvir Adsorption onto Activated Carbon Derived from Argan Shell Residue: Optimization, Kinetic, Thermodynamic and Theoretical Approaches. *J. Mol. Liq.* **2022**, *356*, 119019. [CrossRef]
20. Neme, I.; Gonfa, G.; Masi, C. Activated Carbon from Biomass Precursors Using Phosphoric Acid: A Review. *Heliyon* **2022**, *8*, e11940. [CrossRef]
21. Nahil, M.A.; Williams, P.T. Pore Characteristics of Activated Carbons from the Phosphoric Acid Chemical Activation of Cotton Stalks. *Biomass Bioenergy* **2012**, *37*, 142–149. [CrossRef]
22. Amran, F.; Zaini, M.A.A. Valorization of *Casuarina* Empty Fruit-Based Activated Carbons for Dyes Removal—Activators, Isotherm, Kinetics and Thermodynamics. *Surf. Interfaces* **2021**, *25*, 101277. [CrossRef]
23. Danish, M.; Hashim, R.; Ibrahim, M.N.M.; Sulaiman, O. Optimization Study for Preparation of Activated Carbon from Acacia Mangium Wood Using Phosphoric Acid. *Wood Sci. Technol.* **2014**, *48*, 1069–1083. [CrossRef]
24. Luo, Y.; Li, D.; Chen, Y.; Sun, X.; Cao, Q.; Liu, X. The Performance of Phosphoric Acid in the Preparation of Activated Carbon-Containing Phosphorus Species from Rice Husk Residue. *J. Mater. Sci.* **2019**, *54*, 5008–5021. [CrossRef]
25. Ayalkie Gizaw, B.; Gabbiye Habtu, N. Catalytic Wet Air Oxidation of Azo Dye (Reactive Red 2) over Copper Oxide Loaded Activated Carbon Catalyst. *J. Water Process Eng.* **2022**, *48*, 102797. [CrossRef]
26. Djilani, C.; Zaghdoudi, R.; Modarressi, A.; Rogalski, M.; Djazi, F.; Lallam, A. Elimination of Organic Micropollutants by Adsorption on Activated Carbon Prepared from Agricultural Waste. *Chem. Eng. J.* **2012**, *189–190*, 203–212. [CrossRef]
27. Ofgea, N.M.; Tura, A.M.; Fanta, G.M. Activated Carbon from H_3PO_4-Activated Moringa Stenopetale Seed Husk for Removal of Methylene Blue: Optimization Using the Response Surface Method (RSM). *Environ. Sustain. Indic.* **2022**, *16*, 100214. [CrossRef]

28. Liu, X.; Geng, B.; Du, Q.; Ma, J.; Liu, X. Temperature-Controlled Self-Assembled Synthesis of CuO, Cu$_2$O and Cu Nanoparticles through a Single-Precursor Route. *Mater. Sci. Eng. A* **2007**, *448*, 7–14. [CrossRef]
29. Mishra, S.R.; Ahmaruzzaman, M. CuO and CuO-Based Nanocomposites: Synthesis and Applications in Environment and Energy. *Sustain. Mater. Technol.* **2022**, *33*, e00463. [CrossRef]
30. Chatir, E.M.; El Hadrami, A.; Ojala, S.; Brahmi, R. Production of Activated Carbon with Tunable Porosity and Surface Chemistry via Chemical Activation of Hydrochar with Phosphoric Acid under Oxidizing Atmosphere. *Surf. Interfaces* **2022**, *30*, 101849. [CrossRef]
31. Lee, C.L.; H'ng, P.S.; Paridah, M.T.; Chin, K.L.; Rashid, U.; Maminski, M.; Go, W.Z.; Nazrin, R.A.R.; Rosli, S.N.A.; Khoo, P.S. Production of Bioadsorbent from Phosphoric Acid Pretreated Palm Kernel Shell and Coconut Shell by Two-Stage Continuous Physical Activation via N2 and Air. *R. Soc. Open Sci.* **2018**, *5*, 180775. [CrossRef] [PubMed]
32. De Rose, E.; Bartucci, S.; Poselle Bonaventura, C.; Conte, G.; Agostino, R.G.; Policicchio, A. Effects of Activation Temperature and Time on Porosity Features of Activated Carbons Derived from Lemon Peel and Preliminary Hydrogen Adsorption Tests. *Colloids Surf. Physicochem. Eng. Asp.* **2023**, *672*, 131727. [CrossRef]
33. Khajonrit, J.; Sichumsaeng, T.; Kalawa, O.; Chaisit, S.; Chinnakorn, A.; Chanlek, N.; Maensiri, S. Mangosteen Peel-Derived Activated Carbon for Supercapacitors. *Prog. Nat. Sci. Mater. Int.* **2022**, *32*, 570–578. [CrossRef]
34. Wang, J.; Li, X.; Fang, Y.; Huang, Q.; Wang, Y. Efficient Adsorption of Tetracycline from Aqueous Solution Using Copper and Zinc Oxides Modified Porous Boron Nitride Adsorbent. *Colloids Surf. Physicochem. Eng. Asp.* **2023**, *666*, 131372. [CrossRef]
35. Santos, M.P.F.; Porfírio, M.C.P.; Junior, E.C.S.; Bonomo, R.C.F.; Veloso, C.M. Pepsin Immobilization: Influence of Carbon Support Functionalization. *Int. J. Biol. Macromol.* **2022**, *203*, 67–79. [CrossRef] [PubMed]
36. Mohammadi, M.; Garmarudi, A.B.; Khanmohammadi, M.; Rouchi, M.B. Infrared Spectrometric Evaluation of Carbon Nanotube Sulfonation. *Fuller. Nanotub. Carbon Nanostructures* **2016**, *24*, 219–224. [CrossRef]
37. Ghaedi, A.M.; Karamipour, S.; Vafaei, A.; Baneshi, M.M.; Kiarostami, V. Optimization and Modeling of Simultaneous Ultrasound-Assisted Adsorption of Ternary Dyes Using Copper Oxide Nanoparticles Immobilized on Activated Carbon Using Response Surface Methodology and Artificial Neural Network. *Ultrason. Sonochem.* **2019**, *51*, 264–280. [CrossRef] [PubMed]
38. Pereira, R.G.; Veloso, C.M.; da Silva, N.M.; de Sousa, L.F.; Bonomo, R.C.F.; de Souza, A.O.; da Guarda Souza, M.O.; Fontan, R.D.C.I. Preparation of Activated Carbons from Cocoa Shells and Siriguela Seeds Using H3PO4 and ZnCL2 as Activating Agents for BSA and α-Lactalbumin Adsorption. *Fuel Process. Technol.* **2014**, *126*, 476–486. [CrossRef]
39. Zhang, S.; Chen, Q.; Hao, M.; Zhang, Y.; Ren, X.; Cao, F.; Zhang, L.; Sun, Q.; Wennersten, R. Effect of Functional Groups on VOCs Adsorption by Activated Carbon: DFT Study. *Surf. Sci.* **2023**, *736*, 122352. [CrossRef]
40. Sue-aok, N.; Srithanratana, T.; Rangsriwatananon, K.; Hengrasmee, S. Study of Ethylene Adsorption on Zeolite NaY Modified with Group I Metal Ions. *Appl. Surf. Sci.* **2010**, *256*, 3997–4002. [CrossRef]
41. Zhou, K.; Ma, W.; Zeng, Z.; Ma, X.; Xu, X.; Guo, Y.; Li, H.; Li, L. Experimental and DFT Study on the Adsorption of VOCs on Activated Carbon/Metal Oxides Composites. *Chem. Eng. J.* **2019**, *372*, 1122–1133. [CrossRef]
42. Wang, S.-H.; Hwang, Y.-K.; Choi, S.W.; Yuan, X.; Lee, K.B.; Chang, F.-C. Developing Self-Activated Lignosulfonate-Based Porous Carbon Material for Ethylene Adsorption. *J. Taiwan Inst. Chem. Eng.* **2020**, *115*, 315–320. [CrossRef]
43. OAC. *Official Methods of Analysis of the Association of Official Analytical Chemists: Official Methods of Analysis of AOAC International*, 21st ed.; AOAC: Washington, DC, USA, 2019.
44. Rabemanolontsoa, H.; Ayada, S.; Saka, S. Quantitative Method Applicable for Various Biomass Species to Determine Their Chemical Composition. *Biomass Bioenergy* **2011**, *35*, 4630–4635. [CrossRef]
45. Barrett, E.P.; Joyner, L.G.; Halenda, P.P. The Determination of Pore Volume and Area Distributions in Porous Substances. I. Computations from Nitrogen Isotherms. *J. Am. Chem. Soc.* **1951**, *73*, 373–380. [CrossRef]
46. Kuśmierek, K.; Szala, M.; Świątkowski, A. Adsorption of 2,4-Dichlorophenol and 2,4-Dichlorophenoxyacetic Acid from Aqueous Solutions on Carbonaceous Materials Obtained by Combustion Synthesis. *J. Taiwan Inst. Chem. Eng.* **2016**, *63*, 371–378. [CrossRef]

Disclaimer/Publisher's Note: The statements, opinions and data contained in all publications are solely those of the individual author(s) and contributor(s) and not of MDPI and/or the editor(s). MDPI and/or the editor(s) disclaim responsibility for any injury to people or property resulting from any ideas, methods, instructions or products referred to in the content.

Communication

Optical Properties of Graphene Nanoplatelets on Amorphous Germanium Substrates

Grazia Giuseppina Politano

Department of Environmental Engineering, University of Calabria, 87036 Rende, CS, Italy; grazia.politano@unical.it

Abstract: In this work, the integration of graphene nanoplatelets (GNPs) with amorphous germanium (Ge) substrates is explored. The optical properties were characterized using Variable-Angle Spectroscopic Ellipsometry (VASE). The findings of this study reveal a strong interaction between GNPs and amorphous germanium, indicated by a significant optical absorption. This interaction suggests a change in the electronic structure of the GNPs, implying that amorphous germanium could enhance their effectiveness in devices such as optical sensors, photodetectors, and solar cells. Herein, the use of amorphous germanium as a substrate for GNPs, which notably increases their refractive index and extinction coefficient, is introduced for the first time. By exploring this unique material combination, this study provides new insights into the interaction between GNPs and amorphous substrates, paving the way for the develop of high-performance, scalable optoelectronic devices with enhanced efficiency.

Keywords: thin films; graphene; germanium; ellipsometry; optical properties

Citation: Politano, G.G. Optical Properties of Graphene Nanoplatelets on Amorphous Germanium Substrates. *Molecules* **2024**, *29*, 4089. https://doi.org/10.3390/molecules29174089

Academic Editors: Sake Wang, Nguyen Tuan Hung and Minglei Sun

Received: 10 August 2024
Revised: 24 August 2024
Accepted: 28 August 2024
Published: 29 August 2024

Copyright: © 2024 by the author. Licensee MDPI, Basel, Switzerland. This article is an open access article distributed under the terms and conditions of the Creative Commons Attribution (CC BY) license (https://creativecommons.org/licenses/by/4.0/).

1. Introduction

The scientific community has long been captivated by graphene [1,2], a two-dimensional material known for its extraordinary properties [3,4]. Notably, graphene's exceptional transparency, conductivity, and flexibility [5] have opened new avenues in material science, suggesting a vast potential yet to be fully explored [6].

This material is characterized by several outstanding intrinsic qualities, such as extraordinary carrier mobility, impermeability to atoms, comprehensive optical absorption, and notable flexibility [7]. These properties position graphene as a promising candidate to drive the evolution of microelectronics in the coming years [8].

The properties of graphene, such as mobility and optical characteristics, can be significantly influenced by the substrate on which it is placed [9]. Moreover, it can be noted that graphene can be used not only in sensors but also as a filter with nanopores, contributing to environmental improvements. Nanopores can be made, for example, as demonstrated in Ref. [10], where few-layer graphene films were nanostructured with swift heavy ions, tuning their electronic and transport properties for potential filtration applications. Similarly, the effects of swift ion tracks on suspended graphene were visualized in Ref. [11], further highlighting the substrate's role in modifying graphene's properties.

Recent research focuses on achieving compatibility between graphene and existing silicon-based complementary metal-oxide semiconductor (CMOS) technology [12]. Significant advancements have been made in this direction, particularly with the integration of two-dimensional materials like graphene with three-dimensional semiconductor materials [13]. This combination sis crucial as it improves the interaction between these materials and ensures they work well with existing CMOS technology [14]. CMOS technology primarily uses semiconductors like silicon, germanium, and gallium arsenide, all known for their specific electronic properties at the Fermi level [15]. The successful integration of graphene with these semiconductors highlights the importance of selecting compatible materials in advancing technology, particularly in the semiconductor industry [16].

One notable development in this field is the direct growth of graphene on germanium substrates [17]. This innovation is particularly significant for the semiconductor industry, as it combines graphene's unique properties with the enhanced mobility of charge carriers in germanium, offering a superior alternative to silicon [18,19]. Research in this area has demonstrated that the interface structure between graphene and germanium is crucial for optimizing optoelectronic applications [16]. The exploration of hybrid composites, particularly those combining germanium (Ge) with carbon-based materials like graphene, has addressed several challenges associated with Ge-based anodes [20]. Moreover, the combination of graphene and germanium shows a marked enhancement in charge capacity, stability, and rate capability, which are essential for lithium-ion battery anodes [21].

In terms of production, methods such as micromechanical cleavage of graphite [22] and chemical vapor deposition [23] have been used in graphene synthesis. However, these techniques have limitations in scaling up for industrial applications. In contrast, graphene-based materials [24], including reduced graphene oxide [25], few-layer graphene [26], multilayer graphene [27], and graphene nanoplatelets (GNPs) [28], offer viable alternatives. GNPs [29], in particular, retain several advantageous properties of single-layer graphene and are produced through economically viable processes, making them suitable for widespread applications.

GNPs [30] present an optimal balance in terms of excellent physical characteristics, scalability in mass production, and cost-effectiveness. The production of GNPs [31] is feasible through various scalable industrial techniques, including wet-jet milling [32], microwave irradiation [33], and liquid exfoliation [34]. Recent research on GNPs has significantly advanced our understanding of their multifunctional applications in many fields. For instance, Wu et al. [35] demonstrated how surface-etched GNPs can reinforce magnesium alloys, leading to improved strength and ductility. The integration of GNPs in sustainable solar desalination systems, as explored by Khoei et al. [36] and Lim et al. [37], highlights their utility in enhancing interfacial evaporation processes for efficient water purification. GNPs have been used in developing advanced composites and wearable sensors, as shown in studies by Dong et al. [38] and Zhu et al. [39], where they contribute to improved mechanical properties and superior thermal management. Additionally, noncovalent functionalization of GNPs has opened new avenues for their application in energy storage devices, particularly in supercapacitors, as reported by Haridas et al. [40].

Herein, GNPs on magnetron-sputtered amorphous germanium thin films were studied using Variable-Angle Spectroscopic Ellipsometry (VASE) [41], a technique that enables precise measurement of the optical constants of these materials, specifically the refractive index and thickness.

The optical model reveals a significant alteration in the optical properties resulting from the interaction between GNPs and the amorphous Ge substrate. The resulting composite exhibits an improved refractive index and extinction coefficient, suggesting a stronger light-matter interaction.

The results presented here are a starting point for the comprehension of interactions between GNPs and amorphous germanium, which could facilitate advancements in nanotechnology and materials engineering.

2. Results and Discussion

The optical properties of amorphous germanium substrates were accurately determined by applying the Tauc–Lorentz oscillator [42] using the model implemented in the WVASE32 software. The model applied in this study differs significantly from the Bruggeman approach (see, for example, Ref. [43]). Complete details of the fitting parameters for the substrates are included in the Supplementary Materials.

Figure 1a,b present both the simulated and measured results for the ψ and Δ spectra across various incident angles within the wavelength range of 300 to 1000 nm for amorphous germanium films.

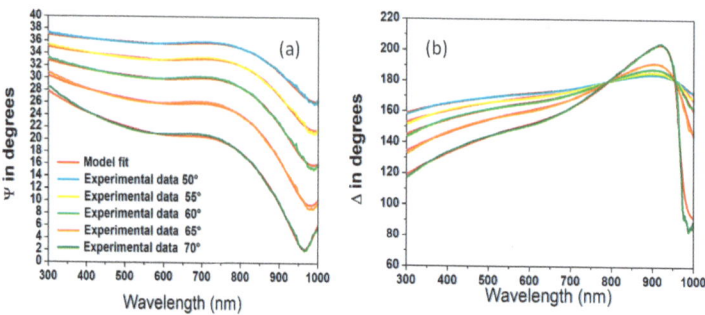

Figure 1. Comparison of experimental and model ψ (**a**) and Δ (**b**) values for germanium substrates at various incidence angles.

Figure 2 presents the estimated dispersion laws of the optical constants for amorphous germanium thin films. The thickness of the films was estimated to be approximately 100 nm.

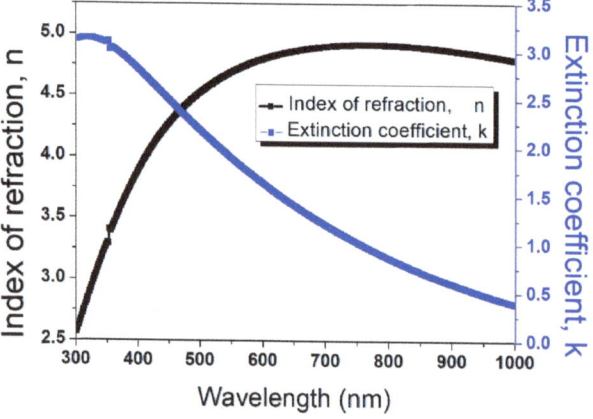

Figure 2. Estimated dispersion laws of germanium substrates. The black curve shows the index of refraction (n), and the blue curve depicts the extinction coefficient (k).

Figure 3a,b display the measured and calculated values of ψ and Δ for GNP thin films on germanium substrates, covering the wavelength range from 300 to 1000 nm.

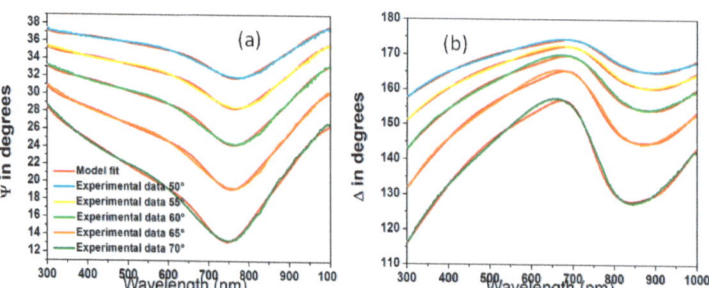

Figure 3. Comparison of experimental and model ψ (**a**) and Δ (**b**) values for GNP films on germanium substrates at various incidence angles.

A comprehensive fit was conducted across the entire wavelength spectrum. This involved the use of a generalized oscillator model [44], incorporating three Gaussian oscillators to represent the imaginary component of the dielectric constant as detailed in the same source. These oscillators are characterized by three fitting parameters: amplitude, energy position, and broadening. The real part of the dielectric constant was derived using the Kramers–Kronig (KK) [45] relations.

The Gaussian oscillators are described by the following formula:

$$\varepsilon_{2,\text{Gauss}} = A\left[\exp\left(-\frac{2\sqrt{\ln 2}(E_{ph}-E_c)}{B}\right)^2 - \exp\left(-\frac{2\sqrt{\ln 2}(E_{ph}+E_c)}{B}\right)^2\right] \quad (1)$$

In this formula, A represents the amplitude in arbitrary units, B denotes the broadening in electronvolts (eV), E_{ph} is the photon energy, and E_c signifies the energy position of the oscillator, also in eV [44].

Figure 4 shows the estimated dispersion laws for GNPs dip-coated on amorphous germanium substrates.

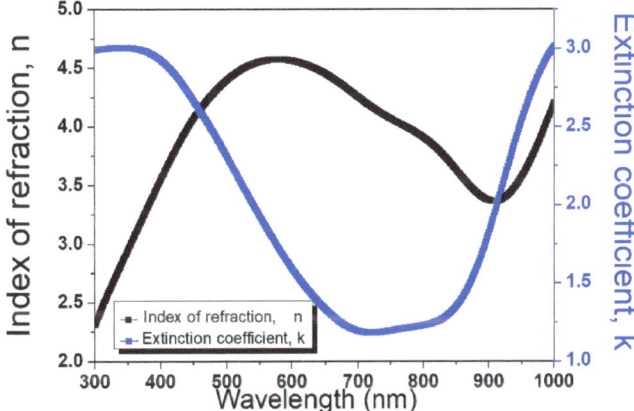

Figure 4. Estimated dispersion laws for GNPs on germanium substrates. The black curve shows the index of refraction (n), and the blue curve depicts the extinction coefficient (k).

The thickness of the GNP thin films was determined to be approximately 55 nm.

The MSE obtained was near 6. Table S2 in the Supplementary Materials lists the parameters derived from the optimal fit for GNPs on silicon (reported in Ref. [29]) and for amorphous germanium (present work).

Figure 5 shows the estimated optical properties of GNPs on silicon as published in the author's previous work [29] for comparative purposes.

As reported in the previous work about GNPs on silicon [29], the oscillator located at 3.7 eV aligns with the surface and interlayer states in graphite's bulk. The oscillator at 2.7 eV is attributed to defect states, and the one around 1.5 eV corresponds with the predicted π* band in graphite [46].

The amplitude values for the oscillators on amorphous germanium are markedly higher than those on silicon, indicating stronger interactions with the germanium substrate. The broadening (B) and energy position (E) parameters also show variations, indicating differences in the electronic band structure influenced by the substrate. There is also a significant variance in film thickness and the high-frequency dielectric constant between the substrates, emphasizing the influence of amorphous material on the GNP films' properties.

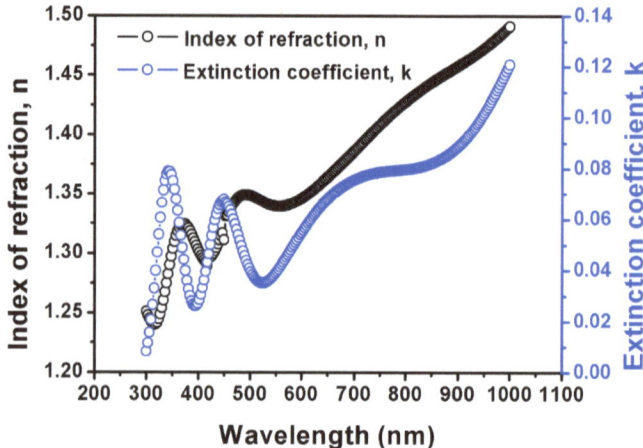

Figure 5. Estimated dispersion laws for GNPs on silicon substrates [29]. The black curve shows the index of refraction (n), and the blue curve depicts the extinction coefficient (k). This content has been reproduced from reference Ref. [29] with permission granted by IOP. Copyright 2019, IOP.

These differences in parameters underscore the significant impact that the substrate material has on the optical properties of graphene nanoplatelets, influencing their potential applications in various optoelectronic devices.

The enhanced refractive index and extinction coefficient of GNPs on germanium (Figure 4) compared to GNPs on silicon (Figure 5) indicate a stronger light–matter interaction, which may be useful in many optoelectronic applications. In particular, the unexpected jump in the extinction coefficient at approximately 350 nm (Figure 4) can be attributed to electronic transitions within the GNPs. This phenomenon occurs when the energy of the incident photons matches the energy difference between electronic states, leading to increased absorption. The interaction between the GNPs and the amorphous germanium substrate might introduce localized states or modify the density of states at the interface, further contributing to this effect.

E. Aktürk et al. [47] provide an in-depth analysis of how germanium atoms interact with graphene, showing that these atoms preferentially bind at the bridge sites of graphene with substantial binding energy. This interaction induces notable changes in graphene's electronic structure, shifting it from semimetallic to metallic and generating a magnetic moment. Such modifications are indicative of a strong interaction between the germanium atoms and the graphene lattice, altering its electronic properties.

Applying these insights to the interaction between GNPs and amorphous germanium, it might be that similar strong binding energies and alterations in electronic structure likely occur at the interface between GNPs and the amorphous germanium substrate. The significant optical absorption observed in GNPs on amorphous germanium can be attributed to these electronic structure modifications, which are critical for enhancing the optoelectronic properties of the composite material.

The coupling between the electronic bands of amorphous germanium and GNPs may also lead to strong absorption effects. This coupling likely results from the interaction of electronic states between germanium and graphene, possibly leading to an increased density of states at the Fermi energy and, consequently, enhanced optical absorption.

Moreover, the non-crystalline nature of amorphous germanium might introduce additional complexity into the interaction. For example, the amorphous nature of the germanium differs from a crystalline structure in its ability to introduce localized energy states or disorders that impact electronic interactions. When GNPs are interfaced with amorphous Ge substrates, the peculiarities of amorphous Ge [48], such as its density variations and optical properties, play a significant role in determining the overall behavior of the composite material. Variations in the germanium substrate's density, porosity, or defect structure may significantly alter the composite material's optical properties.

High optical absorption of GNPs on amorphous germanium substrates makes them particularly suitable for applications such as optical sensors, photodetectors, and solar cells. In these devices, enhanced optical absorption is desirable for improved efficiency and sensitivity.

3. Materials and Methods

The glass substrates underwent a cleaning process using piranha solution, a potent cleaning mixture composed of sulfuric acid (H_2SO_4) and hydrogen peroxide (H_2O_2).

Germanium films with a thickness of 100 nm were deposited onto glass substrates using a DC magnetron sputtering technique [49] (Edwards Auto306 system, West Sussex, UK) at a working pressure of 4.2×10^{-2} mbar, a sputtering power of 40 W, and a sputtering duration of 5 min.

Using X-ray diffraction, no crystalline peaks were observed in the materials, confirming that the Ge layer is amorphous.

The deposition of GNPs onto the prepared germanium films was achieved through a dip-coating process [50]. This procedure was facilitated by a custom-built apparatus, operating at a speed of 0.33 mm/s. The GNPs, with a concentration of 1 mg/mL in water, were purchased from Sigma Aldrich (St. Louis, MO, USA). The composition of the dispersion was 0.1 weight percent graphene and 99.9 weight percent water. The synthesis of the GNPs was achieved through various exfoliation techniques.

In the author's previous work [29], a detailed analysis of the size distribution of the GNPs' major and minor lateral dimensions was provided using scanning transmission electron microscopy (STEM).

The analysis yielded average major and minor lateral sizes of approximately 0.05 µm and 0.02 µm, respectively [29].

Variable-Angle Spectroscopic Ellipsometry (VASE) was used to estimate both the thickness and the optical properties n (refractive index) and k (extinction coefficient) of the GNPs films on germanium samples. Ellipsometric measurements provided precise thickness values and detailed information about the optical properties of the material, enabling a full characterization. The WVASE31 program was used to analyze the ellipsometric data. It employs regression analysis and the Mean Squared Error (MSE) method to fit the model to the experimental data and uses the covariance matrix to provide error bars for the measured values. The ellipsometric parameters, ψ and Δ, were measured using a J.A. M2000 F (Woollam Co., Lincoln, NE, USA) rotating compensator ellipsometer. This measurement spanned a wavelength range of 300–1000 nm, at incident angles varying from 50° to 70° in 5° increments, all conducted at room temperature. The optical model and optimal parameter values for the films were determined using the WVASE32 [44] software from J.A. Woollam, which focuses on minimizing the MSE.

4. Conclusions and Outlook

In this study, GNP films were dip-coated onto magnetron-sputtered amorphous germanium substrates. The optical properties and the thickness of the films were studied using VASE. One of the crucial findings from the ellipsometric data analysis is the observation of a higher refractive index and extinction coefficient of the GNPs on amorphous germanium compared to GNPs on silicon substrates. This indicates a significant change in the optical properties due to the interaction between GNPs and the amorphous germanium substrate. The enhanced refractive index and extinction coefficient are indicative of a stronger light–matter interaction, which may be useful in many optoelectronic applications. Such interaction, which likely involves the merging of electronic states from both materials, might cause a rise in the density of states at the Fermi level, thereby boosting the optical absorption.

These results could be useful for future studies aimed at exploring the full potential of GNP–amorphous germanium composites. The increased absorption and improved light interaction make GNPs on amorphous germanium a suitable candidate for applications requiring high optical sensitivity and efficiency. The potential applications of this graphene–germanium composite are promising for advanced photodetectors, high-efficiency solar cells, and innovative optical sensors.

However, the scalability of the dip-coating process for GNPs and the magnetron sputtering technique for amorphous germanium deposition in industrial applications remains a challenge. Additionally, the long-term stability and durability of the GNP–germanium interface under several operational conditions are aspects that require further investigation.

Supplementary Materials: The following supporting information can be downloaded at: https://www.mdpi.com/article/10.3390/molecules29174089/s1, Figure S1: Scanning transmission electron microscopy image displaying graphene nanoplatelets drop-cast on a gold mesh. Figure S2: Image highlighting the specific marked areas of the graphene nanoplatelets on the gold mesh. Figure S3: Distribution of size (depicted as a histogram and fitted with an exponential function) for the minor (a) and major (b) lateral dimensions of graphene nanoplatelets. Table S1: Tauc-Lorentz oscillators parameters obtained from the best fit of ellipsometric experimental data for germanium/glass substrates. D, A, E_0, C, and E_g are the film thickness, amplitude, peak position, broadening, and optical band gap. Table S2: Parameters of Gaussian oscillators derived from the most accurate fit for graphene nanoplateletes thin films on amorphous germanium and on silicon substrates [24]: amplitude (A), broadening (B), energy position (E), film thickness (d) and high-frequency dielectric constant ε_∞.

Funding: This research received no external funding.

Institutional Review Board Statement: Not applicable.

Informed Consent Statement: Not applicable.

Data Availability Statement: Data are contained within the article and Supplementary Materials.

Conflicts of Interest: The author declares no conflicts of interest.

References

1. Politano, G.G.; Vena, C.; Desiderio, G.; Versace, C. Variable angle spectroscopic ellipsometry characterization of turbostratic CVD-grown bilayer and trilayer graphene. *Opt. Mater.* **2020**, *107*, 110165. [CrossRef]
2. Faggio, G.; Politano, G.G.; Lisi, N.; Capasso, A.; Messina, G. The structure of chemical vapor deposited graphene substrates for graphene-enhanced Raman spectroscopy. *J. Phys. Condens. Matter.* **2024**, *36*, 195303. [CrossRef] [PubMed]
3. Choi, S.H.; Yun, S.J.; Won, Y.S.; Oh, C.S.; Kim, S.M.; Kim, K.K.; Lee, Y.H. Large-scale synthesis of graphene and other 2D materials towards industrialization. *Nat. Commun.* **2022**, *13*, 1484. [CrossRef] [PubMed]
4. Lei, Y.; Zhang, T.; Lin, Y.-C.; Granzier-Nakajima, T.; Bepete, G.; Kowalczyk, D.A.; Lin, Z.; Zhou, D.; Schranghamer, T.F.; Dodda, A.; et al. Graphene and Beyond: Recent Advances in Two-Dimensional Materials Synthesis, Properties, and Devices. *ACS Nanosci. Au* **2022**, *2*, 450–485. [CrossRef] [PubMed]
5. Castriota, M.; Politano, G.G.; Vena, C.; De Santo, M.P.; Desiderio, G.; Davoli, M.; Cazzanelli, E.; Versace, C. Variable Angle Spectroscopic Ellipsometry investigation of CVD-grown monolayer graphene. *Appl. Surf. Sci.* **2019**, *467–468*, 213–220. [CrossRef]
6. Bonaccorso, F.; Sun, Z.; Hasan, T.; Ferrari, A.C. Graphene photonics and optoelectronics. *Nat. Photonics* **2010**, *4*, 611. [CrossRef]
7. Lee, S.-M.; Kim, J.-H.; Ahn, J.-H. Graphene as a flexible electronic material: Mechanical limitations by defect formation and efforts to overcome. *Mater. Today* **2015**, *18*, 336–344. [CrossRef]
8. Ruhl, G.; Wittmann, S.; Koenig, M.; Neumaier, D. The integration of graphene into microelectronic devices. *Beilstein J. Nanotechnol.* **2017**, *8*, 1056–1064. [CrossRef]
9. Tyagi, A.; Mišeikis, V.; Martini, L.; Forti, S.; Mishra, N.; Gebeyehu, Z.M.; Giambra, M.A.; Zribi, J.; Frégnaux, M.; Aureau, D.; et al. Ultra-clean high-mobility graphene on technologically relevant substrates. *Nanoscale* **2022**, *14*, 2167–2176. [CrossRef]
10. Nebogatikova, N.A.; Antonova, I.V.; Erohin, S.V.; Kvashnin, D.G.; Olejniczak, A.; Volodin, V.A.; Skuratov, A.V.; Krasheninnikov, A.V.; Sorokin, P.B.; Chernozatonskii, L.A. Nanostructuring few-layer graphene films with swift heavy ions for electronic application: Tuning of electronic and transport properties. *Nanoscale* **2018**, *10*, 14499–14509. [CrossRef]
11. Nebogatikova, N.A.; Antonova, I.V.; Gutakovskii, A.K.; Smovzh, D.V.; Volodin, V.A.; Sorokin, P.B. Visualization of Swift Ion Tracks in Suspended Local Diamondized Few-Layer Graphene. *Materials* **2023**, *16*, 1391. [CrossRef] [PubMed]
12. Goossens, S.; Navickaite, G.; Monasterio, C.; Gupta, S.; Piqueras, J.J.; Pérez, R.; Burwell, G.; Nikitskiy, I.; Lasanta, T.; Galán, T.; et al. Broadband image sensor array based on graphene–CMOS integration. *Nat. Photonics* **2017**, *11*, 366–371. [CrossRef]

13. Zhao, M.; Zhu, W.; Feng, X.; Yang, S.; Liu, Z.; Tang, S.; Chen, D.; Guo, Q.; Wang, G.; Ding, G. Role of interfacial 2D graphene in high performance 3D graphene/germanium Schottky junction humidity sensors. *J. Mater. Chem. C* **2020**, *8*, 14196–14202. [CrossRef]
14. Liu, C.; Chen, H.; Wang, S.; Liu, Q.; Jiang, Y.-G.; Zhang, D.W.; Liu, M.; Zhou, P. Two-dimensional materials for next-generation computing technologies. *Nat. Nanotechnol.* **2020**, *15*, 545–557. [CrossRef] [PubMed]
15. Radamson, H.H.; Zhu, H.; Wu, Z.; He, X.; Lin, H.; Liu, J.; Xiang, J.; Kong, Z.; Xiong, W.; Li, J.; et al. State of the Art and Future Perspectives in Advanced CMOS Technology. *Nanomaterials* **2020**, *10*, 1555. [CrossRef] [PubMed]
16. Zhao, M.; Xue, Z.; Zhu, W.; Wang, G.; Tang, S.; Liu, Z.; Guo, Q.; Chen, D.; Chu, P.K.; Ding, G.; et al. Interface Engineering-Assisted 3D-Graphene/Germanium Heterojunction for High-Performance Photodetectors. *ACS Appl. Mater. Interfaces* **2020**, *12*, 15606–15614. [CrossRef] [PubMed]
17. Wang, G.; Zhang, M.; Zhu, Y.; Ding, G.; Jiang, D.; Guo, Q.; Liu, S.; Xie, X.; Chu, P.K.; Di, Z.; et al. Direct Growth of Graphene Film on Germanium Substrate. *Sci. Rep.* **2013**, *3*, 2465. [CrossRef]
18. Cavallo, F.; Rojas Delgado, R.; Kelly, M.M.; Sánchez Pérez, J.R.; Schroeder, D.P.; Xing, H.G.; Eriksson, M.A.; Lagally, M.G. Exceptional Charge Transport Properties of Graphene on Germanium. *ACS Nano* **2014**, *8*, 10237–10245. [CrossRef]
19. Kiraly, B.; Jacobberger, R.M.; Mannix, A.J.; Campbell, G.P.; Bedzyk, M.J.; Arnold, M.S.; Hersam, M.C.; Guisinger, N.P. Electronic and Mechanical Properties of Graphene–Germanium Interfaces Grown by Chemical Vapor Deposition. *Nano Lett.* **2015**, *15*, 7414–7420. [CrossRef]
20. Ren, J.-G.; Wu, Q.-H.; Tang, H.; Hong, G.; Zhang, W.; Lee, S.-T. Germanium–graphene composite anode for high-energy lithium batteries with long cycle life. *J. Mater. Chem. A* **2013**, *1*, 1821–1826. [CrossRef]
21. Cheng, J.; Du, J. Facile synthesis of germanium–graphene nanocomposites and their application as anode materials for lithium ion batteries. *CrystEngComm* **2012**, *14*, 397–400. [CrossRef]
22. Sumdani, M.G.; Islam, M.R.; Yahaya, A.N.A.; Safie, S.I. Recent advances of the graphite exfoliation processes and structural modification of graphene: A review. *J. Nanoparticle Res.* **2021**, *23*, 253. [CrossRef]
23. Zhang, Y.; Zhang, L.; Zhou, C. Review of Chemical Vapor Deposition of Graphene and Related Applications. *Acc. Chem. Res.* **2013**, *46*, 2329–2339. [CrossRef]
24. Politano, G.G.; Versace, C. Recent Advances in the Raman Investigation of Structural and Optical Properties of Graphene and Other Two-Dimensional Materials. *Crystals* **2023**, *13*, 1357. [CrossRef]
25. Moon, I.K.; Lee, J.; Ruoff, R.S.; Lee, H. Reduced graphene oxide by chemical graphitization. *Nat. Commun.* **2010**, *1*, 73. [CrossRef] [PubMed]
26. Pirzado, A.A.; Le Normand, F.; Romero, T.; Paszkiewicz, S.; Papaefthimiou, V.; Ihiawakrim, D.; Janowska, I. Few-Layer Graphene from Mechanical Exfoliation of Graphite-Based Materials: Structure-Dependent Characteristics. *ChemEngineering* **2019**, *3*, 37. [CrossRef]
27. Shahil, K.M.F.; Balandin, A.A. Thermal properties of graphene and multilayer graphene: Applications in thermal interface materials. *Solid State Commun.* **2012**, *152*, 1331–1340. [CrossRef]
28. Chen, G.; Yang, M.; Xu, L.; Zhang, Y.; Wang, Y. Graphene Nanoplatelets Impact on Concrete in Improving Freeze-Thaw Resistance. *Appl. Sci.* **2019**, *9*, 3582. [CrossRef]
29. Politano, G.G.; Nucera, A.; Castriota, M.; Desiderio, G.; Vena, C.; Versace, C. Spectroscopic and morphological study of graphene nanoplatelets thin films on Si/SiO$_2$ substrates. *Mater. Res. Express* **2019**, *6*, 106432. [CrossRef]
30. Cataldi, P.; Athanassiou, A.; Bayer, S.I. Graphene Nanoplatelets-Based Advanced Materials and Recent Progress in Sustainable Applications. *Appl. Sci.* **2018**, *8*, 1438. [CrossRef]
31. Moosa, A.; Ramazani, S.A.A.; Ibrahim, M. Mechanical and Electrical Properties of Graphene Nanoplates and Carbon-Nanotubes Hybrid Epoxy Nanocomposites. *Am. J. Mater. Sci.* **2016**, *6*, 157–165.
32. Del Rio Castillo, A.; Pellegrini, V.; Ansaldo, A.; Ricciardella, F.; Sun, H.; Marasco, L.; Buha, J.; Dang, Z.; Gagliani, L.; Lago, E.; et al. High-yield production of 2D crystals by wet-jet milling. *Mater. Horizons* **2018**, *5*, 890–904. [CrossRef]
33. Kumar, D.; Singh, K.; Verma, V.; Bhatti, H.S. Microwave assisted synthesis and characterization of graphene nanoplatelets. *Appl. Nanosci.* **2016**, *6*, 97–103. [CrossRef]
34. Sellathurai, A.J.; Mypati, S.; Kontopoulou, M.; Barz, D.P.J. High yields of graphene nanoplatelets by liquid phase exfoliation using graphene oxide as a stabilizer. *Chem. Eng. J.* **2023**, *451*, 138365. [CrossRef]
35. Wu, X.; Du, X.; Wang, Z.; Li, S.; Liu, K.; Du, W. Surface etched graphene nanoplatelets and their heterogeneous interface to reinforce magnesium alloys for high strength and ductility. *Mater. Sci. Eng. A* **2024**, *913*, 147080. [CrossRef]
36. Khoei, J.K.; Bafqi, M.S.S.; Dericiler, K.; Doustdar, O.; Okan, B.S.; Koşar, A.; Sadaghiani, A. Upcycled graphene nanoplatelets integrated fiber-based Janus membranes for enhanced solar-driven interfacial steam generation. *RSC Appl. Interfaces* **2024**. online ahead of print. [CrossRef]
37. Lim, H.W.; Seung Lee, H.; Joon Lee, S. Laminated chitosan/graphene nanoplatelets aerogel for 3D interfacial solar desalination with harnessing wind energy. *Chem. Eng. J.* **2024**, *480*, 148197. [CrossRef]
38. Dong, P.; Yang, M.; Ma, J.; Zheng, S.; Li, W.; Pi, W. A novel prediction method for nanoplatelets content dependent yield strength of graphene nanoplatelets reinforced metal matrix composites at different temperatures. *Compos. Part A Appl. Sci. Manuf.* **2024**, *179*, 108038. [CrossRef]

39. Zhu, Z.; Tian, Z.; Liu, Y.; Yue, S.; Li, Y.; Wang, Z.L.; Yu, Z.-Z.; Yang, D. Human Nervous System Inspired Modified Graphene Nanoplatelets/Cellulose Nanofibers-Based Wearable Sensors with Superior Thermal Management and Electromagnetic Interference Shielding. *Adv. Funct. Mater.* **2024**, *34*, 2315851. [CrossRef]
40. Haridas, H.; Kader, A.K.A.; Sellathurai, A.; Barz, D.P.J.; Kontopoulou, M. Noncovalent Functionalization of Graphene Nanoplatelets and Their Applications in Supercapacitors. *ACS Appl. Mater. Interfaces* **2024**, *16*, 16630–16640. [CrossRef]
41. Riegler, H. A user's guide to ellipsometry. By Harland G. Tompkins, Academic Press, New York 1993, 260 pp. hardback, ISBN 0-12-603050-0. *Adv. Mater.* **1993**, *5*, 778. [CrossRef]
42. Shahrokhabadi, H.; Bananej, A.; Vaezzadeh, M. Investigation of Cody–Lorentz and Tauc–Lorentz Models in Characterizing Dielectric Function of $(HfO_2)x(ZrO_2)1-x$ Mixed Thin Film. *J. Appl. Spectrosc.* **2017**, *84*, 915–922. [CrossRef]
43. Marin, D.V.; Gorokhov, E.B.; Borisov, A.G.; Volodin, V.A. Ellipsometry of GeO_2 films with Ge nanoclusters: Influence of the quantum-size effect on refractive index. *Opt. Spectrosc.* **2009**, *106*, 436–440. [CrossRef]
44. J.A. Woollam, Co. *WVASE Manual "Guide to Using WVASE32"*; J.A. Woollam Co.: Lincoln, NE, USA, 2010.
45. Kubo, R.; Ichimura, M. Kramers-Kronig Relations and Sum Rules. *J. Math. Phys.* **1972**, *13*, 1454–1461. [CrossRef]
46. Fuchs, H.T.E. Unoccupied Electronic States of a Graphite Surface as Observed by Local Tunnelling Spectroscopy. *Europhys. Lett.* **1987**, *3*, 745. [CrossRef]
47. Aktürk, E.; Ataca, C.; Ciraci, S. Effects of silicon and germanium adsorbed on graphene. *Appl. Phys. Lett.* **2010**, *96*, 123112. [CrossRef]
48. Clark, A.H. Electrical and Optical Properties of Amorphous Germanium. *Phys. Rev.* **1967**, *154*, 750–757. [CrossRef]
49. Kelly, P.J.; Arnell, R.D. Magnetron sputtering: A review of recent developments and applications. *Vacuum* **2000**, *56*, 159–172. [CrossRef]
50. Scriven, L.E. Physics and Applications of DIP Coating and Spin Coating. *MRS Proc.* **1988**, *121*, 717. [CrossRef]

Disclaimer/Publisher's Note: The statements, opinions and data contained in all publications are solely those of the individual author(s) and contributor(s) and not of MDPI and/or the editor(s). MDPI and/or the editor(s) disclaim responsibility for any injury to people or property resulting from any ideas, methods, instructions or products referred to in the content.

Article

Binding Energies and Optical Properties of Power-Exponential and Modified Gaussian Quantum Dots

Ruba Mohammad Alauwaji [1,2], Hassen Dakhlaoui [2,3], Eman Algraphy [2,3], Fatih Ungan [4,5] and Bryan M. Wong [6,*]

1. Physics Department, College of Science, Qassim University, Qassim 51452, Saudi Arabia
2. Physics Department, College of Science of Dammam, Imam Abdulrahman Bin Faisal University, Dammam 34212, Saudi Arabia
3. Nanomaterials Technology Unit, Basic and Applied Scientific Research Center (BASRC), Physics Department, College of Science of Dammam, Imam Abdulrahman Bin Faisal University, Dammam 34212, Saudi Arabia
4. Department of Physics, Faculty of Science, Sivas Cumhuriyet University, Sivas 58140, Turkey
5. Nanophotonics Research and Application Center, Sivas 58070, Turkey
6. Materials Science & Engineering Program, Department of Chemistry, and Department of Physics & Astronomy, University of California-Riverside, Riverside, CA 92521, USA
* Correspondence: bryan.wong@ucr.edu

Abstract: We examine the optical and electronic properties of a GaAs spherical quantum dot with a hydrogenic impurity in its center. We study two different confining potentials: (1) a modified Gaussian potential and (2) a power-exponential potential. Using the finite difference method, we solve the radial Schrodinger equation for the 1s and 1p energy levels and their probability densities and subsequently compute the optical absorption coefficient (OAC) for each confining potential using Fermi's golden rule. We discuss the role of different physical quantities influencing the behavior of the OAC, such as the structural parameters of each potential, the dipole matrix elements, and their energy separation. Our results show that modification of the structural physical parameters of each potential can enable new optoelectronic devices that can leverage inter-sub-band optical transitions.

Keywords: optical absorption coefficient; binding energy; GaAs quantum dot; Schrödinger equation; hydrogenic impurity

Citation: Alauwaji, R.M.; Dakhlaoui, H.; Algraphy, E.; Ungan, F.; Wong, B.M. Binding Energies and Optical Properties of Power-Exponential and Modified Gaussian Quantum Dots. Molecules 2024, 29, 3052. https://doi.org/10.3390/molecules29133052

Academic Editors: Sake Wang, Minglei Sun and Nguyen Tuan Hung

Received: 22 May 2024
Revised: 22 June 2024
Accepted: 22 June 2024
Published: 27 June 2024

Copyright: © 2024 by the authors. Licensee MDPI, Basel, Switzerland. This article is an open access article distributed under the terms and conditions of the Creative Commons Attribution (CC BY) license (https://creativecommons.org/licenses/by/4.0/).

1. Introduction

Quantum structures such as quantum wells, quantum dots (QDs), and nanowires are low-dimensional semiconductors that have enabled several technologies, such as single-electron transistors [1], photovoltaic (PV) devices [2], light-emitting diodes (LEDs) [3], and photodetectors [4–8]. QDs are particularly useful in optoelectronic applications due to quantum confinement effects that enable efficient luminescence, large extinction coefficients, and extensive lifetimes [9–11]. For this reason, QDs are presently employed in various applications, including LEDs, photovoltaics, biomedical imaging, solid-state lighting, QD displays, biosensors, and quantum computing materials [12–19]. QDs can be considered a middle ground between molecules and semiconductor materials that enable quantum mechanical properties that can be tailored by varying their physical features [20–26]. For example, inserting a hydrogenic impurity at the center of a QD center affects the electronic distribution of all energy levels, their separations, and the electronic wavefunctions. This, in turn, affects the electrostatic attraction between the hydrogenic impurity and free carriers, the dipole matrix elements, and the optical absorption coefficient (OAC). There have been several studies that have examined the effects of inserting an impurity in the center of a QD [25,27–33]. The OACs in coupled InAs/GaAs QD systems were studied by Li and Xia, who found that the optical properties in these QD systems were different from QD superlattices [34]. Schrey and coauthors studied the polarization and optical absorption properties in QD-based photodetectors and found that the QD enables large effects on the distribution of minibands in the superlattice [35]. The variation of the OAC and nonlinear refractive index (NRI) as a function of the applied electric field, temperature, and hydrostatic pressure in a Mathieu-like QD potential with a hydrogenic impurity was examined by Bahar et al. [36]. Batra and coauthors also evaluated the effect of a Kratzer-like radial potential on the OAC and NRI of a spherical QD [37]. Bassani and Buczko studied the sensitivity of the optical properties to the impurity of donors and acceptors in spherical QDs [38]. Narvaez and coauthors examined OACs arising from conduction-to-conduction and valence-to-valence bands [39]. A. Ed-Dahmouny et al. studied the effects of electric and magnetic fields on donor impurity electronic states and OACs in a core/shell GaAs/AlGaAs ellipsoidal QD [40]. In their study, they showed that changes in the polarization of light caused blue or red shifts in the inter-sub-band OAC spectra, depending on the orientations of the two external fields and the presence/absence of a hydrogenic impurity. Fakkahi et al. examined the OACs of spherical QDs based on a Kratzer-like confinement potential [41]. In their study, they demonstrated that the OACs and transition energies ($1p$ - $2s$ and $2s$ - $2p$) were strongly influenced by the structural parameters of the Kratzer confinement potential.

In addition, the oscillator strengths in spherical QDs with a hydrogenic impurity were computed by Yilmaz and Safak [42]. Finally, Kirak et al. studied the effect of an applied electric field on OACs in parabolic QDs with a hydrogenic impurity [43]. In recent years, GaAs-based spherical quantum dots have emerged as a subject of intense research due to their unique properties. GaAs has a high electron mobility, good thermal stability, and excellent optical properties. Moreover, GaAs is widely used in thin film production and high-quality epitaxial growth methods. These factors collectively render GaAs quantum dots appealing for advancing high-performance semiconductor devices and facilitating nanoscale optoelectronic applications.

In this work, we compute the two lowest energies, E_{1p} and E_{1s}, in GaAs spherical quantum dots as a function of the structural shape of two confining potentials: (1) a modified Gaussian potential (MGP) and (2) a power-exponential potential (PEP). We then present a complete analysis of OACs and binding energies as a function of energy separation and dipole matrix elements, as the structural parameters of these potentials are varied. The binding energy effectively captures the attractive force between the free electrons in different levels and the inserted impurity. Section 2 provides the mathematical details of our approach, and Section 3 presents our results for each potential.

2. Theoretical Details
2.1. Geometrical Forms of MGP and PEP Potentials

Before calculating the different energy levels and electronic wavefunctions in the QD, we first evaluate the effects of the structural parameters on the geometrical shape of the confining potentials. The spherical symmetry of these potentials introduces a quantization of the angular motion via the angular and magnetic numbers. Within this quantization, the total carrier wavefunction can be expressed by the well-known spherical harmonics. The adjustment and control of electronic transitions in QDs can be attained by varying the size of each layer in the structure or by changing the structural parameters governing the shape of the potentials.

In the present paper, we examine two confining potentials: (1) the power-exponential potential, $V_{\text{PEP}}(r)$, and (2) the modified Gaussian potential, $V_{\text{MGP}}(r)$. These potentials are generated by the application of an external voltage and barriers/wells of the structure with analytical expressions given by the following [44–48]:

$$V_{\text{PEP}}(r) = -V_c \exp\left[-\left(\frac{r}{R_0}\right)^q\right], \qquad (1)$$

$$V_{\text{MGP}}(r) = -V_c \operatorname{sech}\left[\left(\frac{r}{R_0}\right)^q\right], \qquad (2)$$

where V_c and R_0 are the depth and range, respectively, of these potentials, and q is a structural parameter. Figures 1 and 2 plot the two potentials as a function of the radius for a GaAs QD with different values of the structural parameter, q. Figure 1 shows that $V_{\text{PEP}}(r)$ has a global minimum of $-V_c$ at $r = 0$ and increases for higher values of r. For low values of q, the potential has a parabolic shape but gives a square-like confining potential for larger values of q. By increasing q, the potential widens but has the same value at $r = R_0$ regardless of the value of q. These geometrical changes enable us to understand their effect on the desired energy levels and optimize the transitions between the initial and final levels to obtain the desired absorption.

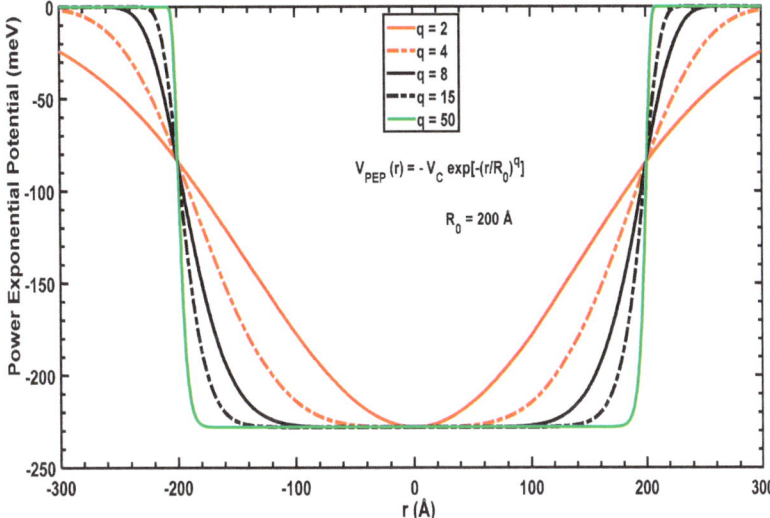

Figure 1. $V_{\text{PEP}}(r)$ for different values of the parameter q with $R_0 = 200$ Å.

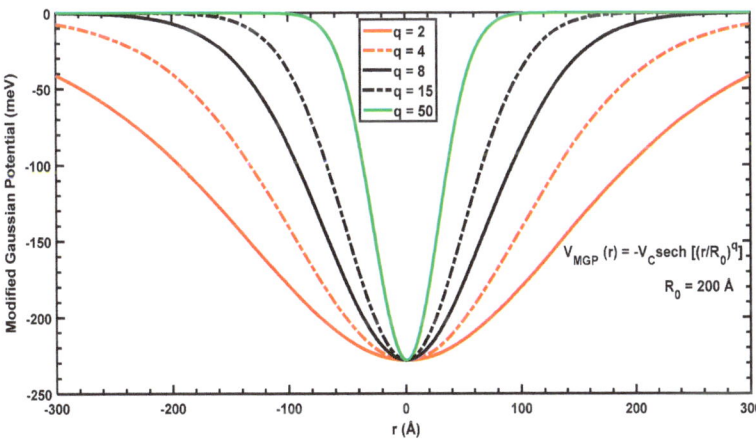

Figure 2. $V_{\text{MGP}}(r)$ for different values of the parameter q with $R_0 = 200$ Å.

Figure 2 plots the modified Gaussian potential as a function of the radius, r. To allow for a straightforward comparison, the same radius of $R_0 = 200$ Å is used in Figure 2 (which was considered in Figure 1). When $q = 2$, the power-exponential and modified Gaussian potentials resemble each other; however, when q is increased, the shape of the potential tends to a negative Dirac-delta function at $r = 0$, which will dramatically affect the confinement of wavefunctions and energy levels of the ground and first-excited states.

2.2. Optical Absorption of the MGP and PEP Potentials

To compute the E_{1s} and E_{1p} energy levels and the $R_{1s}(r)$ and $R_{1p}(r)$ wavefunctions, the radial part of the Schrödinger equation is solved with each of the confining potentials within the effective mass approximation. The Schrödinger equation with the hydrogenic impurity is given by [49,50]:

$$\left[-\frac{\hbar^2}{2} \vec{\nabla}_r \left(\frac{1}{m^*(r)} \vec{\nabla}_r \right) + \frac{\ell(\ell+1)\hbar^2}{2m^*(r)\, r^2} - \frac{Z\, e^2}{\varepsilon\, r} + V_{\text{conf}}(r) \right] R_{n\ell}(r) = E_{n\ell}\, R_{n\ell}(r), \qquad (3)$$

where $\hbar$, ε, and ℓ are the reduced Planck constant, dielectric constant, and angular quantum number, respectively. $V_{\text{conf}}(r)$ is the confining potential, $V_{\text{MGP}}(r)$ or $V_{\text{PEP}}(r)$. In addition, $R_{n\ell}(r)$ and $E_{n\ell}$ denote the radial wavefunction and energy level of the confined electron. Including/neglecting the hydrogenic impurity is controlled by setting $Z = 1$ or $Z = 0$, respectively. To find the values of $E_{n\ell}$ and $R_{n\ell}(r)$, the Schrödinger equation is discretized and transformed to an eigenvalue problem, $Hx = \lambda x$, where H is a tridiagonal matrix, and λ and x represent $E_{n\ell}$ and $R_{n\ell}(r)$, respectively. After discretization, the Schrödinger equation can be written as follows:

$$R_{n\ell}(j+1)\left[-\frac{\hbar^2}{2m^* r_j (\Delta r)} - \frac{\hbar^2}{2m^* (\Delta r)^2} \right] + R_{n\ell}(j)\left[\frac{\hbar^2}{m^* (\Delta r)^2} + \frac{\ell(\ell+1)}{m^* (r_j \cdot \Delta r)^2} + V_{\text{conf}}(j) \right]$$
$$+ R_{n\ell}(j-1)\left[\frac{\hbar^2}{2m^* r_j (\Delta r)} - \frac{\hbar^2}{2m^* (\Delta r)^2} \right] = E_{n\ell} R_{n\ell}(j), \qquad (4)$$

where the elements of H are:

$$H_{ij} = \begin{cases} \frac{\hbar^2}{m^*(\Delta r)^2} + \frac{\ell(\ell+1)}{m^*(r_j \cdot \Delta r)^2} + V_{KP}(j), & \text{if } j = i \\ \frac{\hbar^2}{2m^* r_j (\Delta r)} - \frac{\hbar^2}{2m^*(\Delta r)^2}, & \text{if } j = i-1 \\ -\frac{\hbar^2}{2m^* r(\Delta r)} - \frac{\hbar^2}{2m^*(\Delta r)^2}, & \text{if } j = i+1 \\ 0, & \text{otherwise} \end{cases} \qquad (5)$$

After discretization, the radial coordinate is $r_j = j\Delta r$ with $j = 1, \ldots, N$, where $\Delta r = \frac{R}{N}$ is the width of the radial mesh. As boundary conditions, the ground and first-excited wavefunctions vanish at the external boundary point ($j = N+1$) due to the negligible probability of finding the electron at the edge of the confining potential at $r = R$. In our simulation, we diagonalized the $N \times N$ matrix with $N = 1200$ using the MATLAB (version 9.8) software package.

The OACs of different potentials arise from an electronic transition from the $1s$ to the $1p$ states after the absorption of a photon having an energy of $\hbar\omega = E_f - E_i$. We denote OAC as $\alpha(\hbar\omega)$ and compute it using Fermi's golden rule with the following expression [51]:

$$\alpha(\hbar\omega) = \frac{16\pi^2 \gamma_{\text{FS}} N_{if}}{n_r V_{\text{con}}} \hbar\omega \left| M_{if} \right|^2 \delta\!\left(E_f - E_i - \hbar\omega \right). \qquad (6)$$

The parameters γ_{FS}, V_{con}, n_r, and N_{if} are the fine-structure constant, confinement volume, and refractive index, respectively. The Dirac δ-function in Equation (6) can be replaced with the following Lorentzian function [51]:

$$\delta\!\left(E_f - E_i - \hbar\omega \right) = \frac{\hbar\Gamma}{\pi\left[\left(E_f - E_i - \hbar\omega \right)^2 + (\hbar\Gamma)^2 \right]}. \qquad (7)$$

In our study, the initial $(i = 1)$ and final $(f = 2)$ states are the $1s$ and $1p$ states, respectively. The physical parameters used in this study are: $\gamma_{FS} = 1/137$, $n_r = 3.25$, $\hbar\Gamma = 3$ meV, $m^* = 0.067\, m_0$, and $V_C = 0.228$ eV. Furthermore, we use atomic units $(\hbar = e = m_0 = 1)$ throughout this work, which corresponds to a Rydberg energy and Bohr radius of $1\, R_y \cong 5.6$ meV and $1\, a_B \cong 100$ Å, respectively. In addition, the electromagnetic radiation is polarized along the z-axis, and $|M_{12}|^2$ is given by the following expression [51]:

$$|M_{12}|^2 = \frac{1}{3}\left|\int_0^\infty R_{1s}(r) r^3 R_{1p}(r) dr\right|^2, \tag{8}$$

where the $\frac{1}{3}$ pre-factor arises from the integration of the spherical harmonics.

3. Results and Discussion

3.1. Optical Properties of GaAs Quantum Dot with PEP Potential

In this section, we will discuss the effect of the structural parameter, q, on the E_{1s} and E_{1p} energy levels and the binding energy. We then analyze trends in $|M_{if}|^2$ and the OACs for the transition between these states. Figure 3 plots the energy levels of the ground $(1s)$ and first-excited $(1p)$ states as a function of the structural parameter q with and without the hydrogenic impurity. When q increases, the energy levels decrease rapidly at low values of q and tend toward constant values, which is due to the shape of the confining potential shown in Figure 1 (the energy levels are inversely proportional to the width of the well). Furthermore, in the presence of the hydrogenic impurity $(Z = 1)$, the energy levels are reduced compared to those in the absence of the impurity $(Z = 0)$ due to the strong attraction between the electrons and the impurity at the center of the QD. In addition, we observe a slow decrease in all energy levels for larger values of q, since the width of the potential (see Figure 1) becomes insensitive to the variation of large q values.

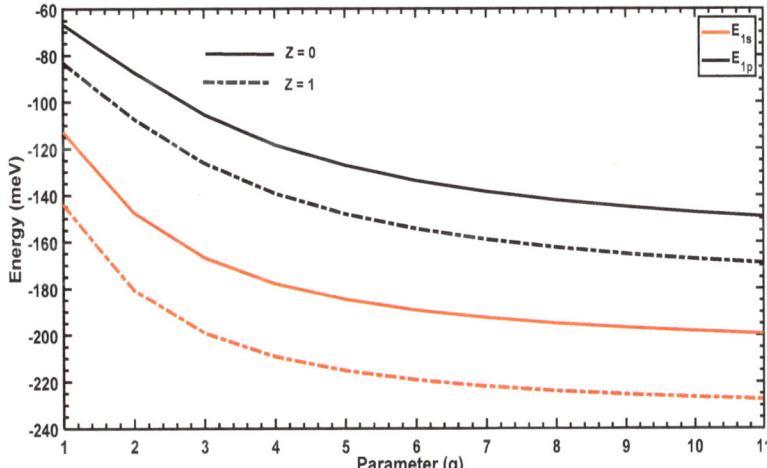

Figure 3. Variations in E_{1s} and E_{1p} as a function of the parameter q. The solid lines are energies without the hydrogenic impurity $(Z = 0)$, and the dashed lines represent energies with the hydrogenic impurity $(Z = 1)$.

The OAC between the ground and first-excited levels depends on the energy separation $\Delta E = E_{1p} - E_{1s}$ and the dipole matrix element, $|M_{12}|^2$. Figure 4 plots these physical quantities as a function of the structural parameter, q. For $Z = 0$, ΔE increases, reaches its maximum at $q = 3$, and subsequently decreases. This arises because E_{1p} and E_{1s} decrease when $q < 3$; however, the decrease in E_{1s} is faster than that of E_{1p}. As such, ΔE shows an increasing variation; however, the opposite trend occurs for $q > 3$, leading to a reduction in ΔE. Consequently, the OAC can undergo a red or blue shift as q increases.

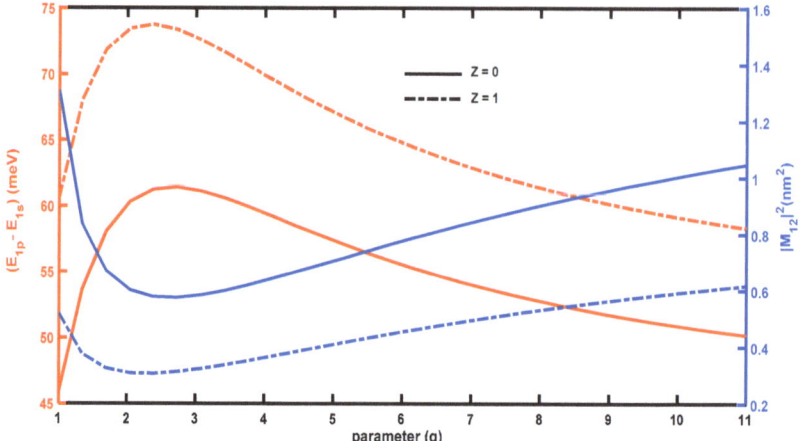

Figure 4. Variations in $E_{1p} - E_{1s}$ and $|M_{12}|^2$ as a function of the parameter q.

Figure 4 shows the variation of the dipole matrix element, $|M_{12}|^2$, which plays a crucial role in controlling the amplitude of the optical absorption. $|M_{12}|^2$ decreases for $q < 3$ and increases for $q > 3$, which is the opposite trend to that of ΔE. For low values of q, the overlap between the ground and first-excited wavefunctions is reduced; however, the overlap increases for larger values of q, resulting in an enhancement of $|M_{12}|^2$.

Figure 5 displays the variation of the OAC as a function of the incident photon energy for three values of the parameter q. The OAC peak moves to the left (redshifts) when q is increased, which arises from the variation of the energy separation shown in Figure 4. Furthermore, the amplitude diminishes for $q = 6$ and subsequently rises again when $q = 11$. The amplitude and position of the OAC is sensitive to q, which affects the geometrical shape of the confining potential and delocalization of the $1s$ and $1p$ wavefunctions.

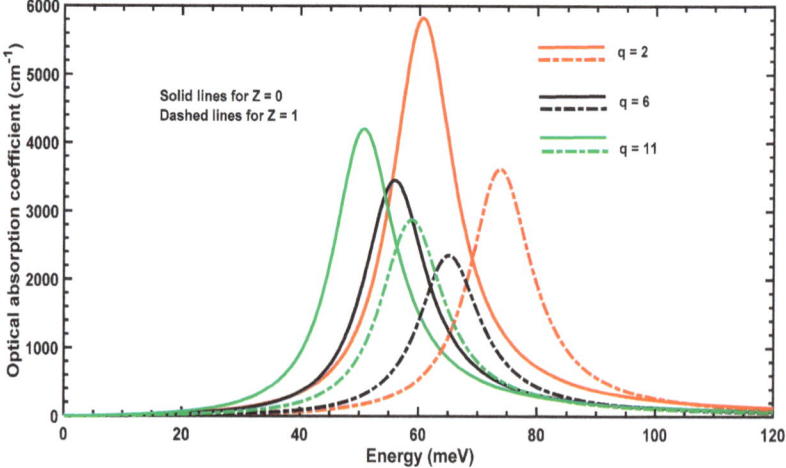

Figure 5. OAC as a function of incident photon energy for different values of the parameter q with ($Z = 1$) and without ($Z = 0$) the impurity.

Figure 6 shows the variation of the binding energy (Eb) as a function of the parameter q. For low values of q, the binding energy increases sharply for both the $1p$ and $1s$ states and subsequently decreases. For all values of q, the binding energy of the $1s$ state is larger than the $1p$ state, which is due to the strong electrostatic attraction between the impurity and the electron in the $1s$ state compared to the $1p$ state. Furthermore, increasing q enlarges the confining potential, as shown in Figure 1, which leads to a reduction in all energy levels of the QD with and without the impurity. Therefore, the binding energy will be influenced by two effects: (1) the electrostatic attraction and (2) the geometrical confinement imposed by the confining potential. For higher values of q, the confining potential becomes too large and dominates the effect of the electrostatic attraction, leading to a reduction in the binding energies, as shown in Figure 6.

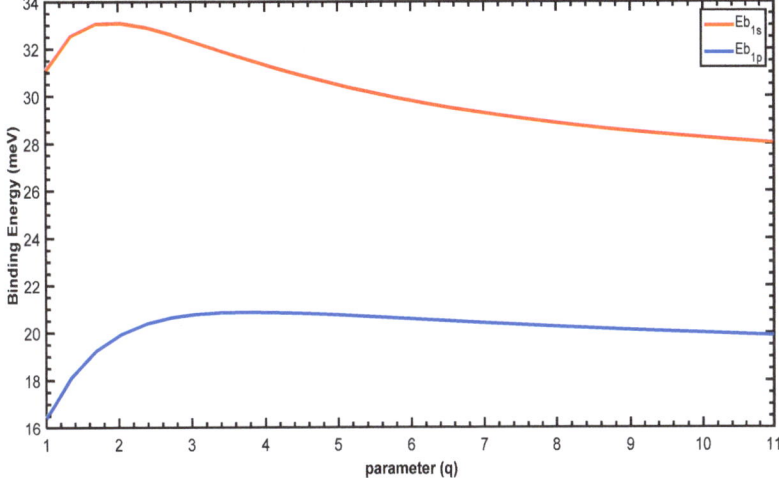

Figure 6. Variation of the binding energy as a function of the parameter q for the $1s$ and $1p$ states.

3.2. Optical Properties of a GaAs Quantum Dot with an MGP Potential

In this section, we examine the effect of the structural parameter q on the E_{1s} and E_{1p} energy levels, their energy separation, and the binding energy. We then discuss the behavior of the dipole matrix elements and the OACs between these states.

Figure 7 plots the energy levels of the ground ($1s$) and first-excited ($1p$) states as a function of the structural parameter q with and without the hydrogenic impurity. When q increases, these energies increase considerably in the presence and absence of the impurity, which is opposite to that observed in the previous section for the PEP confining potential. Increasing q reduces the width of the MGP; for higher values of q, the potential tends to the shape of a negative Dirac-delta potential (Figure 2), which increases the energy levels. In addition, the slope of each energy level is slowly reduced for higher values of q since the confining potential no longer changes for very large values of q.

Comparing Figures 3 and 7, the evolution of the energy levels as a function of the structural parameter are opposite for the PEP and MGP potential. The PEP potential tends to a square-like quantum well, leading to a reduction in energy levels; however, the MGP potential tends to a Dirac-delta form, which shifts all of the energy levels to higher values. Figure 8 plots $|M_{12}|^2$ and $\Delta E = E_{1p} - E_{1s}$ as a function of q, which shows that ΔE increases with q, reaches a maximum, and then diminishes. The maximum of ΔE in the presence of the hydrogenic impurity ($Z = 1$) is slightly different from that in its absence ($Z = 0$), which causes the blue and red shifts observed in the OAC. Furthermore, the amplitude of the OAC is sensitive to the variation of the dipole matrix element $|M_{12}|^2$. Figure 8 shows that $|M_{12}|^2$ first decreases with q, reaches a minimum, and finally increases, which is an opposite trend to that of the energy separation, $\Delta E = E_{1p} - E_{1s}$.

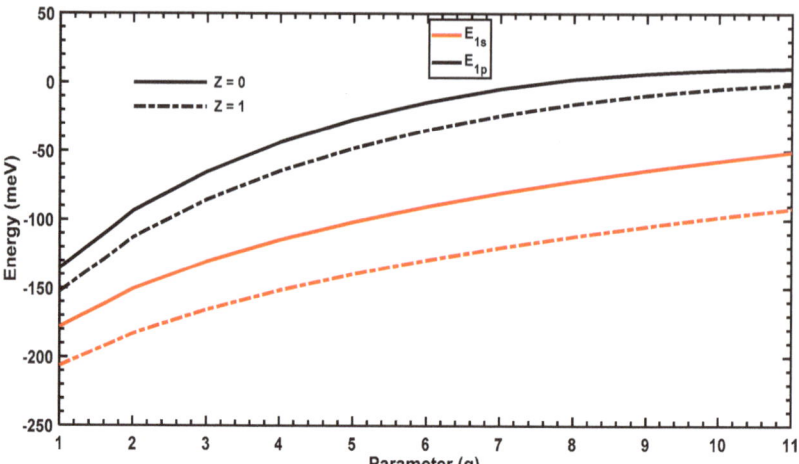

Figure 7. Variation of E_{1s} and E_{1p} as a function of the parameter q. The solid lines are energies without the hydrogenic impurity ($Z = 0$), and the dashed lines represent energies with the hydrogenic impurity ($Z = 1$).

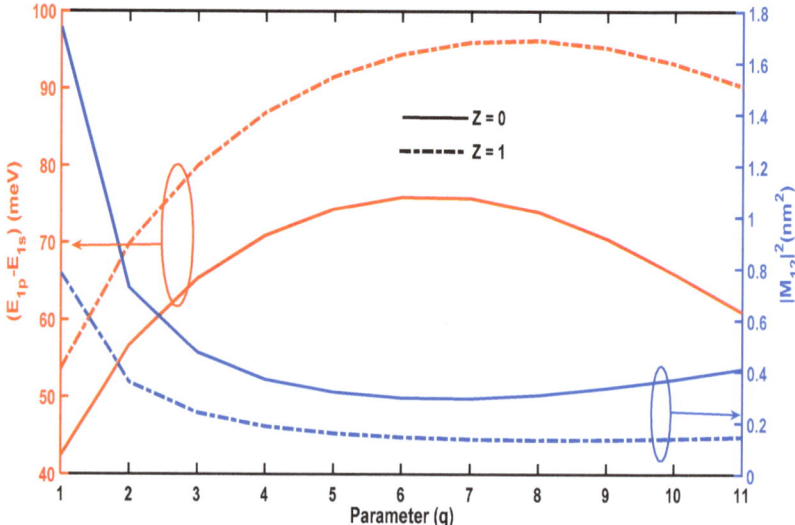

Figure 8. Variations in $E_{1p} - E_{1s}$ and $|M_{12}|^2$ as a function of the parameter q.

Figure 9 displays the variation of the OAC as a function of incident photon energy for three values of the parameter q. The OAC peak moves to the right (blue shifts) when q is increased from 2 to 6; it subsequently moves to the left (redshifts) when q increases from 6 to 11. This arises from the variation of the energy separation shown in Figure 8. Furthermore, the amplitude decreases when q varies between 2 and 6 and rises again when $q = 11$.

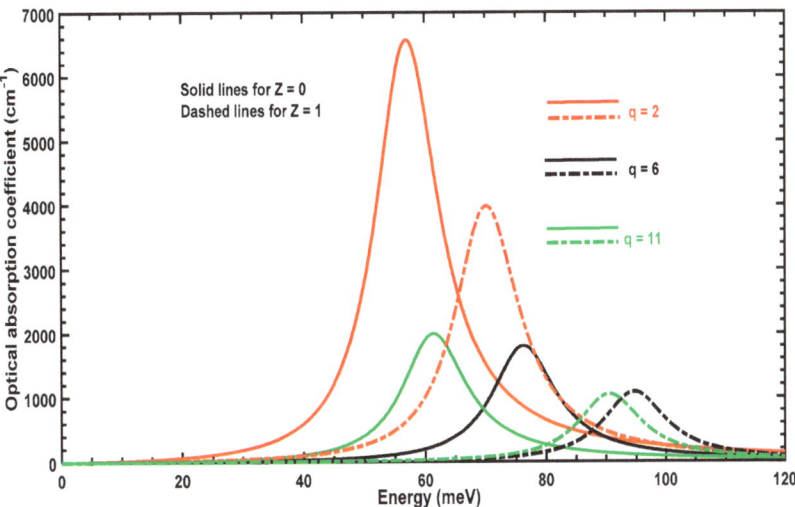

Figure 9. OAC as a function of incident photon energy for different values of parameter q with ($Z = 1$) and without ($Z = 0$) the impurity.

Finally, we plot the binding energy in Figure 10. For low values of q, the binding energy increases gradually for the $1p$ and $1s$ states. For $q > 5$, the binding energy of the $1p$ state starts to decrease, whereas the $1s$ binding energy continues its increase. For all values of the parameter q, the binding energy of the $1s$ state is larger than that of $1p$, which arises from the attraction between the hydrogenic impurity and the free electrons. Furthermore, increasing q subsequently reduces the confining potential (cf. Figure 2), which leads to the enhancement of all energy levels of the QD with and without the presence of the impurity. The difference in the variation of the binding energies in Figures 6 and 10 confirms the effect of the structural parameter q on the PEP and MGP potentials.

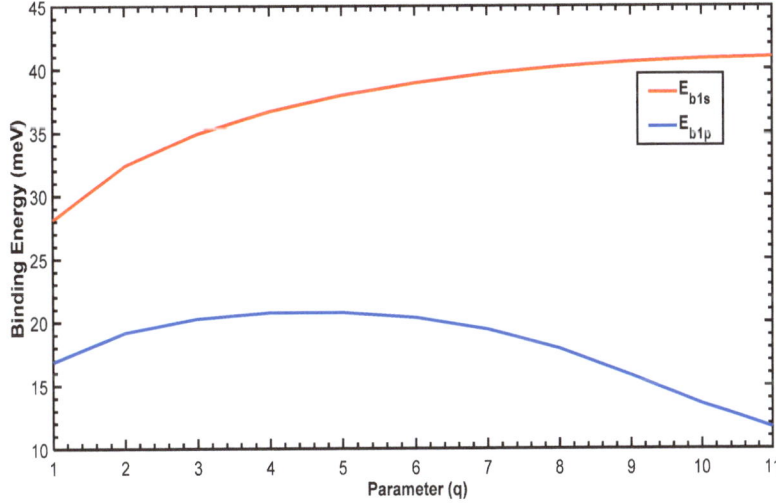

Figure 10. Variation of the binding energy as a function of parameter q for the $1s$ and $1p$ states.

4. Conclusions

In this work, we have examined the optical and electronic characteristics of spherical QDs in PEP and MGP potentials. A finite difference method was used to compute the energy levels, OACs, and binding energies for the two low-lying $1s$ and $1p$ states. Our calculations for the two confining potentials account for a hydrogenic impurity in the center of the QD. We first calculated the energy levels and their corresponding wavefunctions and subsequently evaluated the dipole matrix elements and energy separations between the $1s$ and $1p$ levels. We then examined the behavior of these physical quantities to interpret the blue and red shifts observed in the variation of OAC.

Our findings show that an increase in the structural parameter of the PEP potential produces a red shift in the OAC, which arises from the change in the energy separation due to the widening of the potential. In addition, our findings showed that an increase in the structural parameter of the MGP potential first produces a blue shift in the OAC and, subsequently, a redshift. The trends in the binding energy as a function of the structural parameter of each confining potential were attributed to the attractive force between the free electrons and hydrogenic impurity. Our simulations provide insight into the optical and electronic characteristics of spherical QDs in various confined potentials.

Author Contributions: Writing—original draft, R.M.A., H.D., E.A., F.U. and B.M.W.; Writing—review & editing, B.M.W.; Supervision, B.M.W.; Project administration, B.M.W. All authors have read and agreed to the published version of the manuscript.

Funding: This research received no external funding.

Institutional Review Board Statement: Not applicable.

Informed Consent Statement: Not applicable.

Data Availability Statement: The data that support the findings of this study are available from the corresponding author upon reasonable request.

Conflicts of Interest: The authors declare no conflict of interest.

References

1. Tang, J.; Sargent, E.H. Infrared Colloidal Quantum Dots for Photovoltaics: Fundamentals and Recent Progress. *Adv. Mater.* **2011**, *23*, 12–29. [CrossRef]
2. Kastner, M.A. The single-electron transistor. *Rev. Mod. Phys.* **1992**, *64*, 849. [CrossRef]
3. Nizamoglu, S.; Ozel, T.; Sari, E.; Demir, H.V. White light generation using CdSe/ZnS core–shell nanocrystals hybridized with InGaN/GaN light emitting diodes. *Nanotechnology* **2007**, *18*, 065709. [CrossRef]
4. Levine, B.F. Quantum-well infrared photodetectors. *J. Appl. Phys.* **1993**, *74*, R1–R81. [CrossRef]
5. Liu, H.C.; Gao, M.; McCaffrey, J.; Wasilewski, Z.R.; Fafard, S. Quantum dot infrared photodetectors. *Appl. Phys. Lett.* **2001**, *78*, 79. [CrossRef]
6. Bouzaiene, L.; Alamri, H.; Sfaxi, L.; Maaref, H. Simultaneous effects of hydrostatic pressure, temperature and electric field on optical absorption in InAs/GaAs lens shape quantum dot. *J. Alloys Compd.* **2016**, *655*, 172–177. [CrossRef]
7. Barve, A.V.; Lee, S.K.; Noh, S.K.; Krishna, S. Review of current progress in quantum dot infrared photodetectors. *Laser Photonics Rev.* **2010**, *4*, 738–750. [CrossRef]
8. Guériaux, V.; de l'Isle, N.B.; Berurier, A.; Huet, O.; Manissadjian, A.; Facoetti, H.; Marcadet, X.; Carras, M.; Trinité, V.; Nedelcu, A. Quantum well infrared photodetectors: Present and future. *Opt. Eng.* **2011**, *50*, 061013.
9. Al-Marhaby, F.A.; Al-Ghamdi, M.S. Experimental investigation of stripe cavity length effect on threshold current density for InP/AlGaInP QD laser diode. *Opt. Mater.* **2022**, *127*, 112191. [CrossRef]
10. Al-Sheikhi, A.; Al-Abedi, N.A.A. The luminescent emission and quantum optical efficiency of $Cd_{1-x}Sr_xSe$ QDs developed via ions exchange approach for multicolor-lasing materials and LED applications. *Optik* **2021**, *227*, 166035. [CrossRef]
11. Sargent, E.H. Colloidal quantum dot solar cells. *Nat. Photonics* **2012**, *6*, 133–135. [CrossRef]
12. Nozik, A.J. Quantum dot solar cells. *Physica E* **2002**, *14*, 115–120. [CrossRef]
13. Chung, S.-R.; Chen, S.-S.; Wang, K.-W.; Siao, C.-B. Promotion of solid-state lighting for ZnCdSe quantum dot modified-YAG-based white light-emitting diodes. *RCS Adv.* **2016**, *6*, 51989–51996. [CrossRef]
14. Kim, L.; Anikeeva, P.O.; Coe-Sullivan, S.A.; Steckel, J.S.; Bawendi, M.B.; Bulović, V. Contact printing of quantum dot light-emitting devices. *Nano Lett.* **2008**, *8*, 4513–4517. [CrossRef] [PubMed]
15. Chuang, C.-H.M.; Brown, P.R.; Bulović, V.; Bawendi, M.G. Improved performance and stability in quantum dot solar cells through band alignment engineering. *Nat. Mater.* **2014**, *3*, 796–801. [CrossRef]

16. De Franceschi, S.; Kouwenhoven, L.; Schönenberger, C.; Wernsdorfer, W. Hybrid superconductor–quantum dot devices. *Nat. Nanotechnol.* **2010**, *5*, 703–711. [CrossRef]
17. Gao, X.; Yang, L.; Petros, J.A.; Marshall, F.F.; Simons, J.W.; Nie, S. In vivo molecular and cellular imaging with quantum dots. *Cur. Opin. Biotechnol.* **2005**, *16*, 63–72. [CrossRef]
18. Ben Mahrsia, R.; Choubani, M.; Bouzaiene, L.; Maaref, H. Second-Harmonic Generation in Vertically Coupled InAs/GaAs Quantum Dots with a Gaussian Potential Distribution: Combined Effects of Electromagnetic Fields, Pressure, and Temperature. *Electron. Mater.* **2015**, *44*, 2792–2799.
19. Ameenah, A.N. The anti-crossing and dipping spectral behavior of coupled nanocrystal system under the influence of the magnetic field. *Results Phys.* **2021**, *22*, 103835. [CrossRef]
20. Kastner, M.A. Artificial Atoms. *Phys. Today* **1993**, *46*, 24–31. [CrossRef]
21. Ashoori, R.C. Electrons in artificial atoms. *Nature* **1996**, *379*, 413–419. [CrossRef]
22. Johnson, N.F. Quantum dots: Few-body, low-dimensional systems. *Phys. Condens. Matter* **1995**, *7*, 965. [CrossRef]
23. Akman, N.; Tomak, M. Interacting electrons in a 2D quantum dot. *Phys. B* **1999**, *262*, 317–321. [CrossRef]
24. Bednarek, S.; Szafran, B.; Adamowski, J. Many-electron artificial atoms. *Phys. Rev. B* **1999**, *59*, 13036. [CrossRef]
25. Sahin, M.; Tomak, M. Electronic structure of a many-electron spherical quantum dot with an impurity. *Phys. Rev. B* **2005**, *72*, 125323. [CrossRef]
26. Dakhlaoui, H.; Belhadj, W.; Elabidi, H.; Ungan, F.; Wong, B.M. GaAs Quantum Dot Confined with a Woods–Saxon Potential: Role of Structural Parameters on Binding Energy and Optical Absorption. *Inorganics* **2023**, *11*, 401. [CrossRef]
27. Zhu, J.; Xiong, J.; Gu, B. Confined electron and hydrogenic donor states in a spherical quantum dot of GaAs-Ga$_{1-x}$Al$_x$As. *Phys. Rev. B* **1990**, *41*, 6001. [CrossRef] [PubMed]
28. Porras-Montenegro, N.; Perez-Merchancano, S.T. Hydrogenic impurities in GaAs-(Ga,Al)As quantum dots. *Phys. Rev. B* **1992**, *46*, 9780. [CrossRef] [PubMed]
29. Aktas, S.; Boz, F. The binding energy of hydrogenic impurity in multilayered spherical quantum dot. *Physica E* **2008**, *40*, 753. [CrossRef]
30. Sahin, M. Photoionization cross section and intersublevel transitions in a one- and two-electron spherical quantum dot with a hydrogenic impurity. *Phys. Rev. B* **2008**, *77*, 045317. [CrossRef]
31. Sahin, M. Third-order nonlinear optical properties of a one- and two-electron spherical quantum dot with and without a hydrogenic impurity. *Appl. Phys.* **2009**, *106*, 063710. [CrossRef]
32. Holovatsky, V.A.; Frankiv, I.B. Oscillator strength of quantum transition in multi-shell quantum dots with impurity. *J. Optoelectr. Adv. Mater.* **2013**, *15*, 88–93.
33. Holovatsky, V.; Bernik, I.; Voitsekhivska, O. Oscillator Strengths of Quantum Transitions in Spherical Quantum Dot GaAs/Al$_x$Ga$_{1-x}$As/GaAs/Al$_x$Ga$_{1-x}$As with On-Center Donor Impurity. *Acta Phys. Pol. A* **2014**, *125*, 93. [CrossRef]
34. Li, S.-S.; Xia, J.-B. Intraband optical absorption in semiconductor coupled quantum dots. *Phys. Rev. B* **1997**, *55*, 15434. [CrossRef]
35. Schrey, F.F.; Rebohle, L.; Muller, T.; Strasser, G.; Unterrainer, K.; Nguyen, D.P.; Regnault, N.; Ferreira, R.; Bastard, G. ntraband transitions in quantum dot superlattice heterostructures. *Phys. Rev. B* **2005**, *72*, 155310. [CrossRef]
36. Bahar, M.K.; Baser, P. Nonlinear optical characteristics of thermodynamic effects- and electric field-triggered Mathieu quantum dot. *Micro Nanostruct.* **2022**, *170*, 207371. [CrossRef]
37. Batra, K.; Prasad, V. Spherical quantum dot in Kratzer confining potential: Study of linear and nonlinear optical absorption coefficients and refractive index changes. *Eur. Phys. J. B* **2018**, *91*, 298. [CrossRef]
38. Buczko, R.; Bassani, F. Bound and resonant electron states in quantum dots: The optical spectrum. *Phys. Rev. B* **1996**, *54*, 2667. [CrossRef]
39. Narvaez, G.A.; Zunger, A. Calculation of conduction-to-conduction and valence-to-valence transitions between bound states in (In,Ga)As/GaAs quantum dots. *Phys. Rev. B* **2007**, *75*, 085306. [CrossRef]
40. Ed-Dahmouny, A.; Arraoui, R.; Jaouane, M.; Fakkahi, A.; Sali, A.; Es-Sbai, N.; El-Bakkari, K.; Zeiri, N.; Duque, C.A The influence of the electric and magnetic fields on donor impurity electronic states and optical absorption coefficients in a core/shell GaAs/AlGaAs ellipsoidal quantum dot. *Eur. Phys. J. Plus* **2023**, *138*, 774. [CrossRef]
41. Fakkahi, A.; Jaouane, M.; Kirak, M.; Khordad, R.; Sali, A.; Arraoui, R.; El-bakkari, K.; Ed-Dahmouny, A.; Azmi, H. Optical absorption coefficients of a single electron in a multilayer spherical quantum dot with a Kratzer-like confinement potential. *Results Opt.* **2023**, *13*, 100553. [CrossRef]
42. Yilmaz, S.; Safak, H. Oscillator strengths for the intersubband transitions in a CdS–SiO$_2$ quantum dot with hydrogenic impurity. *Physica E* **2007**, *36*, 40–44. [CrossRef]
43. Kirak, M.; Yilmaz, S.; Sahin, M.; Gencaslan, M.J. The electric field effects on the binding energies and the nonlinear optical properties of a donor impurity in a spherical quantum dot. *Appl. Phys.* **2011**, *109*, 094309. [CrossRef]
44. Khordad, R. Use of modified Gaussian potential to study an exciton in a spherical quantum dot. *Superlattices Microstruct.* **2013**, *54*, 7–15. [CrossRef]
45. Ciurla, M.; Adamowski, J.; Szafran, B.; Bednarek, S. Modelling of confinement potentials in quantum dots. *Physica E* **2002**, *15*, 261. [CrossRef]
46. Bednarek, S.; Szafran, B.; Lis, K.; Adamowski, J. Modeling of electronic properties of electrostatic quantum dots. *Phys. Rev. B* **2003**, *68*, 155333. [CrossRef]

47. Szafran, B.; Bednarek, S.; Adamowski, J. Parity symmetry and energy spectrum of excitons in coupled self-assembled quantum dots. *Phys. Rev. B* **2001**, *64*, 125301. [CrossRef]
48. Gharaati, A.; Khordad, R. A new confinement potential in spherical quantum dots: Modified Gaussian potential. *Superlattices Microstruct.* **2010**, *48*, 276–287. [CrossRef]
49. Harrison, P. *Quantum Wells, Wires and Dots: Theoretical and Computational Physics of Semiconductor Nanostructures*; Wiley: Chichester, UK, 2009.
50. Ben Daniel, D.J.; Duke, C.B. Space-Charge Effects on Electron Tunneling. *Phys. Rev.* **1996**, *152*, 683. [CrossRef]
51. Sahin, M.; Koksal, K. The linear optical properties of a multi-shell spherical quantum dot of a parabolic confinement for cases with and without a hydrogenic impurity. *Semicond. Sci. Technol.* **2012**, *27*, 125011. [CrossRef]

Disclaimer/Publisher's Note: The statements, opinions and data contained in all publications are solely those of the individual author(s) and contributor(s) and not of MDPI and/or the editor(s). MDPI and/or the editor(s) disclaim responsibility for any injury to people or property resulting from any ideas, methods, instructions or products referred to in the content.

molecules

Article

Magnetic Exchange Mechanism and Quantized Anomalous Hall Effect in Bi$_2$Se$_3$ Film with a CrWI$_6$ Monolayer

He Huang [1], Fan He [1], Qiya Liu [2], You Yu [2] and Min Zhang [1,*]

[1] School of Physics and Astronomy, China West Normal University, Nanchong 637002, China; huangh@stu.cwnu.edu.cn (H.H.); hefan@cwnu.edu.cn (F.H.)
[2] College of Optoelectronic Engineering, Chengdu University of Information Technology, Chengdu 610225, China; liuqiya@cuit.edu.cn (Q.L.); yy2012@cuit.edu.cn (Y.Y.)
* Correspondence: zmzmi1987@cwnu.edu.cn

Abstract: Magnetizing the surface states of topological insulators without damaging their topological features is a crucial step for realizing the quantum anomalous Hall (QAH) effect and remains a challenging task. The TI–ferromagnetic material interface system was constructed and studied by the density functional theory (DFT). A two-dimensional magnetic semiconductor CrWI$_6$ has been proven to effectively magnetize topological surface states (TSSs) via the magnetic proximity effect. The non-trivial phase was identified in the Bi$_2$Se$_3$ (BS) films with six quantum layers (QL) within the CrWI$_6$/BS/CrWI$_6$ heterostructure. BS thin films exhibit the generation of spin splitting near the TSSs, and a band gap of approximately 2.9 meV is observed at the Γ in the Brillouin zone; by adjusting the interface distance of the heterostructure, we increased the non-trivial band gap to 7.9 meV, indicating that applying external pressure is conducive to realizing the QAH effect. Furthermore, the topological non-triviality of CrWI$_6$/6QL-BS/CrWI$_6$ is confirmed by the nonzero Chern number. This study furnishes a valuable guideline for the implementation of the QAH effect at elevated temperatures within heterostructures comprising two-dimensional (2D) magnetic monolayers (MLs) and topological insulators.

Keywords: topological insulator; first principles; magnetic proximity effect; spin

1. Introduction

The discovery of topological insulators (TIs) has spurred significant research interest in condensed matter physics, particularly in exploring exotic phenomena induced by breaking time-reversal symmetry (TRS) and opening surface band gaps [1,2]. In particular, magnetic interactions with topological properties lead to exotic quantum states in materials, including Majorana fermions [3] (non-Abelian statistics) and the quantum anomalous Hall (QAH) effect (dissipationless chiral edge states) [4,5]. Of the two widely employed approaches to break the TRS in TIs, the magnetic proximity effect, achieved through interface formation with a magnetic insulator, has advantages over doping with transition metal atoms into TIs. This is commonly associated with magnetic doping, leading to the formation of defects, including lattice defects, microscopic phase segregation, and impurity bands [6,7]. Following theoretical predictions, the QAH effect was first confirmed in magnetically Cr-doped (Bi, Sb)$_2$Te$_3$ [5]. Subsequently, researchers observed the QAH effect in various systems, such as Cr-doped (Bi, Sb)$_2$Te$_3$/Cr$_2$O$_3$ [8] and a Cr-doped (Bi, Sb)$_2$Te$_3$/(Bi, Sb)$_2$Te$_3$ sandwich structure [9]. Despite some work reports on introducing magnetism into Bi$_2$Se$_3$ (BS), it is still a challenge to achieve complete resistance to the QAH effect due to the influence of the carrier concentration [10–12]. Therefore, it is necessary to explore new magnetic introduction methods.

Recently, some theoretical studies have reported on heterojunctions employing magnetic insulator (MI)/TI heterojunctions [13–16]. Regrettably, in many cases, the hybridization at the interface between an MI and TI is excessively strong, leading to the disruption or misalignment of the topological surface states (TSSs) concerning the Fermi level [17,18]. Theoretical work has proposed that in the heterostructure of three-dimensional topological insulators and magnetic insulators, the magnetic order introduced by the magnetic proximity effect breaks the TRS and exhibits gapped surface states [18–20]. Nevertheless, achieving the QAH effect at high temperatures remains a considerable challenge. Ordinarily, the existence of the Mermin–Wagner theorem [21] has led to the belief that establishing a stable magnetic order in two-dimensional materials is impossible. It has been reported that magnetic anisotropy can alleviate this limitation and promote phenomena such as the emergence of two-dimensional (2D) Ising ferromagnetism [22]. Consequently, researchers have unveiled that CrI_3 is an out-of-plane spin-oriented Ising ferromagnet via magneto-optical Kerr effect microscopy [23], confirming the existence of a 2D Ising magnetic monolayer (ML) in CrI_3. Due to the weak magnetic coupling [24–26] established by super-exchange, ML CrI_3 exhibits a low Curie temperature ($T_c \approx 45$ K). Doping metal elements to alloy CrI_3 is beneficial to enhance its magnetic coupling effect, thereby increasing the Curie temperature. Using alloyed CrI_3 as a magnetic substrate for topological insulator materials is more conducive to achieving a high-temperature QAH effect.

In our study, a 2D-vdW magnetic semiconductor, $CrWI_6$, was created through doping. A $CrWI_6$ ML as a semiconductor can maintain the transport properties of TIs, while showing a highly stable magnetic order and significant magnetic anisotropy [23]. Therefore, a $CrWI_6$ ML provides an optimal way to magnetize the TSSs of TIs. Additionally, the $CrWI_6/BS/CrWI_6$ heterostructures with different thicknesses of BS thin films were constructed. Based on the density functional theory (DFT), the feasible route to enhance the non-trivial gaps in the heterostructures of utilizing a $CrWI_6$ ML to magnetize the TSSs of BS is reported. It is observed that the TSSs of BS can be effectively magnetized by the $CrWI_6$ ML. Furthermore, the topological non-triviality of the $CrWI_6/6QL-BS/CrWI_6$ heterostructures was confirmed by the presence of nonzero Chern numbers (C_N). The experimental observation and practical application of the QAH effect based on the BS/2D-vdW magnetic semiconductor interface system can be further facilitated by our work.

2. Model and Computational Details

The DFT calculations used the Vienna Ab Initio Simulation Package (VASP) code [27,28] to study the geometrical and electronic properties of a system. The Perdew–Burke–Ernzerhof (PBE) [29] of the Generalized Gradient Approximation (GGA) [30] exchange–correlation functional method was employed in this study. To better investigate the geometry of heterojunctions, the vdW interaction form of DFT-D3 [31] with Becke–Jonson damping was employed for calculations, as it has been demonstrated to provide a more accurate description for various systems. The plane-wave cutoff energy was set to 520 eV, and the convergence criteria for the total energy and atomic forces were set to be less than 10^{-6} eV and 0.02 eV/Å, respectively. Dudarev's method [32] was used in all calculations, while the effective Hubbard Ueff 3 eV and 1 eV were added to the Cr-d and W-d orbits, respectively. Spin–orbit coupling (SOC) was also considered, and Brillouin zone (BZ) integration was performed using Gamma-centered $3 \times 3 \times 1$ k-point grids. To avoid spurious interactions due to periodic boundary conditions, a vacuum space of more than 15 Å was used along the Z direction.

The present work introduces a theoretical investigation of the magnetic proximity effect in the interface of $CrWI_6/BS$ (See Figure 1). To check the thermal stability of the interface of the heterojunction, molecular dynamics simulations were carried out using the CP2K 2024.1 software package [33,34], a $4 \times 4 \times 1$ supercell (included 424 atoms) was used, the temperature was maintained at 300 K by using the Nose–Hoover thermostat [35], a time step of 1 femto-second (fs) was used to integrate the equations of motion, and the 5 pico-second (ps) trajectory was generated and used for analysis. There are three positions with high symmetry within the $CrWI_6/BS/CrWI_6$ heterostructure. Among them, the configuration where Cr ions are positioned above Se ions exhibits the lowest energy (see Figure S1). Therefore, unless otherwise specified in the following sections, we adopt the heterostructure model where Cr ions are positioned above Se ions. The interlayer distance between the heterojunctions and calculation parameters were determined through an energy test (see Figure S2); the results indicated that the optimal interlayer distance, denoted as d_0, for the $CrWI_6/BS/CrWI_6$ heterojunction is 3.18 Å. Further stability calculations confirmed that d_0 remains unchanged after the relaxation of the structure, demonstrating its structural stability.

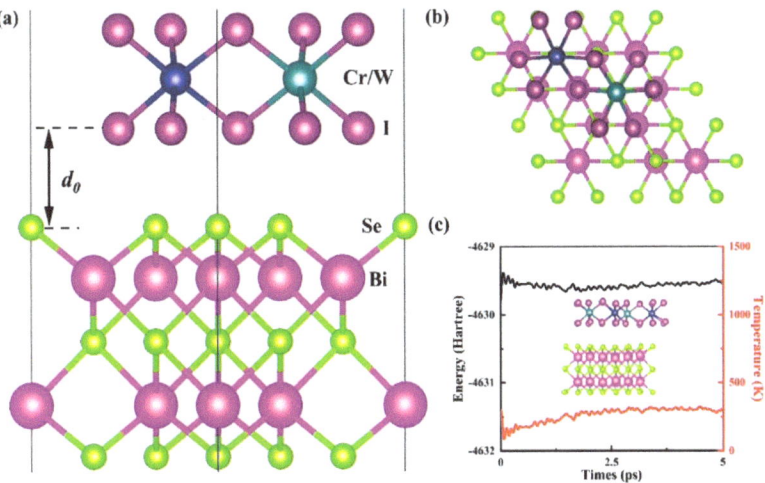

Figure 1. (a) Side view of the $CrWI_6/BS$ interface, d_0 denotes the optimized interlayer distance. The blue atom (left) represent Cr and the green atom (right) represent W. (b) Top view of the $CrWI_6/BS$ interface. (c) Energy and temperature fluctuations observed in molecular dynamics simulations of heterojunction interfaces, the inset depicting the final structure.

3. Results and Discussion

3.1. Magnetism and Structure of $CrWI_6$

To check the dynamical stabilities of the $CrWI_6$ ML, the phonon spectra are computed using density functional perturbation theory (DFPT) [36]. The phonon dispersion relation is obtained by processing the data with the PHONOPY 2.26.0 software package [37], as illustrated in Figure S3. After structural optimization, it was observed that the lattice constant of the $CrWI_6$ ML measures 6.94 Å, representing a diminution of merely 3.9% in comparison to the lattice constant of the $\sqrt{3} \times \sqrt{3}$ BS supercell (7.22 Å). Additionally, the application of biaxial strain (−5~5%) revealed that the magnetic properties of the $CrWI_6$ ML remain unaffected in the presence of minor lattice stretching (see Figure 2).

The presence of magnetic anisotropy is a crucial requirement for the existence of long-range magnetic ordering in two-dimensional systems [38]. To assess the magnetic anisotropy energy (MAE) in the single-layer $CrWI_6$ with spin orientation along different directions, the angular dependence of the MAE was determined using the following equation [39]:

$$MAE(\theta, \varphi) = E(\theta, \varphi) - E(\theta = 0, \varphi = 0) \quad (1)$$

Here, θ represents the angle between the magnetization direction and the Z-axis, while φ denotes the angle between the projection of the magnetization direction on the XY-plane and the X-axis. The calculation results reveal that the magnetic easy axis of $CrWI_6$ ML is perpendicular to the XY-plane, which means that the easy magnetization axis consistently aligns with the Z-axis, and it can effectively magnetize the BS film peeled from the BS 001 crystal plane. In two-dimensional magnetic materials, the magnetic anisotropy of $CrWI_6$ is relatively more significant [40]. This unique characteristic is predicted to play a crucial role in stabilizing long-range magnetic coupling by offering resistance against thermal perturbations (See Figure S4). The substantial MAE observed in $CrWI_6$ is indicative of a remarkably robust magnetic coupling interaction within the material. This is further highlighted by its ability to maintain magnetic order even at elevated temperatures. To quantify this behavior, the T_c of CrWI$_6$ was determined through Monte Carlo simulations applied to the Heisenberg model:

$$H = \sum_{ij} J_{ij} S_i S_j \quad (2)$$

Generally, considering only the exchange interaction between the nearest and next-nearest neighbors is sufficient to describe the magnetic interactions in magnetic systems [41], where J_{ij} represents the nearest and the next-nearest-neighbor exchange interactions, and S_i represents the spin at site i, respectively. For CrWI$_6$ ML, our results demonstrate that the establishment of the ferromagnetic (FM) ground state is mainly governed by the nearest-neighbor exchange interaction (J_1). Additionally, the next-nearest-neighbor interactions (J_2) also favor FM ordering; notably, due to an order of magnitude difference, we exclusively considered the magnetic coupling constants J_1 and J_2, excluding J_3. The T_c is taken as the critical point of the specific heat. As shown in Figure 3, the simulation results indicate that the T_c transition for CrI$_3$ is at 43.3 K (very close to the experimental value ~45 K [42]), thereby validating the reliability of our simulation. In contrast, CrWI$_6$ ML exhibits the T_c of 182.7 K, FM super-exchange interactions in CrWI$_6$ ML via the e_g-(p_x, p_y, p_z)-t_{2g} orbitals, and the Cr-I-W bonding angle is close to 90 degrees, which explains the occurrence of FM coupling according to the well-known Goodenough–Kanamori–Anderson (GKA) rules [24–26] of the super-exchange theorem. Given the elevated T_c observed in CrWI$_6$, the formation of a heterojunction with the BS becomes a feasible prospect, potentially enabling the realization of the QAH effect at a higher temperature.

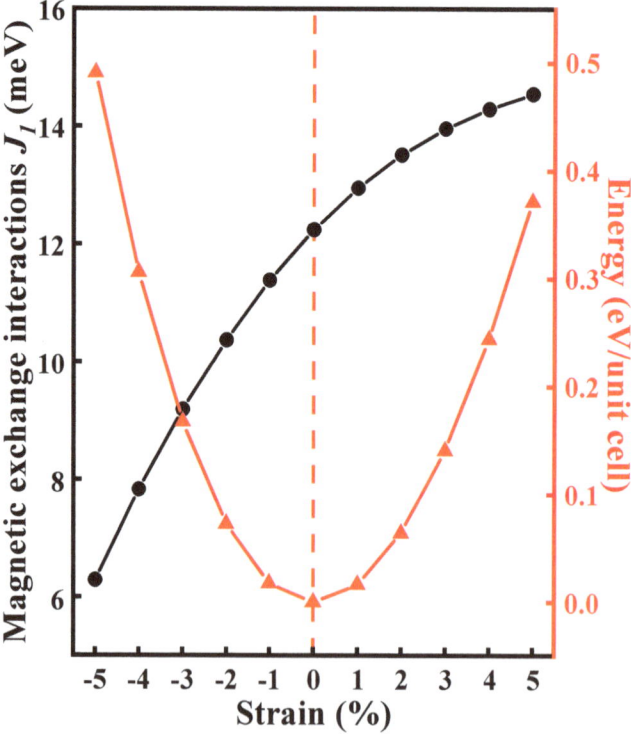

Figure 2. The influence of biaxial strain on the nearest-neighbor magnetic exchange coupling effect and energy transformation trend of CrWI$_6$.

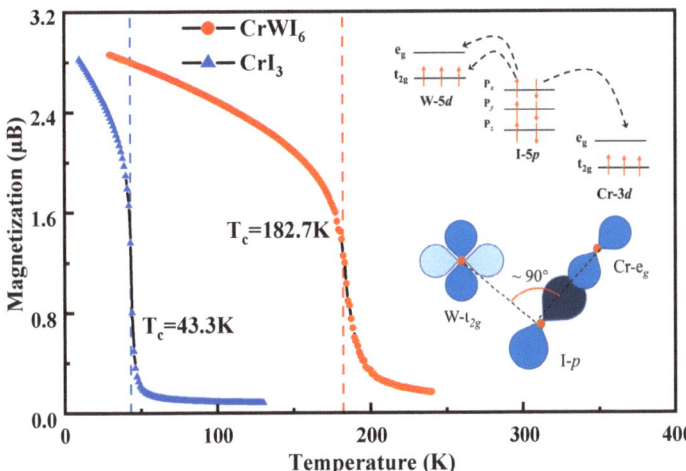

Figure 3. Based on the Heisenberg model, the T_c of CrWI$_6$ and CrI$_3$ are computed through the application of Monte Carlo simulation. Schematic diagrams of the super-exchange interaction and FM coupling in CrI$_3$ and CrWI$_6$ ML.

3.2. Electronic Properties of Heterojunction

To clarify the interactions at the interface, the spin density $\Delta\sigma$ and charge density difference $\Delta\rho$ at the CrWI$_6$/6QL-BS interface were plotted, which is defined as follows

$$\Delta\sigma = \sigma_\uparrow - \sigma_\downarrow \quad (3)$$

$$\Delta\rho = \rho_{total} - \rho_{CrWI6} - \rho_{BS} \quad (4)$$

Here, $\sigma_\uparrow$ and $\sigma_\downarrow$ denote the spin-up electrons and spin-down electrons, and ρ_{total}, ρ_{CrWI6}, and ρ_{BS} were the charge densities of the interface of the heterojunction, CrWI$_6$ ML, and the BS film, respectively, as shown in Figure 4. Similar to prior studies [20,43], the magnetic proximity effect is prominent in the QL nearest to the interface. The Se atomic layer of BS closest to the interface acquires a small but non-negligible magnetic moment, aligning with the Cr/W ion spin in the CrWI$_6$ ML. To further discuss the origin of the magnetic proximity effect, the charge differential density at the interface of the heterojunction was analyzed. The charge transfer of the interface of the heterojunction predominantly takes place at the interface between CrWI$_6$ and BS; the positive values (yellow area) indicate electron accumulation, and the negative values (cyan area) represent electron depletion. Additionally, the plane-averaged charge density difference at the interface is illustrated. Regions where $\Delta\rho$ is less than zero indicate electron dissipation, whereas values greater than zero signify electron accumulation. Consequently, it is evident that within the CrWI$_6$ ML, electrons are transferred from I to Cr/W atoms. At the interface, the electron transfer occurs from I to Se-p orbit, providing an explanation for the spin-up magnetic moment of Se atoms.

For BS, a reduction in the number of QLs results in the hybridization and annihilation of TSSs on both the upper and lower surfaces of the TI [44]. Hence, the heterojunction consists of 4, 5, and 6 QLs of BS and CrWI$_6$. Through the projected band structure analysis (See Figure 5), the band contributions in proximity to the Fermi level at the heterojunction were ascertained. The energy band near the Dirac cone is primarily composed of BS, while the contribution from CrWI$_6$ is concentrated around 0.5 eV and −0.2 eV, the same as with the energy band structure of intrinsic CrWI$_6$ (See Figure S5), and the TSSs near the Dirac cone are almost unaffected by CrWI$_6$. These results show that 2D-vdW magnetic semiconductors are more suitable for the magnetization of TSS than other ferromagnetic or antiferromagnetic films.

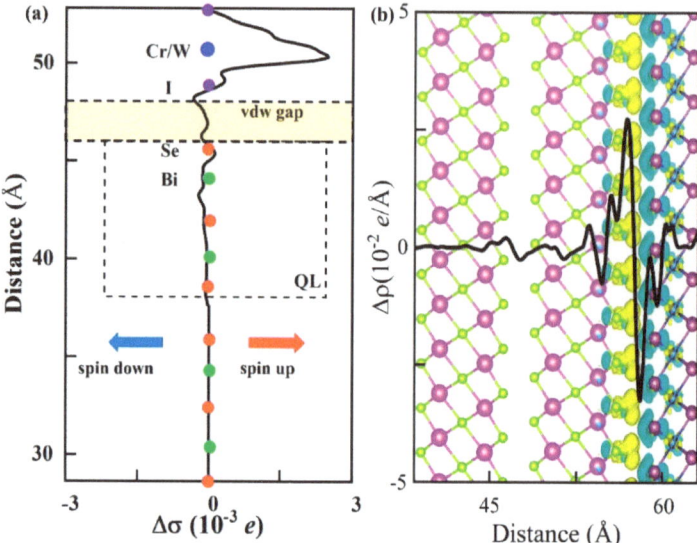

Figure 4. (a) Planar-averaged spin density $\Delta\sigma$ in the interfacial region of the interface of the heterostructure. (b) The distribution of the charge difference $\Delta\rho$.

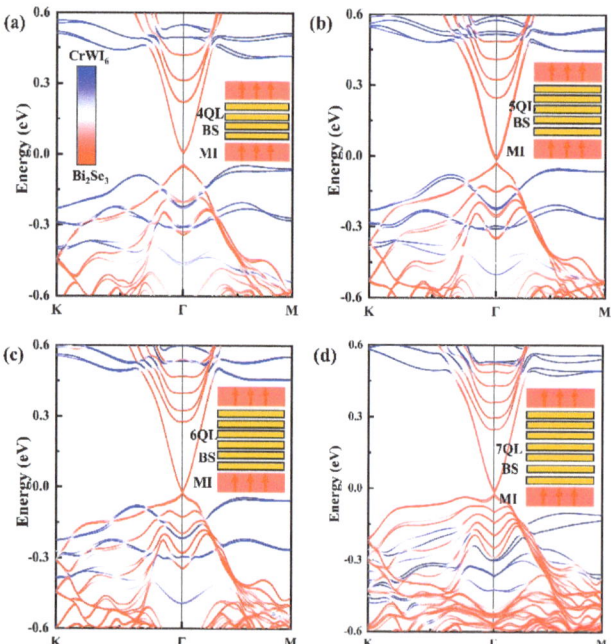

Figure 5. Band structures of $CrWI_6/BS/CrWI_6$: (a) $CrWI_6/4QL\text{-}BS/CrWI_6$, (b) $CrWI_6/5QL\text{-}BS/CrWI_6$, (c) $CrWI_6/6QL\text{-}BS/CrWI_6$, and (d) $CrWI_6/7QL\text{-}BS/CrWI_6$. Colors in the main panels indicate the weights of bands from BS (red) and $CrWI_6$ (yellow).

To reveal the successful magnetization of TSSs, the spin projections of the band structure near the Dirac cone in the 4–6 QLs energy band are plotted, with the red arrow indicating spin-up electrons and the blue arrow indicating spin-down electrons. It is observed that the spin degeneracy of TSSs near the Fermi level is disrupted in all cases (See Figure 6). Upon reaching a 6QL of the BS film, the band composition near the Fermi level undergoes a change. There is a band gap of approximately 2.9 meV near the Dirac cone. Specifically, in the 6QL-BS/CrWI6 system, the valence band is composed of spin-up components, while the conduction band is composed of spin-down components. In contrast, for BS systems with fewer than 5QLs, both the valence band and conduction band contain spin-up and spin-down components. The significant difference in the energy band composition at this point indicates that the system is transitioning from a normal insulator to a Chern insulator state, signifying that the system is topologically non-trivial [43].

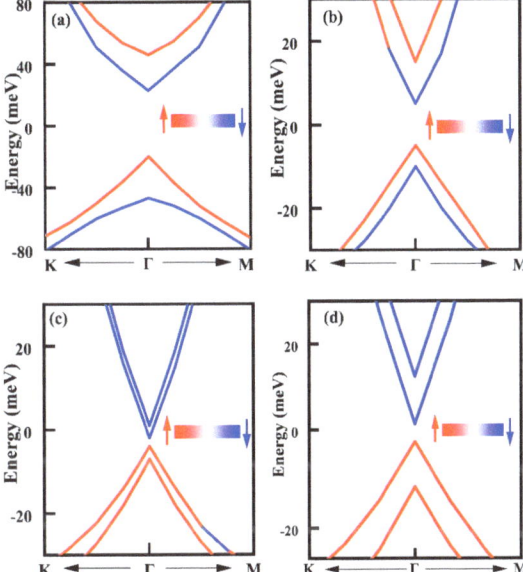

Figure 6. Spin projections of $CrWI_6/BS/CrWI_6$ band structure: (a) $CrWI_6/4QL\text{-}BS/CrWI_6$, (b) $CrWI_6/5QL\text{-}BS/CrWI_6$, (c) $CrWI_6/6QL\text{-}BS/CrWI_6$, and (d) $CrWI_6/7QL\text{-}BS/CrWI_6$. The red arrow indicates spin-up electrons and the blue arrow indicates spin-down electrons.

To verify the potential influence of different interface stacking manners on the results, we computed energy bands for two additional stacking manners. The spin projections for the two alternative stacking manners in $CrWI_6/6QL\text{-}BS/CrWI_6$ are shown in Figure S6. The results indicate that different stacking manners do indeed affect the band gaps. The stacking configuration of Cr/W atoms above Se atoms exhibits the largest non-trivial band gaps, possibly due to the lowest energy associated with this stacking mode. However, due to the six QLs of BS, the TSSs of this system remain stable, and thus, the non-trivial band structure is preserved. The energy gap data can be found in Table 1.

Table 1. Energy gaps E_g of various thickness $CrWI_6/BS/CrWI_6$ heterostructures.

Thickness of BS	Energy Gaps E_g (meV)
4QLs	23.5 (Se site)
5QLs	9.8 (Se site)
6QLs	2.9 (Se site) 1.4 (Bi site) 1.1 (Hole site)
7QLs	3.2 (Se site)

3.3. Topological Properties

To assess the topological features of the magnetized TSSs of $CrWI_6/6QL\text{-}BS/CrWI_6$, a tight-binding model system based on Wannier functions was constructed. Using this model, the band structure and Berry phase by integrating the Berry curvature over the BZ were calculated. The VASPBERRY package [45] was used to perform calculations for the Berry curvature and Chern numbers of BZ, and the Berry curvature $\Omega(k)$ is defined by the following equation [46,47]:

$$\Omega(k) = \sum_n f_n \Omega_n(k) = -\sum_{n' \neq n} 2\text{Im} \frac{<\psi_{nk}|v_x|\psi_{n'k}><\psi_{n'k}|v_y|\psi_{nk}>}{(\varepsilon_{n'k} - \varepsilon_{nk})} \quad (5)$$

$$C_N = \sum_{n \in \{0\}} C_n = \frac{1}{2\pi} \int \sum_{n \in \{0\}} \Omega_n(k) d^2k = \sum_{n \in \{0\}} (C_{n,\uparrow} + C_{n,\downarrow}) \quad (6)$$

Here, f_n is the Fermi distribution, ψ_{nk} is the eigenstate of the wave function, v_x and v_y are the velocity operator, and $\{0\}$ represents the occupied state. The anomalous Hall conductance around the Fermi level is defined by the following equation:

$$\sigma_{xy} = \frac{e^2}{h} C_N \quad (7)$$

Firstly, utilizing the above formula, the distribution of the Berry curvature for the valence band of the $CrWI_6/6QL\text{-}BS/CrWI_6$ system was computed, as depicted in Figure 7a. By integrating the Berry curvature, a non-trivial Chern number $C_N = 1$ was determined. Within the energy window of the SOC gap, one can observe a quantized Hall plateau at a value of e^2/h, as illustrated in Figure 7b.

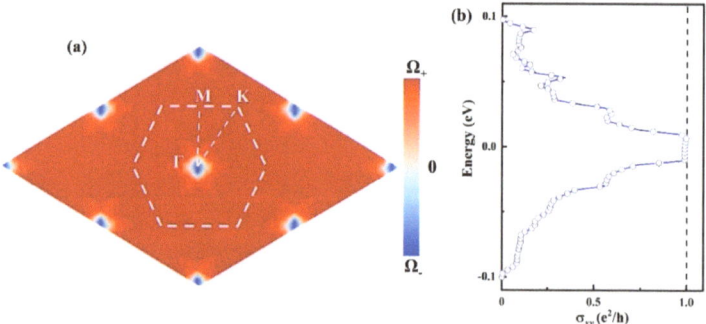

Figure 7. (a) Berry curvature within the SOC gap in reciprocal space of $CrWI_6/6QL\text{-}BS/CrWI_6$. (b) Hall conductivity of $CrWI_6/6QL\text{-}BS/CrWI_6$.

Secondly, by computing the $CrWI_6/BS/CrWI_6$ heterostructure with different layer thicknesses of BS, we summarized the trends of the band gaps and Chern numbers for the $CrWI_6/BS/CrWI_6$ heterostructure at different layers, as shown in Figure 8a. The calculated gaps of $CrWI_6/6QL\text{-}BS/CrWI_6$ and $CrWI_6/7QL\text{-}BS/CrWI_6$ are 2.3 meV (26.7 K) and 3.2 meV (37.1 K), respectively. Therefore, it can be concluded that a $CrWI_6/BS/CrWI_6$ heterostructure with fewer than six QLs of BS behaves as a normal insulator. Considering that the T_c of $CrWI_6$ ML is 183 K, and that the heterostructure system does not involve other complex factors such as uncontrollable doping distribution and local magnetic ordering, the QAH effect should be observed in the heterostructure at temperatures as high as several tens of Kelvin.

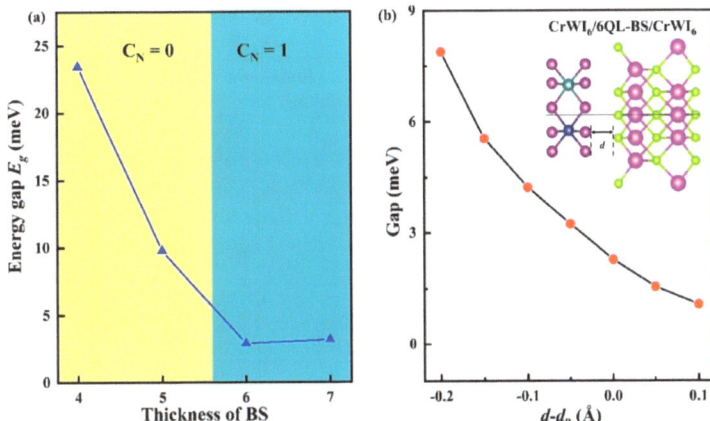

Figure 8. (a) Dependence of Chern numbers and gaps of CrWI$_6$/BS/CrWI$_6$ on the QL of BS film. (b) The impact of the interlayer distance on the gap of CrWI$_6$/6QL-BS/CrWI$_6$.

Finally, although the non-trivial band gaps in the CrWI$_6$/6QL-BS/CrWI$_6$ heterostructure can reach 2.9 meV, they are still relatively small compared to similar systems with the same number of BS layers, such as Cr$_2$Ge$_2$Te$_6$/Bi$_2$Se$_3$/Cr$_2$Ge$_2$Te$_6$ 19.5 meV [48] and MnBi$_2$Te$_4$ 33 meV [49]. To increase the system's non-trivial band gaps and enhance the temperature at which the QAHE is observed, we simulated external pressure by adjusting the interface distance d of CrWI$_6$/6QL-BS/CrWI$_6$ to compress the heterostructure. The results are shown in Figure 8b. The results indicate that reducing the interlayer distance can effectively widen the band gaps, facilitating the realization of the QAH effect. When reducing the interface distance to simulate external compression in the vertical direction, the band gaps of the CrWI$_6$/6QL-BS/CrWI$_6$ heterostructure can reach 7.9 meV (91.6 K); it is indicated that it is beneficial to apply external pressure to reduce *d* for the realization of the QAH effect.

4. Conclusions

In summary, this study has explored the feasibility of achieving the QAH effect in a TI–ferromagnetic material interface system composed of BS and CrWI$_6$ at higher temperatures. Through DFT to investigate the electron features, we have demonstrated that the 2D-vdW, a 2D-vdW magnetic semiconductor, can effectively magnetize the TSSs of TIs while preserving their topological features around the Fermi level. The topological non-trivial features of the CrWI$_6$/6QL-BS/CrWI$_6$ interface were verified by nonzero Chern numbers. There was a band gap of approximately 2.9 meV near the Dirac cone. By adjusting the interface distance of the heterostructure, we increased the non-trivial band gap to 7.9 meV, indicating that applying external pressure is conducive to realizing the QAH effect. This strongly indicates the significant possibility of detecting an induced QAH effect in the experiment. Consequently, this research has the potential to advance the experimental observation and practical utilization of the QAH effect in the TI-MI interface system.

Supplementary Materials: The following supporting information can be downloaded at: https://www.mdpi.com/article/10.3390/molecules29174101/s1.

Author Contributions: H.H.: Conceptualization, data curation, formal analysis, methodology, writing—original draft. F.H.: Writing—review and editing, methodology, supervision. M.Z.: Funding acquisition, investigation, project administration, supervision, writing—review and editing. Q.L.: Writing—review and editing, investigation, supervision. Y.Y.: Software, resources. All authors have read and agreed to the published version of the manuscript.

Funding: This work was supported by the Sichuan Science and Technology Program (No. 2023YFG0086, 2023ZYD0175).

Institutional Review Board Statement: Not applicable.

Informed Consent Statement: Not applicable.

Data Availability Statement: No new data were created or analyzed in this study. Data sharing is not applicable to this article.

Conflicts of Interest: The authors declare that they have no known competing financial interests or personal relationships that could have appeared to influence the work reported in this paper.

References

1. Liu, J.; Hesjedal, T. Magnetic Topological Insulator Heterostructures: A Review. *Adv. Mater.* **2023**, *35*, 2102427. [CrossRef]
2. Li, X.-G.; Zhang, G.-F.; Wu, G.-F.; Chen, H.; Culcer, D.; Zhang, Z.-Y. Proximity effects in topological insulator heterostructures. *Chin. Phys. B* **2013**, *22*, 097306. [CrossRef]
3. Qi, X.-L.; Zhang, S.-C. Topological insulators and superconductors. *Rev. Mod. Phys.* **2011**, *83*, 1057–1110. [CrossRef]
4. Yu, R.; Zhang, W.; Zhang, H.J.; Zhang, S.C.; Dai, X.; Fang, Z. Quantized Anomalous Hall Effect in Magnetic Topological Insulators. *Science* **2010**, *329*, 61–64. [CrossRef]
5. Chang, C.-Z.; Zhang, J.; Feng, X.; Shen, J.; Zhang, Z.; Guo, M.; Li, K.; Ou, Y.; Wei, P.; Wang, L.-L.; et al. Experimental Observation of the Quantum Anomalous Hall Effect in a Magnetic Topological Insulator. *Science* **2013**, *340*, 167–170. [CrossRef] [PubMed]
6. Zhang, Z.; Feng, X.; Wang, J.; Lian, B.; Zhang, J.; Chang, C.; Guo, M.; Ou, Y.; Feng, Y.; Zhang, S.-C.; et al. Magnetic quantum phase transition in Cr-doped $Bi_2(Se_xTe_{1-x})_3$ driven by the Stark effect. *Nat. Nanotechnol.* **2017**, *12*, 953–957. [CrossRef]
7. Wei, P.; Katmis, F.; Assaf, B.A.; Steinberg, H.; Jarillo-Herrero, P.; Heiman, D.; Moodera, J.S. Exchange-Coupling-Induced Symmetry Breaking in Topological Insulators. *Phys. Rev. Lett.* **2013**, *110*, 186807. [CrossRef]
8. Pan, L.; Grutter, A.; Zhang, P.; Che, X.; Nozaki, T.; Stern, A.; Street, M.; Zhang, B.; Casas, B.; He, Q.L.; et al. Observation of Quantum Anomalous Hall Effect and Exchange Interaction in Topological Insulator/Antiferromagnet Heterostructure. *Adv. Mater.* **2020**, *32*, 2001460. [CrossRef] [PubMed]
9. Jiang, J.; Xiao, D.; Wang, F.; Shin, J.-H.; Andreoli, D.; Zhang, J.; Xiao, R.; Zhao, Y.-F.; Kayyalha, M.; Zhang, L.; et al. Concurrence of quantum anomalous Hall and topological Hall effects in magnetic topological insulator sandwich heterostructures. *Nat. Mater.* **2020**, *19*, 732–737. [CrossRef] [PubMed]
10. Kanagaraj, M.; Yizhe, S.; Ning, J.; Zhao, Y.; Tu, J.; Zou, W.; He, L. Topological quantum weak antilocalization limit and anomalous Hall effect in semimagnetic $Bi_{2-x}Cr_xSe_3/Bi_2Se_{3-y}$ Tey heterostructure. *Mater. Res. Express* **2020**, *7*, 016401. [CrossRef]
11. Zhang, L.; Zhao, D.; Zang, Y.; Yuan, Y.; Jiang, G.; Liao, M.; Zhang, D.; He, K.; Ma, X.; Xue, Q.; et al. Ferromagnetism in vanadium-doped Bi_2Se_3 topological insulator films. *APL Mater.* **2017**, *5*, 076106. [CrossRef]
12. Chang, C.-Z.; Tang, P.; Wang, Y.-L.; Feng, X.; Li, K.; Zhang, Z.; Wang, Y.; Wang, L.-L.; Chen, X.; Liu, C.; et al. Chemical-Potential-Dependent Gap Opening at the Dirac Surface States of Bi_2Se_3 Induced by Aggregated Substitutional Cr Atoms. *Phys. Rev. Lett.* **2014**, *112*, 056801. [CrossRef] [PubMed]
13. Bhattacharyya, S.; Akhgar, G.; Gebert, M.; Karel, J.; Edmonds, M.T.; Fuhrer, M.S. Recent Progress in Proximity Coupling of Magnetism to Topological Insulators. *Adv. Mater.* **2021**, *33*, 2007795. [CrossRef] [PubMed]
14. Mathimalar, S.; Sasmal, S.; Bhardwaj, A.; Abhaya, S.; Pothala, R.; Chaudhary, S.; Satpati, B.; Raman, K.V. Signature of gate-controlled magnetism and localization effects at Bi_2Se_3/EuS interface. *Npj Quantum Mater.* **2020**, *5*, 64. [CrossRef]
15. Eremeev, S.V.; Men, V.N.; Tugushev, V.V.; Chulkov, E.V. Interface induced states at the boundary between a 3D topological insulator Bi_2Se_3 and a ferromagnetic insulator EuS. *J. Magn. Magn. Mater.* **2015**, *383*, 30–33. [CrossRef]
16. Eremeev, S.V.; Otrokov, M.M.; Chulkov, E.V. New Universal Type of Interface in the Magnetic Insulator/Topological Insulator Heterostructures. *Nano Lett.* **2018**, *18*, 6521–6529. [CrossRef]
17. Lee, A.T.; Han, M.J.; Park, K. Magnetic proximity effect and spin-orbital texture at the Bi_2Se_3/EuS interface. *Phys. Rev. B* **2014**, *90*, 155103. [CrossRef]
18. Eremeev, S.V.; Men'shov, V.N.; Tugushev, V.V.; Echenique, P.M.; Chulkov, E.V. Magnetic proximity effect at the three-dimensional topological insulator/magnetic insulator interface. *Phys. Rev. B* **2013**, *88*, 144430. [CrossRef]
19. Luo, W.; Qi, X.-L. Massive Dirac surface states in topological insulator/magnetic insulator heterostructures. *Phys. Rev. B* **2013**, *87*, 085431. [CrossRef]
20. Hou, Y.; Kim, J.; Wu, R. Magnetizing topological surface states of Bi_2Se_3 with a CrI_3 monolayer. *Sci. Adv.* **2019**, *5*, eaaw1874. [CrossRef]
21. Mermin, N.D.; Wagner, H. Absence of Ferromagnetism or Antiferromagnetism in One- or Two-Dimensional Isotropic Heisenberg Models. *Phys. Rev. Lett.* **1966**, *17*, 1133–1136. [CrossRef]
22. Gong, C.; Zhang, X. Two-dimensional magnetic crystals and emergent heterostructure devices. *Science* **2019**, *363*, eaav4450. [CrossRef] [PubMed]
23. Huang, C.; Feng, J.; Wu, F.; Ahmed, D.; Huang, B.; Xiang, H.; Deng, K.; Kan, E. Toward Intrinsic Room-Temperature Ferromagnetism in Two-Dimensional Semiconductors. *J. Am. Chem. Soc.* **2018**, *140*, 11519–11525. [CrossRef] [PubMed]

24. Kanamori, J. Superexchange Interaction and Symmetry Properties of Electron Orbitals. *J. Phys. Chem. Solids* **1959**, *10*, 87–98. [CrossRef]
25. Goodenough, J.B. Theory of the Role of Covalence in the Perovskite-Type Manganites [La, M(II)]MnO$_3$. *Phys. Rev.* **1955**, *100*, 564. [CrossRef]
26. Anderson, P.W. Antiferromagnetism. Theory of Superexchange Interaction. *Phys. Rev.* **1950**, *79*, 350–356. [CrossRef]
27. Kresse, G.; Hafner, J. Ab initio molecular dynamics for open-shell transition metals. *Phys. Rev. B* **1993**, *48*, 13115–13118. [CrossRef]
28. Kresse, G.; Furthmüller, J. Efficiency of ab-initio total energy calculations for metals and semiconductors using a plane-wave basis set. *Comput. Mater. Sci.* **1996**, *6*, 15–50. [CrossRef]
29. Perdew, J.P.; Burke, K.; Ernzerhof, M. Generalized Gradient Approximation Made Simple. *Phys. Rev. Lett.* **1996**, *77*, 3865–3868. [CrossRef]
30. Blöchl, P.E. Projector augmented-wave method. *Phys. Rev. B* **1994**, *50*, 17953–17979. [CrossRef]
31. Grimme, S.; Antony, J.; Ehrlich, S.; Krieg, H. A consistent and accurate ab initio parametrization of density functional dispersion correction (DFT-D) for the 94 elements H-Pu. *J. Chem. Phys.* **2010**, *132*, 154104. [CrossRef] [PubMed]
32. Dudarev, S.L.; Botton, G.A.; Savrasov, S.Y.; Humphreys, C.J.; Sutton, A.P. Electron-energy-loss spectra and the structural stability of nickel oxide: An LSDA+U study. *Phys. Rev. B* **1998**, *57*, 1505–1509. [CrossRef]
33. Hutter, J.; Iannuzzi, M.; Schiffmann, F.; VandeVondele, J. CP2K: ATOMISTIC simulations of condensed matter systems. *WIREs Comput. Mol. Sci.* **2014**, *4*, 15–25. [CrossRef]
34. VandeVondele, J.; Krack, M.; Mohamed, F.; Parrinello, M.; Chassaing, T.; Hutter, J. Quickstep: Fast and accurate density functional calculations using a mixed Gaussian and plane waves approach. *Comput. Phys. Commun.* **2005**, *167*, 103–128. [CrossRef]
35. Martyna, G.J.; Tobias, D.J.; Klein, M.L. Constant pressure molecular dynamics algorithms. *J. Chem. Phys.* **1994**, *101*, 4177–4189. [CrossRef]
36. Baroni, S.; de Gironcoli, S.; Corso, A.D.; Giannozzi, P. Phonons and related crystal properties from density-functional perturbation theory. *Rev. Mod. Phys.* **2001**, *73*, 515–562. [CrossRef]
37. Togo, A.; Tanaka, I. First principles phonon calculations in materials science. *Scr. Mater.* **2015**, *108*, 1–5. [CrossRef]
38. Gong, C.; Li, L.; Li, Z.; Ji, H.; Stern, A.; Xia, Y.; Cao, T.; Wang, Y.; Qiu, Z.Q.; Cava, R.J.; et al. Discovery of Intrinsic Ferromagnetism in Two-Dimensional van Der Waals Crystals. *Nature* **2017**, *546*, 265–269. [CrossRef]
39. Wang, X.; Wu, R.; Wang, D.; Freeman, A.J. Torque method for the theoretical determination of magnetocrystalline anisotropy. *Phys. Rev. B* **1996**, *54*, 61–64. [CrossRef] [PubMed]
40. Zhang, W.-B.; Qu, Q.; Zhu, P.; Lam, C.-H. Robust intrinsic ferromagnetism and half semiconductivity in stable two-dimensional single-layer chromium trihalides. *J. Mater. Chem. C* **2015**, *3*, 12457–12468. [CrossRef]
41. Wang, H.; Fan, F.; Zhu, S.; Wu, H. Doping enhanced ferromagnetism and induced half-metallicity in CrI$_3$ monolayer. *EPL* **2016**, *114*, 47001. [CrossRef]
42. Huang, B.; Clark, G.; Navarro-Moratalla, E.; Klein, D.R.; Cheng, R.; Seyler, K.L.; Zhong, D.; Schmidgall, E.; McGuire, M.A.; Cobden, D.H.; et al. Layer-dependent ferromagnetism in a van der Waals crystal down to the monolayer limit. *Nature* **2017**, *546*, 270–273. [CrossRef]
43. Chen, L.; Shi, C.; Jiang, H.; Liu, H.; Cui, G.; Wang, D.; Li, X.; Gao, K.; Zhang, X. Realization of quantized anomalous Hall effect by inserting CrI$_3$ layer in Bi$_2$Se$_3$ film. *New J. Phys.* **2020**, *22*, 073005. [CrossRef]
44. Zhang, Y.; He, K.; Chang, C.-Z.; Song, C.-L.; Wang, L.-L.; Chen, X.; Jia, J.-F.; Fang, Z.; Dai, X.; Shan, W.-Y.; et al. Crossover of the three-dimensional topological insulator Bi$_2$Se$_3$ to the two-dimensional limit. *Nat. Phys.* **2010**, *6*, 584–588. [CrossRef]
45. Kim, S.-W.; Kim, H.-J.; Cheon, S.; Kim, T.-H. Circular dichroism of emergent chiral stacking orders in quasi-one-dimensional charge density waves. *Phys. Rev. Lett.* **2022**, *128*, 046401. [CrossRef]
46. Yao, Y.; Kleinman, L.; MacDonald, A.H.; Sinova, J.; Jungwirth, T.; Wang, D.; Wang, E.; Niu, Q. First Principles Calculation of Anomalous Hall Conductivity in Ferromagnetic bcc Fe. *Phys. Rev. Lett.* **2004**, *92*, 037204. [CrossRef]
47. Qi, X.-L.; Wu, Y.-S.; Zhang, S.-C. Topological quantization of the spin Hall effect in two-dimensional paramagnetic semiconductors. *Phys. Rev. B* **2006**, *74*, 085308. [CrossRef]
48. Li, P.; You, Y.; Huang, K.; Luo, W. Quantum Anomalous Hall Effect in Cr$_2$Ge$_2$Te$_6$/Bi$_2$Se$_3$/Cr$_2$Ge$_2$Te$_6$ Heterostructures. *J. Phys.: Condens. Matter* **2021**, *33*, 465003. [CrossRef]
49. Li, P.; Yu, J.; Wang, Y.; Luo, W. Electronic Structure and Topological Phases of the Magnetic Layered Materials MnBi$_2$Te$_4$. MnBi$_2$Se$_4$ and MnSb$_2$Te$_4$. *Phys. Rev. B* **2021**, *103*, 155118. [CrossRef]

Disclaimer/Publisher's Note: The statements, opinions and data contained in all publications are solely those of the individual author(s) and contributor(s) and not of MDPI and/or the editor(s). MDPI and/or the editor(s) disclaim responsibility for any injury to people or property resulting from any ideas, methods, instructions or products referred to in the content.

Communication

Complex Formation of Ag⁺ and Li⁺ with Host Molecules Modeled on Intercalation of Graphite

Yuriko Uetake and Hiroyuki Takemura *

Department of Chemical and Biological Sciences, Faculty of Science, Japan Women's University, Mejirodai 2-8-1, Bunkyo-ku, Tokyo 112-8681, Japan; yuriko.uetake@gmail.com
* Correspondence: takemurah@fc.jwu.ac.jp; Tel.: +81-3-5981-3664

Abstract: Pi-stacked and box-shaped host molecules with xanthene as the basis and pyrene as the π-plane were synthesized to verify cation–π interactions between graphene and metal cations. Since crystal structure analysis was not available, DFT calculations were performed to determine the optimized structure, and the π-planes were found to have a slipped parallel structure, with average distances of 456.2–581.0 pm for the stacked compound and 463.4–471.4 pm for the box-shaped compound. Li⁺ and Ag⁺ were chosen as acceptors for complexation with metal ions, and their interactions with the π-plane were clarified by NMR titration. Clearly, the interaction with metal ions increased when pyrene π-planes were stacked rather than the pyrene itself. In the stacked compound, the association constants of Ag⁺ and Li⁺ were similar; however, in the box-shaped host molecule, only Ag⁺ had moderate coordination ability, but the interaction with Li⁺ was very weak, comparable to the interaction with pyrene. As a result, intercalation is more likely to occur in stacked host compound **1**, which has some degree of freedom in the pyrene rings, than in the box-shaped compound.

Keywords: intercalation; cation intercaland; cation–Pi; Pi-ligand; pyrene

Citation: Uetake, Y.; Takemura, H. Complex Formation of Ag⁺ and Li⁺ with Host Molecules Modeled on Intercalation of Graphite. *Molecules* **2024**, *29*, 3987. https://doi.org/10.3390/molecules29173987

Academic Editors: Sake Wang, Minglei Sun and Nguyen Tuan Hung

Received: 2 August 2024
Revised: 21 August 2024
Accepted: 21 August 2024
Published: 23 August 2024

Copyright: © 2024 by the authors. Licensee MDPI, Basel, Switzerland. This article is an open access article distributed under the terms and conditions of the Creative Commons Attribution (CC BY) license (https:// creativecommons.org/licenses/by/ 4.0/).

1. Introduction

Since the interaction between π-electrons and alkali metal cations in a vacuum has been measured [1,2], the cation–π interaction has been investigated in detail by gas-phase experiments and theoretical calculations [3–6]. Many models have been designed and synthesized and their interactions with cations have been investigated in solutions and solids [7–10]. On the other hand, it is well known that graphite intercalates with metals, metal salts, acids, etc. Complexes of graphene, its constituent unit, with metal ions are currently the subject of basic research on drug delivery [11]. Graphene is also a basic research topic for improving the capacities of Li batteries [12–14]. In addition, the permeation of alkali and alkaline earth metal ions through graphene oxide membranes is also being investigated [15,16]. Graphene, the structural unit of graphite, is rich in π-electrons and is a very interesting target for studying cation–π interactions. We previously designed an intramolecular cation–π recognition system and synthesized compounds in which the cation-binding moiety is a crown ether unit, and pyrene and/or coronene are used as the π-planes [17,18]. Using this system, cation–π interactions in a solution can be visualized. The measurement of the association constants of the complexes and comparison of the conformational and energetic changes by DFT calculations revealed that the two units come closer together as a result of the complexation of the cation in the crown part. In the next stage, the authors focused on graphene as the π-plane. Two molecules were designed and synthesized as models for the intercalation of cations in graphite, and the magnitude of cation–π interactions was estimated. Such studies have been reported only in the case of theoretical calculations, and are few in number [19–21]. Pyrene is a large π-conjugated aromatic compound with a high fluorescence quantum yield that readily exhibits strong excimer emission. The excimer luminescence of pyrene and its derivatives have been widely used for the analytical detection of various compounds and as a common emission source for sensors, organic light-emitting diodes (OLEDs), and organic photovoltaic cells (OPVs). Although many basic studies have reported the fluorescent properties of pyrene for the development of functional materials, its application as an intercaland model has not been reported. However, several types of layered compounds have been reported, as shown in Figure 1. The effect of the molecular configuration of 1,2-di(pyrenyl)benzene on singlet fission (SF) dynamics was investigated by steady-state and time-resolved spectroscopy [22]. Singlet fission (SF) is a process in which two excited triplet states are formed by a spin-allowed transition from an excited singlet state and has been studied for applications in solar cell photovoltaics. A stacked compound with two pyrenes bonded to the 1- and 8-positions of the naphthalene ring, 1,8-bis(pyren-2-yl)naphthalene (BPyN), has been reported to exhibit remarkable single-molecule excimer emission in solution and films [23]. Li et al. synthesized a compound, **X2P**, in which 1-pyrene units were introduced at the 4- and 5-positions of xanthene, and the distance between the pyrene units was kept constant, and described the π-π interaction of the intramolecular excimer in the crystal [24]. We designed stacked compound **1** and boxed compound **2** as model molecules for a metal-ion sandwich complex formation, referring to these types of stacked compounds (Figure 2).

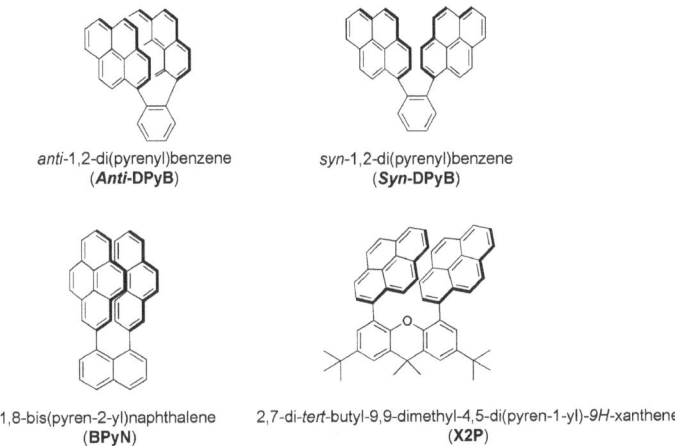

Figure 1. Structures of previously reported pyrene-layered compounds.

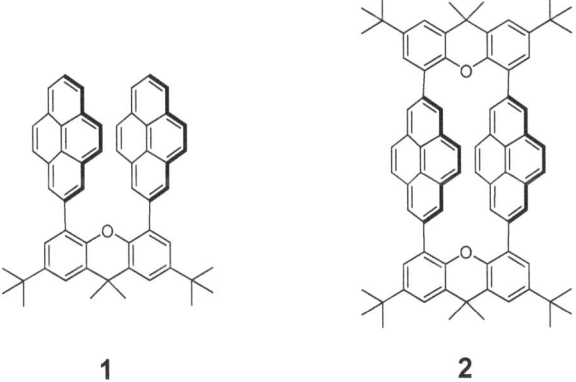

Figure 2. Intercalate model compounds **1** and **2**.

2. Results and Discussion

2.1. Preparation of Compounds

Compound **1** is a known compound and was synthesized according to a previously reported method [25]. Pyrene, bis-pinacolato diboron, [Ir(OMe)COD]$_2$, and dtbpy were added as catalysts to cyclohexane under Ar and heated and stirred at 80 °C for 14 h. After separation by silica gel chromatography, 2-pyreneboronic acid pinacol ester (2-pyrene-BPin) and pyrene-2,7-diboronic acid pinacol ester (2,7-pyrene-diBPin) were simultaneously obtained in a 30.7% and 22.5% yield, respectively (Scheme 1) [26]. Two pyrene boronic acid derivatives obtained from this reaction were used to synthesize compounds **1** and **2**. In the synthesis of **1**, the reaction did not proceed with 2-pyrene-BPin; however, when converted to potassium 2-pyrenyltriolborate and reacted in the presence of potassium carbonate and Pd(PPh$_3$)$_4$, compound **1** was obtained in 24.3% yield [27]. Compound **2** was obtained in 23.0% yield by the reaction of 2,7-pyrene-diBPin with 4,5-dibromo-2,7-di-*tert*-butyl-9,9-dimethyl-9*H*-xanthene in the presence of K$_2$CO$_3$ and Pd(PPh$_3$)$_4$, without using high-dilution conditions.

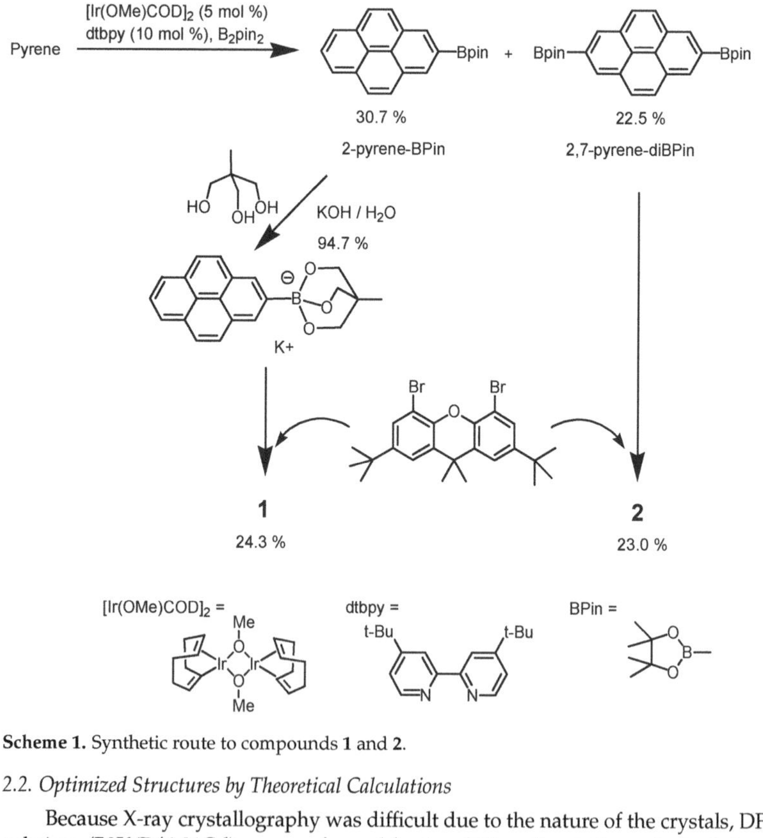

Scheme 1. Synthetic route to compounds **1** and **2**.

2.2. Optimized Structures by Theoretical Calculations

Because X-ray crystallography was difficult due to the nature of the crystals, DFT calculations (B3LYP/6-31Gd) were performed for **1** and **2** to estimate the optimized structures. *tert*-Butyl and methyl groups were omitted from the calculations.

In the case of compound **1**, the π-planes of the optimized structures were found to be slipped parallel (Figure 3). The intersection angles of the xanthene and pyrene rings were 130.5° and 134.1°, respectively. Therefore, the two pyrene rings were not parallel. The distance between the pyrene–pyrene rings was 456.2–581.0 pm, which was equivalent to the spacing of graphite intercalates. Moreover, pyrene rings are flexible, allowing the inclusion of guests of various sizes. The pyrene ring in compound **2** also had a slipped parallel structure, and the distance between the pyrene rings was 463.4–471.4 pm. The angle between the xanthene skeleton and pyrene ring was 130.1°, which was similar to that of compound **1**, although the pyrene ring was more rigidly fixed than that in compound **1**.

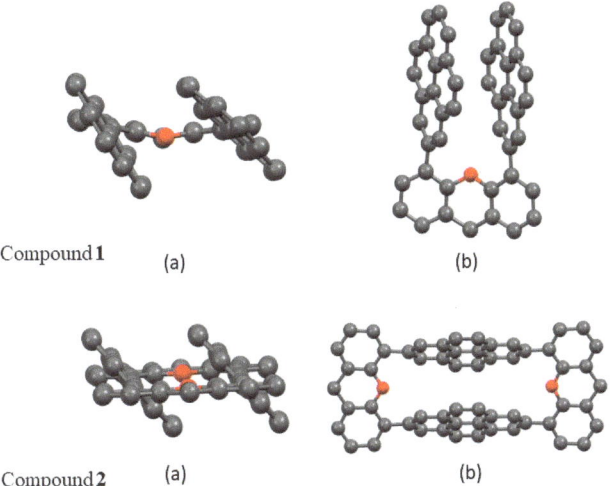

Figure 3. Optimized structures of the main skeletons of compounds **1** and **2** (B3LYP/6-31Gd): (**a**) top view and (**b**) side view.

2.3. Complexation of Li$^+$ and Ag$^+$ Ions

Li$^+$ and Ag$^+$ were chosen as the guest metal cations. Among the alkali metals, Li$^+$ was chosen because it is the hardest, has the smallest ionic radius, and is known to interact strongly with π-electrons [3–6]. Among the transition metals, Ag$^+$ was chosen because it is the most popular metal ion and has long been shown experimentally to interact with π electrons [28–31].

NMR titration (THF-d_8) was performed to investigate the complexation with metal ions (Ag$^+$ and Li$^+$). The data were analyzed using a nonlinear least-squares fitting method with a variant of the Benesi–Hildebrand equation [32]. K_a values were calculated using the change in the chemical shifts in the pyrene ring proton signal. When the host and guest form a 1:1 complex, the chemical shift changes induced by the complexation ($\Delta\delta_{obs}$) can be expressed by Equation (1). For each plot, the binding constants were obtained by curve fitting using Kaleida Graph™ by applying Equation (1).

$$\Delta\delta_{obs} = \frac{\Delta\delta}{2K[H]_0}\left[1 + K[G]_0 + K[H]_0 - \left\{(1 + K[G]_0 + K[H]_0)^2 - 4K^2[G]_0[H]_0\right\}^{\frac{1}{2}}\right] \quad (1)$$

In Equation (1), fitting is possible only when the [Host]/[Guest] ratio of the complex is 1:1. For reconfirmation, a job plot was also constructed for the complexation of **1** with Ag$^+$ ions, and indeed a 1:1 complexation was observed. The results are shown in Table 1. Fitting of the data points obtained as a result of titration yielded good R values (see Figures S5–S11).

When Ag$^+$ was added to the pyrene solution, all the signals of the aromatic rings shifted to a lower field. With compound **1** and Ag$^+$, some signals were high-field-shifted and some were low-field-shifted depending on the proton. In the case of compound **2** and Ag$^+$, the chemical shift was very small, but shifted to higher fields. However, when Li$^+$ was added to pyrene or host molecules **1** and **2**, the proton signal of the host shifted to higher fields in both cases. These phenomena are probably due to the conformational changes in hosts **1** and **2** upon the inclusion of Li$^+$ and Ag$^+$ ions, the different interaction sites, and the different nature of their interactions. The main interaction of Li$^+$ with the π-plane is a cation–π interaction; however, in the case of Ag$^+$, in addition to the cation–π interaction, the d-π* interaction caused by the overlap between the d orbital of Ag$^+$ and the π* orbital of the π-plane is considered to be significant. The titration results clearly showed that the interaction with metal ions increased for pyrene-ring-layered compounds **1** and **2** rather than for pyrene itself. In stacked compound **1**, the K$_a$ values for Ag$^+$ and Li$^+$ were similar, but the ability of **1** to form complexes with Li$^+$ was much higher than that of compound **2** or pyrene. In contrast, box-shaped host molecule **2** had a relatively strong coordination ability only for Ag$^+$, and the interaction with Li$^+$ was as weak as that of pyrene. In the case of compound **1**, this was probably due to the π-electron clustering effect of the two pyrene rings and the fact that the host could flexibly change its interlayer distance to sandwich the guest (Ag$^+$ and Li$^+$). Conversely, the π-plane of the box-shaped host molecule **2** is rigidly fixed, and the solvated Li$^+$ ion, which is bulky, has a large solvation force; therefore, extra energy is required for the Li$^+$ ion to desolvate and then interact with **2**, which causes the pyrene rings to be less likely to interact with the Li$^+$ ion. In contrast, in the case of Ag$^+$, the system may be stabilized by sandwiching between two π-planes that are maintained at a distance suitable for dπ-pπ* interactions. Furthermore, the strength and selectivity of the interaction between Li$^+$, Ag$^+$, and π electrons can be explained based on the HSAB theory. Information on the crystal structures of the complexes of Li$^+$ and Ag$^+$ with **1** or **2** is of great interest for understanding the results of NMR experiments, but all attempts to isolate the complexes in the presence of large excesses of metal ions with compounds **1** and **2** have failed. Yáñez et al. used theoretical calculations to determine the position of Li$^+$ ion in complexes of PAHs and Li$^+$ and showed that there are several structures for the energy minima of Li$^+$-pyrene complexes [33] In the case of our ligand, the system is more complex, and calculations could not be performed; however, it can be inferred that there are several stable structures for Li$^+$-**1** and Li$^+$-**2**.

Table 1. Association constants of pyrene, **1**, and **2** with cations.

Host · Guest	K$_a$ (L·mol^{-1})
Pyrene·Ag$^+$	4.19 ± 1.6
Pyrene·Li$^+$	0.30 ± 0.02
1·Ag$^+$	39.3 ± 9.7
1·Li$^+$	57.6 ± 12.4
2·Ag$^+$	108.1 ± 15.1
2·Li$^+$	0.45 ± 0.09

3. Materials and Methods

3.1. Experimental Section

3.1.1. General Procedure

Melting points were obtained using a Yanaco MP-500D apparatus in Ar-sealed tubes and were uncorrected. NMR spectra were collected on a JEOL AL-300 spectrometer (300.4 MHz for ^{1}H, 75.6 MHz for ^{13}C) with TMS or solvents as internal references. FAB-MS data were collected on a JEOL JMS-SX/SX102A instrument. Silica gel PTLC was performed using PF$_{254}$ (Merck). All reagents were commercially available and used as supplied without further purification.

3.1.2. Synthesis of Compound **2** from 2,7-bis(4,4,5,5-tetramethyl-1,3,2-dioxaborolan-2-yl)pyrene

In a 100 mL three-necked flask, toluene (40 mL), 2 M aq. K$_2$CO$_3$ (4 mL), ethanol (10 mL), 2,7-pyrene-diBPin (0.20 g, 0.44 mmol), 4,5-dibromo-2,7-di-*tert*-butyl-9,9-dimethyl-9*H*-xanthene (0.21 g, 0.46 mmol), and Pd(PPh$_3$)$_4$ (0.050 g, 0.043 mmol) were added and bubbled using Ar gas for 20 min. The mixture was then heated and stirred at 90 °C for two days. The precipitates produced were filtered by suction and dissolved in CH$_2$Cl$_2$, and the solution was filtered through a celite pad. Toluene (10 mL) was added to the filtrate and concentrated to obtain pure **2** as a white powder (0.052 g, 23.0%). M.p. > 460 °C. ^{1}H NMR (300 Hz, CDCl$_3$) δ 1.44 (s, 36H), 1.89 (s, 12H), 7.07 (s, 8H), 7.45 (s, 4H), 7.56 (s, 4H), 7.93 (s, 8H). ^{13}C NMR (300 Hz, CDCl$_3$) δ 31.67, 33.21, 34.68, 34.97, 76.59, 77.02, 77.22, 77.44, 122.35, 123.49, 126.04, 126.44, 126.98, 129.48, 129.61, 130.52, 135.46, 145.36, 145.57. HRMS, calcd for C$_{78}$H$_{73}$O$_2$ (M+H)$^+$ = 1041.5614. Found, 1041.5599.

3.1.3. Silver Ion Complexation with **1**: Preparation of Solutions for ^{1}H NMR Titration

- Preparation of host solution (**1**): Dissolve **1** (0.72 mg, 2×10^{-3} mmol) in 1.0 mL of THF-d_8 to make a solution of 2.0×10^{-3} mol/L.
- Preparation of guest (AgClO$_4$) solution (**2**): AgClO$_4$ (12.4 mg, 0.06 mmol) was dissolved in 0.30 mL of THF-d_8 to make a 0.20 mol/L solution.
- Preparation of guest (AgClO$_4$) solution (**3**): AgClO$_4$ (0.25 g, 1.2 mmol) was dissolved in 0.60 mL THF-d_8 to make a solution of 2.0 mol/L.

Solution (1) (0.30 mL) was placed in NMR tubes, and solution (2) was added to a constant volume (0–80 μL) using a microsyringe. In other NMR tubes, solution (3) was added in 10 μL increments using a microsyringe to prepare a mixed solution. THF-d_8 was then added until the total volume was 0.60 mL and mixed well. The solution was prepared such that the [H]/[G] ratio was 0–120. ^{1}H NMR was performed, and the chemical shifts (pyrene) were recorded. Equation (1), a modification of the Benesi–Hilderand method, was used for the analysis.

The obtained titration curve was analyzed by the nonlinear least square method using Kaleida Graph^{TM} for equation to obtain the binding constant (39.3 $\pm$ 9.7 L mol^{-1}). The following complexation experiments between Ag$^+$, Li$^+$, and pyrene, compounds **1** and **2** were performed in the same way.

3.1.4. Job Plot

A THF-d_8 solution of compound **1** (0.02 mol/L) and a solution of AgClO$_4$ (0.02 mol/L) were prepared. These solutions were added to the NMR tubes in 0.02–0.18 mL increments to give [host] + [guest] = 0.20 mL.

4. Conclusions

Compounds **1** and **2** were synthesized to investigate the complexation of metal cations with layered compounds as a model for the intercalation of graphite and cations. Through computer-assisted optimization, compound **1** was found to have a structure in which the π-planes are in a slipped parallel structure, and the distance between the π-planes is relatively flexible. In the case of compound **2**, the pyrene rings can move in parallel, but the distance is not variable. Pi-plane dimerization is generally more favorable in the slipped parallel form than in the fully overlapped form. Therefore, the misalignment of the pyrene rings in **1** and **2** may be caused by the repulsion of the π-electron dispersion forces. Interactions with the metal ions (Li$^+$ and Ag$^+$) were measured by ^{1}H NMR titration. When AgClO$_4$ and LiClO$_4$ were mixed, the NMR titration results showed 1:1 complexation, which led to the conclusion that the metal ions were encapsulated between the two pyrene rings. Another feature is that the association constant varies depending on the structure of the compound. These results indicate that not only does metal intercalate, which is a process already in practical use, but cations can also intercalate between the graphite layers.

Supplementary Materials: The following supporting information can be downloaded at: https://www.mdpi.com/article/10.3390/molecules29173987/s1.

Author Contributions: Y.U. and H.T.: conceived and designed the project. H.T.: wrote the manuscript. All authors have read and agreed to the published version of the manuscript.

Funding: This research received no external funding.

Institutional Review Board Statement: Not applicable.

Informed Consent Statement: Not applicable.

Data Availability Statement: Data are contained within the article and Supplementary Materials.

Acknowledgments: The authors would like to thank Katsuya Sako of the Nagoya Institute of Technology for the mass spectra measurements.

Conflicts of Interest: The authors declare no conflicts of interest.

References

1. Woodin, R.L.; Beauchamp, J.L. Binding of Li$^+$ to Lewis Bases in the Gas Phase. Reversals in Methyl Substituent Effects for Different Reference Acids. *J. Am. Chem. Soc.* **1978**, *100*, 501–508. [CrossRef]
2. Sunner, J.; Nishizawa, K.; Kebarle, P. Ion-solvent molecule interactions in the gas phase. *Potassium Ion Benzene. J. Phys. Chem.* **1981**, *85*, 1814–1820. [CrossRef]
3. Gal, J.-F.; Maria, P.-C.; Decouzon, M.; Mó, O.; Yáñez, M. Gas-phase lithium-cation basicities of some benzene derivatives: An experimental and theoretical study, Int. *J. Mass Spectrom.* **2002**, *219*, 445–456. [CrossRef]
4. Hoyau, S.; Norrman, K.; McMahon, T.B.; Ohanessian, G.A. Quantitative Basis for a Scale of Na$^+$ Affinities of Organic and Small Biological Molecules in the Gas Phase. *J. Am. Chem. Soc.* **1999**, *121*, 8864–8875. [CrossRef]
5. Armentrout, P.B.; Rodgers, M.T. An Absolute Sodium Cation Affinity Scale: Threshold Collision-Induced Dissociation Experiments and ab Initio Theory. *J. Phys. Chem. A* **2000**, *104*, 2238–2247. [CrossRef]
6. Mó, O.; Yáñez, M.; Gal, J.-F.; Maria, P.-C.; Decouzon, M. Enhanced Li$^+$ Binding Energies in Alkylbenzene Derivatives: The Scorpion Effect. *Chem. Eur. J.* **2003**, *9*, 4330–4338. [CrossRef]
7. Gross, J.; Harder, G.; Siepen, A.; Harren, J.; Vögtle, F.; Stephan, H.; Gloe, K.; Ahlers, B.; Cammann, K.; Rissanen, K. Concave Hydrocarbons. *Chem. Eur. J.* **1996**, *2*, 1585–1595. [CrossRef]
8. Gokel, G.W.; Wall, S.L.; Meadows, E.S. Experimental Evidence for Alkali Metal Cation-π Interactions. *Eur. J. Org. Chem.* **2000**, *17*, 2967–22978. [CrossRef]
9. Choi, H.S.; Kim, D.; Tarakeshwar, P.; Suh, S.B.; Kim, K.S. A New Type of Ionophore Family Utilizing the Cation-Olefinic π Interaction: Theoretical Study of [n]Beltenes. *J. Org. Chem.* **2002**, *67*, 1848–1851. [CrossRef]
10. Takemura, H.; Nagaoka, M.; Kawasaki, C.; Tokumoto, K.; Tobita, N.; Takano, Y.; Iwanaga, T. Synthetic study and structure of cage-type cyclophane $C_{36}H_{36}S_6$. *Tetrahedron Lett.* **2017**, *58*, 1066–1070. [CrossRef]
11. Zaboli, A.; Raissi, H.; Farzad, F.; Hashemzadeh, H.; Fallahi, F. Cation-pi interaction: A strategy for enhancing the performance of graphene-based drug delivery systems. *Inorg. Chem. Commun.* **2022**, *141*, 109542. [CrossRef]
12. Kucinskis, G.; Bajars, G.; Kleperis, J. Graphene in lithium ion battery cathode materials: A review. *J. Power Sources* **2013**, *240*, 66–79. [CrossRef]
13. Kheirabadi, N.; Shafiekhani, A. Graphene/Li-ion battery Crossmark: Check for Updates. *J. Appl. Phys.* **2012**, *112*, 124323. [CrossRef]
14. Cai, X.; Lai, L.; Shen, Z.; Lin, J. Graphene and graphene-based composites as Li-ion battery electrode materials and their application in full cells. *J. Mater. Chem. A* **2017**, *5*, 15423–15446. [CrossRef]
15. Sun, P.; Zheng, F.; Zhu, M.; Song, Z.; Wang, K.; Zhong, M.; Wu, D.; Little, R.B.; Xu, Z.H. Selective Trans-Membrane Transport of Alkali and Alkaline Earth Cations through Graphene Oxide Membranes Based on Cation-π Interactions. *ACS Nano* **2014**, *8*, 850–859. [CrossRef]
16. Zhao, G.; Zhu, H. Cation–π Interactions in Graphene-Containing Systems for Water Treatment and Beyond. *Adv. Mater.* **2020**, *32*, 1905756. [CrossRef] [PubMed]
17. Takemura, H.; Nakamichi, H.; Sako, K. Pyrene−azacrown ether hybrid: Cation−π interaction. *Tetrahedron Lett.* **2005**, *46*, 2063–2066. [CrossRef]
18. Takemura, H. and Sako, K. Li$^+$···π interaction in coronene−azacrown ether system. *Tetrahedron Lett.* **2005**, *46*, 8169–8172. [CrossRef]
19. Li, Y.; Lu, Y.; Adelhelm, P.; Titirici, M.-M.; Hu, Y.-S. Intercalation chemistry of graphite: Alkali metal ions and beyond. *Chem Soc Rev.* **2019**, *48*, 4655–4687. [CrossRef]
20. Kim, Y.-O.; Park, S.-M. Intercalation Mechanism of Lithium Ions into Graphite Layers Studied by Nuclear Magnetic Resonance and Impedance Experiments. *J. Electrochem. Soc.* **2001**, *148*, A194–A199. [CrossRef]
21. Maeda, Y.; Sugimori, D.; Inagaki, M. Electrochemical Intercalation of Alkali Metal Ions into Graphite. *Tanso* **1991**, *149*, 244–247. [CrossRef]
22. Choi, J.; Kim, S.; Ahn, M.; Kim, J.; Cho, D.; Kim, D.; Eom, S.; Im, D.; Kim, Y.; Kim, S.H.; et al. Singlet fission dynamics modulated by molecular configuration in covalently linked pyrene dimers, *Anti*- and *Syn*-1,2-di(pyrenyl)benzene. *Communs. Chem.* **2023**, *6*, 1–11. [CrossRef]
23. Hu, J.-Y.; Pu, Y.-J.; Nakata, G.; Kawata, S.; Sasabe, H.; Kido, J. A single-molecule excimer-emitting compound for highly efficient fluorescent organic light-emitting devices. *Chem. Commun.* **2012**, *48*, 8434–8436. [CrossRef]
24. Wang, J.; Dang, Q.; Gong, Y.; Liao, Q.; Song, G.; Li, Q.; Li, Z. Precise Regulation of Distance between Associated Pyrene Units and Control of Emission Energy and Kinetics in Solid State. *CCS Chem.* **2021**, *3*, 274–286. [CrossRef]

25. Li, G.-Q.; Yamamoto, Y.; Miyaura, N. Double-coupling of dibromo arenes with aryltriolborates for synthesis of diaryl-substituted planar frameworks. *Tetrahedron* **2011**, *67*, 6804–6811. [CrossRef]
26. Coventry, D.N.; Batsanov, A.S.; Goeta, A.E.; Howard, J.A.K.; Marder, T.B.; Perutz, R.N. Selective Ir-catalysed borylation of polycyclic aromatic hydrocarbons: Structure of naphthalene-2,6-bis(boronate), pyrene-2,7-bis(boronate) and perylene-2,5,8,11-tetra(boronate) esters. *Chem. Commun.* **2005**, *16*, 2172–2174. [CrossRef] [PubMed]
27. Akula, M.R.; Yao, M.-L.; Kabalka, G.W. Triolborates: Water-soluble complexes of arylboronic acids as precursors to iodoarenes. *Tetrahedron Lett.* **2010**, *51*, 1170–1171. [CrossRef]
28. Kammermeier, S.; Jones, P.G.; Dix, I.; Herges, R. A Silver(I) Complex of a Tube-Shaped Hydrocarbon. *Acta Cryst.* **1998**, *C54*, 1078–1081. [CrossRef]
29. Seppälä, T.; Wegelius, E.; Rissanen, K. [2.2.2]m,p,p- and [2.2.2]m,m,p-Cyclophane-Ag-triflate: New p-prismand complexes. *New J. Chem.* **1998**, *22*, 789–791. [CrossRef]
30. McMurry, J.E.; Haley, G.J.; Matz, J.R.; Clardy, J.C.; Mitchell, J. Pentacyclo[12.2.2.2^{2,5}.2^{6,9}.2^{10,13}]-1,5,9,13-tetracosatetraene and Its Reaction with AgOTf. Synthesis of a Square-Planar d^{10} Organometallic Complex. *J. Am. Chem. Soc.* **1986**, *108*, 515–516. [CrossRef]
31. Lindeman, S.V.; Rathore, R.; Kochi, J.K. Silver(I) Complexation of (Poly)aromatic Ligands. Structural Criteria for Depth Penetration into *cis*-Stilbenoid Cavities. *Inorg. Chem.* **2000**, *39*, 5707–5716. [CrossRef] [PubMed]
32. *Experimental Methods in Biofunctional Chemistry*, The Chemical Society of Japan, Division of Biofunctional Chemistry, ed.; The Chemical Society of Japan: Tokyo, Japan, 2003.
33. Gal, J.-F.; Maria, P.-C.; Decouzon, M.; Mó, O.; Yáñez, M.; Abboud, J.L.M. Lithium-Cation/π Complexes of Aromatic Systems. The Effect of Increasing the Number of Fused Rings. *J. Am. Chem. Soc.* **2003**, *125*, 10394–10401. [CrossRef] [PubMed]

Disclaimer/Publisher's Note: The statements, opinions and data contained in all publications are solely those of the individual author(s) and contributor(s) and not of MDPI and/or the editor(s). MDPI and/or the editor(s) disclaim responsibility for any injury to people or property resulting from any ideas, methods, instructions or products referred to in the content.

Article

Density Functional Theory-Based Indicators to Estimate the Corrosion Potentials of Zinc Alloys in Chlorine-, Oxidizing-, and Sulfur-Harsh Environments

Azamat Mukhametov [1,†], Insaf Samikov [1,†], Elena A. Korznikova [1,2] and Andrey A. Kistanov [1,*]

1 The Laboratory of Metals and Alloys Under Extreme Impacts, Ufa University of Science and Technology, 450076 Ufa, Russia; elena.a.korznikova@gmail.com (E.A.K.)
2 Polytechnic Institute (Branch) in Mirny, North-Eastern Federal University, 678170 Mirny, Russia
* Correspondence: andrei.kistanov.ufa@gmail.com
† These authors contributed equally to this work.

Abstract: Nowadays, biodegradable metals and alloys, as well as their corrosion behavior, are of particular interest. The corrosion process of metals and alloys under various harsh conditions can be studied via the investigation of corrosion atom adsorption on metal surfaces. This can be performed using density functional theory-based simulations. Importantly, comprehensive analytical data obtained in simulations including parameters such as adsorption energy, the amount of charge transferred, atomic coordinates, etc., can be utilized in machine learning models to predict corrosion behavior, adsorption ability, catalytic activity, etc., of metals and alloys. In this work, data on the corrosion indicators of Zn surfaces in Cl-, S-, and O-rich harsh environments are collected. A dataset containing adsorption height, adsorption energy, partial density of states, work function values, and electronic charges of individual atoms is presented. In addition, based on these corrosion descriptors, it is found that a Cl-rich environment is less harmful for different Zn surfaces compared to an O-rich environment, and more harmful compared to a S-rich environment.

Keywords: dataset; adsorption energy; work function; surface; single-atom adsorption

Citation: Mukhametov, A.; Samikov, I.; Korznikova, E.A.; Kistanov, A.A. Density Functional Theory-Based Indicators to Estimate the Corrosion Potentials of Zinc Alloys in Chlorine-, Oxidizing-, and Sulfur-Harsh Environments. *Molecules* **2024**, *29*, 3790. https://doi.org/10.3390/molecules29163790

Academic Editors: Sake Wang, Minglei Sun and Nguyen Tuan Hung

Received: 30 July 2024
Revised: 8 August 2024
Accepted: 8 August 2024
Published: 10 August 2024

Copyright: © 2024 by the authors. Licensee MDPI, Basel, Switzerland. This article is an open access article distributed under the terms and conditions of the Creative Commons Attribution (CC BY) license (https://creativecommons.org/licenses/by/4.0/).

1. Introduction

Surface adsorption is one of the fundamental processes in many fields, including catalysis, the environment, energy, and medicine. In medicine, biodegradable metals and alloys and their corrosion behavior are of particular interest. In turn, the corrosion process of metals and alloys is associated with the adsorption of corrosive atoms to their surfaces. The corrosion process of biodegradable metals and alloys can be studied in an immersion test. In that case, biodegradable metals and alloys are placed in special solutions, such as Hanks' solution, compositionally similar to blood plasma. These solutions are rich in various minerals, such as sodium, calcium, magnesium carbonates, and bicarbonates [1]. In addition, it is also crucial to understand the corrosion behavior of biodegradable metals and alloys in sulfur- and oxygen-enriched atmospheres [2–4]. Nowadays, various computational approaches have become trustable and utilize comparably fast and cheap tools to assess the corrosion behavior of biodegradable metals and alloys [5–8]. For instance, using density functional theory (DFT)-based modeling, it has been shown that in α-Al$_2$O$_3$(0001), the insertion of a Cl atom in an aluminum vacancy is an endothermic process and the activation energy for the Cl ingress exceeds 2 eV, while the insertion a Cl atom in an oxygen vacancy is an exothermic process [9]. DFT-based simulations have uncovered the fundamental mechanism of the interaction between pure and doped TiO$_2$ with S and O species [10]. Another theoretical investigation of anticorrosion properties of doped Ni-based alloys in Br-rich and O-rich environments can be enhanced via adsorption inhibition [11]. Pure metal surfaces have also been actively studied in terms of corrosion behaviors in different corrosion mediums. Experimental scanning tunneling microscope results combined with DFT calculations have revealed a dynamic process of chlorine adsorption on the Au surface, where Au atoms form a complex superlattice of a Au–Cl surface compound [12]. In the framework of DFT, the coverage of the Fe(100) surface by Cl atoms has been studied [13]. Adsorption energies and buckling distances have been calculated for various coverages as a percentage. Notably, a limited number of works are available on corrosive atom adsorption on Zn surfaces. The DFT results have shown that the S species can chemically interact with Zn atoms at the smithsonite surface and repel water molecules from it [14]. In this study [15], among 22 considered single atoms, Al, Ag, Cd, In, Sn, Au, Hg, Tl, and Bi atoms have been sorted out as doping atoms that possess a weak adsorption energy and low diffusion activation energy on the Zn surface.

Meanwhile, comprehensive analytical data, such as adsorption energy, the amount of charge transferred, atomic coordinates (bond lengths, bond angles, and interatomic distances), etc., can be used in machine learning models as indicators to predict corrosion behavior, adsorption ability, catalytic activity, etc., of materials [16]. For example, the adsorption characteristics of graphene modified using single-atom adsorption has been investigated via DFT-based methods [17]. Various adsorption features, such as atomic and electronic structures, magnetic properties, and adsorption energies of single-atom-modified graphene have been collected and analyzed. These descriptors can be used in machine learning models for the development and design of graphene-based single-atom catalysts. In the work [18], the adsorption of single atoms on the graphene surface of a graphene/aluminum composite was studied using DFT methods supplied with a machine learning approach. A dataset was created containing basic information on atoms, such as atomic radius, ionic radius, etc., as well as adsorption energy and interatomic distances of about thirty atoms at the graphene surface. As a result, it has been shown that single atoms of individual elements, such as Zr, Ti, Sc, and Si can affect the reaction in the interfacial region of graphene and aluminum. Furthermore, the hydrogen storage capacity of MXene materials was tested based on the machine learning models using hydrogen adsorption energy as a main descriptor [19]. Accordingly, several bilayer MXenes with excellent hydrogen adhesion and storage capacities have been designed. Another recent study [20] has introduced the hierarchically interacting particle neural network to predict the energies of molecules based on their physical principles available in the QM9 dataset. Such a method can also work with the data obtained from ab initio molecular dynamics calculations; thus, a dataset can be created based on a specific task. A similar model based on the hierarchically interacting particle neural network has later been used to predict single-atom adsorption on metal and bimetal surfaces [21]. Adsorption energy, adsorption height, and buckling of the surface has been predicted for H, N, and O atoms adsorbed on clean FCC metal surfaces. Some advanced machine learning-based models can also predict the dissociative adsorption energy of single molecules to metal nanoparticles [22] and surfaces [23]. Therefore, the collection of data that will facilitate the development of an adsorption model is a long-term goal in surface and interface science.

This work aims to address two scientific challenges: assessing the corrosion behavior of Zn surfaces in the Cl-, O-, and S-rich harsh environments, and collecting data on the corrosion indicators of Zn surfaces in Cl-, S-, and O-rich harsh environments. A dataset containing the adsorption height d, adsorption energy E_{ads}, partial density of states (PDOS), work function (WF) values, and electronic charges of individual atoms for the single Cl, O, and S atoms on the Zn(111), Zn(110), and Zn(100) surfaces was created. Consequently, based on the data obtained, the corrosion ability of Zn surfaces was evaluated in relation to their tendency to adsorb corrosive Cl, O, and S atoms.

2. Results

The interaction of corrosive atoms with Zn surfaces is considered based on various adsorption characteristics. These characteristics are calculated and collected in Table 1 and Table S1. First, the lowest energy configurations of Cl, O, and S atoms on the Zn(111),

Zn(110), and Zn(100) surfaces are studied. For that, studied adsorbates are placed at the high-symmetry adsorption sites [24,25] as shown in Figure 1. According to Figure 2 and Table 1, the hcp site is the most favorable for a Cl atom on the Zn(111) and Zn(110) surfaces, while the bridge site is the preferable site for a Cl atom on the Zn(100) surface. In the case of O (Figure S1) and S (Figure S2) atoms, the lowest energy positions on the Zn(111), Zn(110), and Zn(100) surfaces are the bridge, hcp, and bridge, respectively.

Table 1. Results for the lowest-energy configurations of Cl, O, and S atoms on the Zn(111), Zn(110), and Zn(100) surfaces. The distance d between the atom and the surface, adsorption energy E_a, and the amount of charge transfer Δq to/from the atom on the surface. A positive (negative) Δq indicates a loss (gain) of electrons.

Structure	Position	d, Å	E_a, eV	Doping Nature	Δq, e
Zn(111) + Cl	hcp	1.54	−2.88	acceptor	0.598
Zn(110) + Cl	hcp	1.99	−2.16	acceptor	0.595
Zn(100) + Cl	bridge	1.68	−2.38	acceptor	0.601
Zn(111) + O	bridge	0.27	−7.30	acceptor	1.231
Zn(110) + O	hcp	0.85	−6.81	acceptor	1.196
Zn(100) + O	bridge	0.42	−7.08	acceptor	1.195
Zn(111) + S	bridge	1.44	−4.69	acceptor	0.823
Zn(110) + S	hcp	1.56	−4.39	acceptor	0.820
Zn(100) + S	bridge	0.81	−4.96	acceptor	0.825

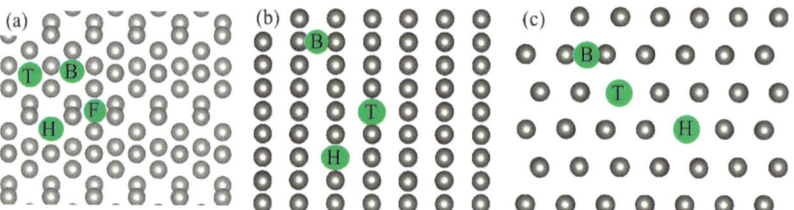

Figure 1. High-symmetry adsorption sites. FCC (F), hcp (H), bridge (B), and top (T) on the (**a**) Zn(111), (**b**) Zn(110), and (**c**) Zn(100) surfaces. Zn atoms and high-symmetry adsorption sites are indicated by gray balls and green balls, respectively.

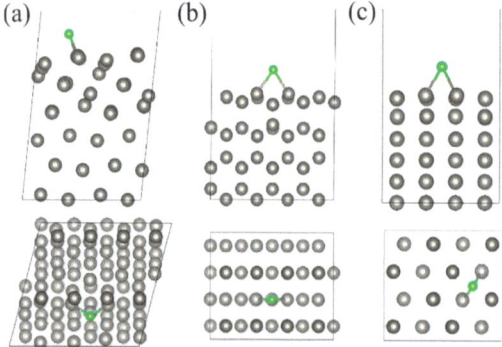

Figure 2. The side (the upper panel) and top (the lower panel) views of the lowest-energy configuration of the Cl atom adsorbed on the (**a**) Zn(111), (**b**) Zn(110), and (**c**) Zn(100) surfaces. Zn and Cl atoms are indicated by gray and green balls, respectively.

Adsorption height d and E_{ads} are other representative markers of adsorption. The lower the distance from the corrosive adsorbate to the surface atom and the more negative the E_{ads} value is, the stronger the adsorbate binds to the surface and, consequently, the lower the resistance of the surface to the corrosion [11]. According to Table 1, the Cl atom possesses the shortest d = 1.54 Å and the lowest E_{ads} = −2.88 eV on the Zn(111) surface. For comparison, an E_{ads} of −4.44 eV has been previously reported for the Cl atom on the Fe(100) surface [13]; thus, Zn may be more stable than Fe in Cl-rich mediums, while both will corrode rapidly. Intercalation of the O atom to the surface with the lowest E_{ads} of ~−7.30 eV is observed in the case of the Zn(111) surface. The S atom adsorbed on the Zn surface has the shortest d of 0.81 Å and the lowest E_{ads} of ~−4.96 eV at the bridge site of the Zn(100) surface.

The binding of the adsorbate atom to the surface can cause a remarkable charge redistribution on the reacting atoms. These changes can also be valuable descriptors of the corrosion process. Differential charge density (DCD) graphs and a Bader analysis are utilized to visualize and to quantify the charge transfer upon the atoms' adsorption on Zn surfaces. Figure 3 shows the DCD plots for the Cl atom adsorbed on the Zn(111), Zn(110), and Zn(100) surfaces. The accumulation of electrons is observed in the Cl atom, while electron depletion is found on the surface Zn atoms surrounding the Cl atom. The Bader analysis (Table 1) proves the charge transfers of 0.598 e, 0.595 e, and 0.601 e, respectively, from Zn atoms on the Zn(111), Zn(110), Zn(100) surfaces to the adsorbed Cl atom. Similarly, the DCD plots in Figure S3 suggest the charge transfer occurs from the surface Zn atoms to the adsorbed O atom. The amount of the charge transferred from the Zn(111), Zn(110), and Zn(100) surfaces to the O atom is 1.231 e, 1.196 e, and 1.195 e, respectively (Table 1). According to Figure S4 and Table 1, there is a charge transferred from the surface Zn atoms to the S atom. This is confirmed by the Bader analysis, which suggests the S atom is an acceptor to the Zn(111), Zn(110), and Zn(100) surfaces with the amount of the charge transferred being 0.823 e, 0.820 e, and 0.825 e per atom, respectively.

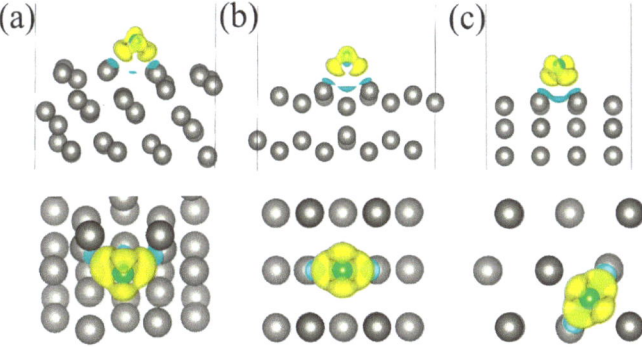

Figure 3. The DCD isosurface plots (0.005 Å^{-3}) of the Cl atom adsorbed on (**a**) Zn(111), (**b**) Zn(110), and (**c**) Zn(100) surfaces. The yellow (blue) color represents an accumulation (depletion) of electrons.

PDOS diagrams of a Cl atom after adsorption on the Zn(111), Zn(110), and Zn(100) surfaces is shown in Figure 4a–c. Energy level-splitting of Cl-p orbitals produced by the spin–orbit interaction is visible in Figure 4a,c. In the case of Cl adsorbed on the Zn(111) surface (Figure 4a), the S-p orbital is splitting and broadening because of a strong coupling to the Zn-d orbital. For the Cl atom adsorbed on the Zn(110) surface (Figure 4b), there is no spin decomposition, while the Cl-p orbital is significantly broadened, which can be due to Cl atoms having the highest E_{ads} on the Zn(110) surface compared to the other surfaces considered. In the case of the Cl atom on the Zn (100) surfaces (Figure 4c), a decomposition and significant broadening of the Cl-p orbital is observed, indicating a strong hybridization between the Cl-p and the Zn-d orbitals. PDOS graphs for the O atom adsorbed on the Zn(111), Zn(110), and Zn(100) surfaces is shown in Figure S5a–c. A strong degeneracy of the O-p orbital of the O atom due to the spin–orbit interaction with the Zn-d orbital of the surface Zn atoms can be seen in the case of the Zn(111) (Figure S5a), Zn(110) (Figure S5b), and Zn(100) (Figure S5c) surfaces. Figure S6a–c present PDOS diagrams of the S atom adsorbed on the Zn(111), Zn(110), and Zn(100) surfaces, respectively. Energy level-splitting for the S-p orbital produced by the spin–orbit interaction is visible in Figure S6. In the case of S adsorbed on the Zn(111) surface (Figure S6a), the S-p orbital is split and broadened because of a strong coupling to the Zn-d orbital. For S adsorbed on the Zn(110) (Figure S6b) and Zn(100) (Figure S6c) surfaces, the S-p orbital is broadened, signifying coupling to the Zn-d orbital, while no spin decomposition is found. The PDOS plots can be useful for predicting the broadening of inner molecular orbitals of adsorbed atoms (molecules, etc.); thus, they can facilitate the prediction of selectivity and corrosion resistance of metal surfaces [26,27].

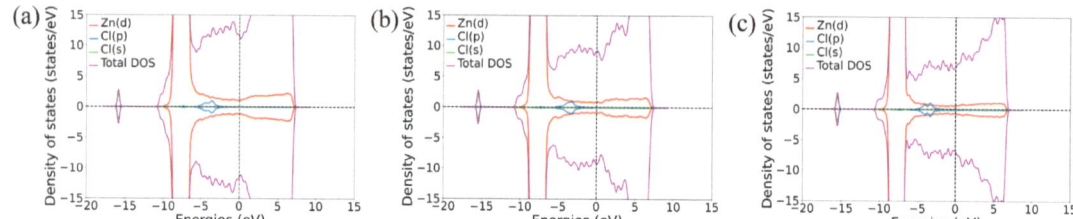

Figure 4. PDOS diagrams of a Cl atom adsorbed on (**a**) Zn(111), (**b**) Zn(110), and (**c**) Zn(100) surfaces.

Another characteristic feature of the surface that can be changed by adsorbed atoms is the WF value [28–30]. The WF values for the pure and Cl-, O-, and S-adsorbed Zn(111), Zn(110), and Zn(100) surfaces are calculated and summarized in Table 2. The presence of adsorbed atoms on the Zn surface leads to a noticeable increase in the WF value. The highest WF values of 4.46 eV and 4.36 are found for the S-adsorbed Zn(111) surface and for the Cl-adsorbed Zn(110) surface, respectively. This can be explained based on the above-mentioned strong charge flows (Figures 3, S3 and S4) from the surface toward the adsorbate, which can lead to the dipole formation on the Zn surface. Such a dipole can be attributed to the WF modifications [31–33].

Table 2. WF (eV) values for the cases of the lowest-energy configurations of Cl, O, and S atoms on the Zn(111), Zn(110), and Zn(100) surfaces.

Structure	Pure	Cl	O	S
Zn(111)	4.18	4.35	4.20	4.46
Zn(110)	4.15	4.36	4.25	4.34
Zn(100)	4.03	4.27	4.10	4.16

Figure 5a–d below present the descriptors collected in this work, such as adsorption height, adsorption energy, Bader charges of individual atoms, and work function values. A clear correlation between these physical characteristics can be seen. Specifically, the lower adsorption height (Figure 5a), the lower adsorption energy (Figure 5b), and the higher charge redistribution (Figure 5c) between the Cl, O, and S adsorbates and the Zn surfaces. The highest sensitivity of the Zn surfaces is attributed to the O atom, while the lowest is to the Cl atom. Notably, the change in the work function values of the Zn surfaces has an inverse relationship with the amount of charge transferred from the surface to the specific adsorbate (Figure 5d).

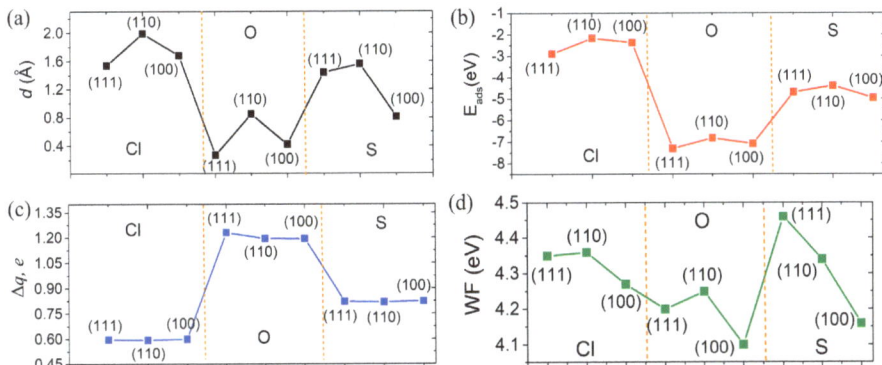

Figure 5. (**a**) The distance d between the atom and the surface, (**b**) adsorption energy E_a, (**c**) Bader charge per adsorbed atom, and (**d**) WF value for the lowest-energy configurations of Cl, O, and S atoms on the Zn(111), Zn(110), and Zn(100) surfaces.

3. Conclusions

The development of an analytical model based on physical parameters characterizing the local atomic environment, such as structural parameters and electronic features can open the door to fast predictions of the adsorption energy landscape, selectivity, and chemical activity of metal surfaces toward various atomic species. However, a proper selection and comprehensive descriptions of these structural parameters and electronic features are needed. This work presents an atomic-scale consideration of corrosive Cl, O, and S atoms interacting with Zn surfaces. A strong correlation between adsorption height, adsorption energy, Bader charges of individual atoms, and work function value change is shown. Based on the descriptors presented in this work, it is concluded that a Cl-rich environment is less harmful for the Zn surface compared to an O-rich environment, and more harmful compared to a S-rich environment. Furthermore, the obtained data is collected and presented in the form of dataset. Following the collected data, one can check the sensitivity of a given Zn surface toward Cl-, O-, and S-rich corrosive media. Moreover, the presented PDOS plots can be used to evaluate the selectivity and corrosion resistance of Zn surfaces. A potential future research direction is to collect corrosion descriptors for various metal surfaces to expand the dataset for predicting their corrosion behavior.

4. Materials and Methods

Spin-polarized simulations were conducted based on the DFT using the Vienna Ab Initio Simulation Package (VASP) [34]. The projector augmented plane–wave method [35] was used to treat the ion–electron interactions. An energy cut-off of 540 eV was adopted for the plane–wave expansion of the electronic wave functions. The exchange correlation functional based on the general gradient approximation of Perdew–Burke–Ernzerh [36] was used in conjunction with Grimme's DFT-D3 method [37] to accurately describe the dispersion correction of long-range van der Waals interactions between atoms and surfaces. All considered structures were fully optimized until the maximum force acting on each atom was less than 0.02 eV/Å. The change in the total energy was less than 10^4 eV in the case of structure optimization and less than 10^8 eV in the case of electronic self-consistent simulations. The Brillouin zone was sampled with a $4 \times 4 \times 1$ and $8 \times 8 \times 1$ centered k-mesh grid for the bulk Zn and for the surfaces of Zn, respectively. The functional and k-point sampling choice was based on previous work [7].

The unit cell of zinc was optimized using variable–cell relaxation, where all degrees of freedom, such as volume, shape, and internal atomic positions were allowed to relax for structural optimization. The calculated lattice parameters of zinc of $a = b = 2.64$ and $c = 4.71$ Å was found to be in good agreement with the literature. The surfaces of Zn were supposed to have different reactivities. For instance, the 100 surface has the lowest surface energy, thus it may be more favorable for adsorption, while the 111 and 110 surfaces possess a much larger surface energy, thus they may be less favorable for adsorption [7,38]. Therefore, these 111, 110, and 100 surfaces were considered. The size of the created Zn slabs was selected to avoid self-interaction of the replicated cells. A vacuum space of 20 Å was used in the out-of-plane direction relative to the surface plane.

The strength of adsorption of an atom to a Zn slab surface is measured based on the E_{ads}, which is determined as follows:

$$E_{ads} = E_{slab+a} - E_a - E_{slab}, \tag{1}$$

where E_{slab+a} is the energy of the atom adsorbed on the slab surface, E_a is the energy of the pure atom, and E_{slab} is the energy of the pure slab surface.

The DCD $\Delta\rho(r)$ is determined as follows:

$$\Delta\rho(r) = \rho_{E slab+a}(r) - \rho_a(r) - \rho_{E slab}(r), \tag{2}$$

where, $\rho_{E slab+a}(r)$, $\rho_a(r)$, and $\rho_{E slab}(r)$ are the charge densities of the atom-adsorbed slab surface, the pure atom, and the pure slab surface.

The WF is determined as follows:

$$WF = E_{vac} - E_{Fermi} \tag{3}$$

where E_{vac} is the energy level of a stationary electron in the vacuum and E_{Fermi} corresponds to the Fermi level of the system.

The amount of charge transferred between the atom and the slab surface due to adsorption was quantitatively calculated using Bader charge analysis [39]. Structural and DCD analysis and plotting were conducted using VESTA 3 programs [40], while the PDOS analysis and plotting were conducted utilizing an open-source programming language (Python) [41].

Supplementary Materials: The following supporting information can be downloaded at: https://www.mdpi.com/article/10.3390/molecules29163790/s1, Table S1: The distance d between the atom and the surface, adsorption energy E_a, and the amount of charge transfer Δq to/from the atom on the surface. A positive (negative) Δq indicates a loss (gain) of electrons.; Figure S1: The side (the upper panel) and top (the lower panel) views of the lowest-energy configuration of the O atom adsorbed on (a) Zn(111), (b) Zn(110), and (c) Zn(100) surfaces.; Figure S2: The side (the upper panel) and top (the lower panel) views of the lowest-energy configuration of the S atom adsorbed on (a) Zn(111), (b) Zn(110), and (c) Zn(100) surfaces.; Figure S3: The DCD isosurface plots (0.005 Å^{-3}) of the O atom adsorbed on (a) Zn(111), (b) Zn(110), and (c) Zn(100) surfaces. The green (blue) color represents an accumulation (depletion) of electrons.; Figure S4: The DCD isosurface plots (0.005 Å^{-3}) of the S atom adsorbed on (a) Zn(111), (b) Zn(110), and (c) Zn(100) surfaces. The green (blue) color represents an accumulation (depletion) of electrons.; Figure S5: PDOS diagrams of O atom (a) before interaction and adsorbed on (b) Zn(111), (c) Zn(110), and (d) Zn(100) surfaces.; Figure S6: PDOS diagrams of (a) S atom before interaction and S atom adsorbed on (b) Zn(111), (c) Zn(110), and (d) Zn(100) surfaces.

Author Contributions: Conceptualization, A.A.K.; methodology, A.A.K.; validation, E.A.K. and A.A.K.; formal analysis, A.A.K.; investigation, A.M. and I.S.; resources, E.A.K.; data curation, A.A.K.; writing—original draft preparation, A.M., I.S. and A.A.K.; writing—review and editing, A.M., I.S., E.A.K. and A.A.K.; visualization, A.M., I.S. and A.A.K.; supervision, E.A.K. and A.A.K.; project administration, A.A.K.; funding acquisition, E.A.K. All authors have read and agreed to the published version of the manuscript.

Funding: The work was supported by the Ministry of Science and Higher Education of the Russian Federation within the framework of the state task of the Ufa University of Science and Technologies (No. 075-03-2024-123/1) of the youth research laboratory "Metals and Alloys under Extreme Impacts".

Institutional Review Board Statement: Not applicable.

Informed Consent Statement: Not applicable.

Data Availability Statement: The raw data supporting the conclusions of this article will be made available by the authors upon request.

Conflicts of Interest: The authors declare no conflicts of interest.

References

1. Wątroba, M.; Mech, K.; Bednarczyk, W.; Kawałko, J.; Marciszko-Wiąckowska, M.; Marzec, M.; Shepherd, D.E.T.; Bała, P. Long-term in vitro corrosion behavior of Zn-3Ag and Zn-3Ag-0.5Mg alloys considered for biodegradable implant applications. *Mater. Des.* **2022**, *213*, 110289. [CrossRef]
2. Shao, L.; Xie, G.; Zhang, C.; Liu, X.; Lu, W.; He, G.; Huang, J. Combustion of metals in oxygen-enriched atmospheres. *Metals* **2020**, *10*, 128. [CrossRef]
3. Langman, J.B.; Ali, J.D.; Child, A.W.; Wilhelm, F.M.; Moberly, J.G. Sulfur species, bonding environment, and metal mobilization in mining-impacted lake sediments: Column experiments replicating seasonal anoxia and deposition of algal detritus. *Minerals* **2020**, *10*, 849. [CrossRef]
4. O'Donnell, J.A.; Carey, M.P.; Koch, J.C.; Baughman, C.; Hill, K.; Zimmerman, C.E.; Sullivan, P.F.; Dial, R.; Lyons, T.; Cooper, D.J.; et al. Metal mobilization from thawing permafrost to aquatic ecosystems is driving rusting of Arctic streams. *Commun. Earth Environ.* **2024**, *5*, 268. [CrossRef]
5. Dong, H.; Lin, F.; Boccaccini, A.R.; Virtanen, S. Corrosion behavior of biodegradable metals in two different simulated physiological solutions: Comparison of Mg, Zn and Fe. *Corr. Sci.* **2021**, *182*, 109278. [CrossRef]
6. Barzegari, M.; Mei, D.; Lamaka, S.V.; Geris, L. Computational modeling of degradation process of biodegradable magnesium biomaterials. *Corr. Sci.* **2021**, *190*, 109674. [CrossRef]
7. Bryzgalov, V.; Kistanov, A.A.; Khafizova, E.; Polenok, M.; Izosimov, A.; Korznikova, E.A. Experimental study of corrosion rate supplied with an ab-initio elucidation of corrosion mechanism of biodegradable implants based on Ag-doped Zn alloys. *Appl. Surf. Sci.* **2024**, *652*, 159300. [CrossRef]
8. Chiter, F.; Bonnet, M.L.; Lacaze-Dufaure, C.; Tang, H.; Pebere, N. DFT studies of the bonding mechanism of 8-hydroxyquinoline and derivatives on the (111) aluminum surface. *Phys. Chem. Chem. Phys.* **2018**, *20*, 21474–21486. [CrossRef] [PubMed]
9. Liu, M.; Jin, Y.; Leygraf, C. A DFT-Study of Cl Iingress into α-Al2O3(0001) and Al(111) and its possible influence on localized corrosion of Al. *J. Electrochem. Soc.* **2019**, *166*, 3124–3130. [CrossRef]
10. Pan, Y.; Guan, W. Origin of enhanced corrosion resistance of Ag and Au doped anatase TiO_2. *Int. J. Hydrogen Energy* **2019**, *44*, 10407–10414. [CrossRef]
11. Li, J.; Zhao, J.; Shao, H.; Zhang, Y.; Dong, H.; Xia, L.; Hu, S. Anticorrosion mechanism of Cr-doped nickel-base alloy in Br/O environment: A DFT study. *Mol. Simul.* **2019**, *45*, 1506–1514. [CrossRef]
12. Gao, W.; Baker, T.A.; Zhou, L.; Pinnaduwage, D.S.; Kaxiras, E.; Friend, C.M. Chlorine adsorption on Au(111): Chlorine overlayer or surface chloride? *J. Am. Chem. Soc.* **2008**, *130*, 3560–3565. [CrossRef]
13. Saraireh, S.A.; Altarawneh, M.; Tarawneh, M.A. Nanosystem's density functional theory study of the chlorine adsorption on the Fe(100) surface. *Nanotech. Rev.* **2021**, *10*, 719–727. [CrossRef]
14. Liu, J.; Zeng, Y.; Ejtemaei, M.; Nguyen, A.V.; Wang, Y.; Wen, S. DFT simulation of S-species interaction with smithsonite (001) surface: Effect of water molecule adsorption position. *Results Phys.* **2019**, *15*, 102575. [CrossRef]
15. Kang, M.; Huang, K.; Li, J.; Liu, H.; Lian, C. Theoretical design of dendrite-free zinc anode through intrinsic descriptors from symbolic regression. *J. Mater. Inf.* **2024**, *4*, 6. [CrossRef]
16. Giulimondi, V.; Mitchell, S.; Pérez-Ramírez, J. Challenges and opportunities in engineering the electronic structure of single-atom catalysts. *ACS Catal.* **2023**, *13*, 2981–2997. [CrossRef]
17. Xie, T.; Wang, P.; Tian, C.; Zhao, G.; Jia, J.; He, C.; Zhao, C.; Wu, H. Adsorption characteristics of gas molecules adsorbed on graphene doped with Mn: A first principle study. *Molecules* **2022**, *27*, 2315. [CrossRef] [PubMed]
18. Huang, J.; Chen, M.; Xue, J.; Li, M.; Cheng, Y.; Lai, Z.; Hu, J.; Zhou, F.; Qu, N.; Liu, Y.; et al. A Study of the adsorption properties of individual atoms on the graphene surface: Density functional theory calculations assisted by machine learning techniques. *Materials* **2024**, *17*, 1428. [CrossRef] [PubMed]
19. Tian, W.; Ren, G.; Wu, Y.; Lu, S.; Huan, Y.; Peng, T.; Liu, P.; Sun, J.; Su, H.; Cui, H. Machine-learning-assisted hydrogen adsorption descriptor design for bilayer MXenes. *J. Clean. Prod.* **2024**, *450*, 141953. [CrossRef]
20. Lubbers, N.; Smith, J.S.; Barros, K. Hierarchical modeling of molecular energies using a deep neural network. *J. Chem. Phys.* **2018**, *148*, 241715. [CrossRef]
21. Malone, W.; Kara, A. Predicting adsorption energies and the physical properties of H, N, and O adsorbed on transition metal surfaces: A machine learning study. *Surf. Sci.* **2023**, *731*, 122252. [CrossRef]
22. Dean, J.; Taylor, M.G.; Mpourmpakisu, G. Unfolding adsorption on metal nanoparticles: Connecting stability with catalysis. *Sci. Adv.* **2019**, *5*, eaax5101. [CrossRef]

23. Restuccia, P.; Ahmada, E.A.; Harrison, N.M. A transferable prediction model of molecular adsorption on metals based on adsorbate and substrate properties. *Phys. Chem. Chem. Phys.* **2022**, *24*, 16545–16555. [CrossRef] [PubMed]
24. Zeng, Z.-H.; Da Silva, J.L.F.; Deng, H.-Q.; Li, W.-X. Density functional theory study of the energetics, electronic structure, and core-level shifts of NO adsorption on the Pt(111) surface. *Phys. Rev. B* **2009**, *79*, 205413. [CrossRef]
25. Gallego, S.; Sanchez, N.; Martin, S.; Muñoz, M.C.; Szunyogh, L. Reversible enhancement of the magnetism of ultrathin Co films by H adsorption. *Phys. Rev. B* **2010**, *82*, 085414. [CrossRef]
26. Liu, W.; Jiang, Y.; Dostert, K.H.; O'Brien, C.P.; Riedel, W.; Savara, A.; Schauermann, S.; Tkatchenko, A. Catalysis beyond frontier molecular orbitals: Selectivity in partial hydrogenation of multi-unsaturated hydrocarbons on metal catalysts. *Sci. Adv.* **2017**, *3*, e1700939. [CrossRef] [PubMed]
27. Korostelev, V.; Wagnera, J.; Klyukin, K. Simple local environment descriptors for accurate prediction of hydrogen absorption and migration in metal alloys. *J. Mater. Chem. A* **2023**, *11*, 23576–23588. [CrossRef]
28. Zhang, Y.; Wang, D.; Wei, G.; Li, B.; Mao, Z.; Xu, S.-M.; Tang, S.; Jiang, J.; Li, Z.; Wang, X.; et al. Engineering spin polarization of the surface-adsorbed fe atom by intercalating a transition metal atom into the MoS_2 bilayer for enhanced nitrogen reduction. *JACS Au* **2024**, *4*, 1509–1520. [CrossRef] [PubMed]
29. Ungerer, M.J.; Santos-Carballal, D.; Cadi-Essadek, A.; van Sittert, C.G.C.E.; de Leeuw, N.H. Interaction of H_2O with the platinum Pt (001), (011), and (111) surfaces: A density functional theory study with long-range dispersion corrections. *J. Phys. Chem. C* **2019**, *123*, 27465–27476. [CrossRef]
30. Romero, M.A.; Bonetto, F.; García, E.A. Influence of single adsorbed atoms on charge exchange during ion-surface collisions. *Phys. Rev. A* **2023**, *107*, 032803. [CrossRef]
31. Hofmann, O.T.; Egger, D.A.; Zojer, E. Work-Function Modification beyond Pinning: When Do Molecular Dipoles Count? *Nano Lett.* **2010**, *10*, 4369–4374. [CrossRef] [PubMed]
32. Rusu, P.C.; Brocks, G. Surface dipoles and work functions of alkylthiolates and fluorinated alkylthiolates on Au(111). *J. Phys. Chem. B* **2006**, *110*, 22628–22634. [CrossRef] [PubMed]
33. Kistanov, A.A. Atomic insights into the interaction of N_2, CO_2, NH_3, NO, and NO_2 gas molecules with Zn_2(V, Nb, Ta)N_3 ternary nitride monolayers. *Phys. Chem. Chem. Phys.* **2024**, *26*, 13719. [CrossRef] [PubMed]
34. Kresse, G. Furthmuller, Efficient iterative schemes for ab initio total-energy calculations using a plane-wave basis set. *J. Phys. Rev. B Condens. Matter Mater. Phys.* **1996**, *54*, 11169. [CrossRef] [PubMed]
35. Joubert, D. From ultrasoft pseudopotentials to the projector augmented-wave method. *Phys. Rev. B* **1999**, *59*, 1758–1775.
36. Perdew, J.P.; Burke, K.; Ernzerhof, M. Generalized gradient approximation made simple. *Phys. Rev. Lett.* **1996**, *77*, 3865–3868. [CrossRef] [PubMed]
37. Grimme, S.; Antony, J.; Ehrlich, S.; Krieg, S. A consistent and accurate ab initio parametrization of density functional dispersion correction (DFT-D) for the 94 elements H-Pu. *J. Chem. Phys.* **2010**, *132*, 154104. [CrossRef] [PubMed]
38. Kabalan, L.; Kowalec, I.; Catlow, C.R.A.; Logsdail, A.J. A computational study of the properties of low- and high-index Pd, Cu and Zn surfaces. *Phys. Chem. Chem. Phys.* **2021**, *23*, 14649–14661. [CrossRef]
39. Bader, R.F.W. *Atoms in Molecules—A Quantum Theory*; Oxford University Press: New York, NY, USA, 1990.
40. Momma, K.; Izumi, F. VESTA 3 for three-dimensional visualization of crystal, volumetric and morphology data. *J. Appl. Crystallogr.* **2011**, *44*, 1272–1276. [CrossRef]
41. Python Software Foundation. Python Language Reference, Version 3.7. Available online: http://www.python.org (accessed on 25 January 2020).

Disclaimer/Publisher's Note: The statements, opinions and data contained in all publications are solely those of the individual author(s) and contributor(s) and not of MDPI and/or the editor(s). MDPI and/or the editor(s) disclaim responsibility for any injury to people or property resulting from any ideas, methods, instructions or products referred to in the content.

Article

Polyethyleneimine Modified Two-Dimensional GO/MXene Composite Membranes with Enhanced Mg^{2+}/Li^{+} Separation Performance for Salt Lake Brine

Jun Wang [1], Andong Wang [2,*], Jiayuan Liu [2], Qiang Niu [1], Yijia Zhang [1], Ping Liu [1], Chengwen Liu [3], Hongshan Wang [3], Xiangdong Zeng [3] and Guangyong Zeng [3,4,*]

1. College of Biological and Chemical Engineering, Panzhihua University, Panzhihua 617000, China; enjoygreenlife@126.com (J.W.); pzh_niuqiang@163.com (Q.N.); 17764944212@163.com (Y.Z.); 15982680425@163.com (P.L.)
2. The 4th Geological Brigade of Sichuan, Chengdu 611130, China; pzhljy1990@163.com
3. College of Materials and Chemistry & Chemical Engineering, Chengdu University of Technology, Chengdu 610059, China; chengwenliu@stu.cdut.edu.cn (C.L.); hongshanwang@stu.cdut.edu.cn (H.W.); zengxiandong17@cdut.edu.cn (X.Z.)
4. Tianfu Yongxing Laboratory, Chengdu 610213, China
* Correspondence: 13882330145@163.com (A.W.); wuwu5125@163.com (G.Z.); Tel./Fax: +86-(0)28-8407-3864 (G.Z.)

Citation: Wang, J.; Wang, A.; Liu, J.; Niu, Q.; Zhang, Y.; Liu, P.; Liu, C.; Wang, H.; Zeng, X.; Zeng, G. Polyethyleneimine Modified Two-Dimensional GO/MXene Composite Membranes with Enhanced Mg^{2+}/Li^{+} Separation Performance for Salt Lake Brine. *Molecules* **2024**, *29*, 4326. https://doi.org/10.3390/molecules29184326

Academic Editors: Sake Wang, Minglei Sun and Nguyen Tuan Hung

Received: 3 August 2024
Revised: 8 September 2024
Accepted: 8 September 2024
Published: 12 September 2024

Copyright: © 2024 by the authors. Licensee MDPI, Basel, Switzerland. This article is an open access article distributed under the terms and conditions of the Creative Commons Attribution (CC BY) license (https://creativecommons.org/licenses/by/4.0/).

Abstract: As global demand for renewable energy and electric vehicles increases, the need for lithium has surged significantly. Extracting lithium from salt lake brine has become a cutting-edge technology in lithium resource production. In this study, two-dimensional (2D) GO/MXene composite membranes were fabricated using pressure-assisted filtration with a polyethyleneimine (PEI) coating, resulting in positively charged PEI-GO/MXene membranes. These innovative membranes, taking advantage of the synergistic effects of interlayer channel sieving and the Donnan effect, demonstrated excellent performance in Mg^{2+}/Li^{+} separation with a mass ratio of 20 (Mg^{2+} rejection = 85.3%, Li^{+} rejection = 16.7%, $S_{Li,Mg}$ = 5.7) in simulated saline lake brine. Testing on actual salt lake brine in Tibet, China, confirmed the composite membrane's potential for effective Mg^{2+}/Li^{+} separation. In the actual brine test with high concentration, Mg^{2+}/Li^{+} after membrane separation is 2.2, which indicates that the membrane can significantly reduce the concentration of Mg^{2+} in the brine. Additionally, the PEI-GO/MXene composite membrane demonstrated strong anti-swelling properties and effective divalent ion rejection. This research presents an innovative approach to advance the development of 2D membranes for the selective removal of Mg^{2+} and Li^{+} from salt lake brine.

Keywords: GO/MXene nanosheets; 2D nanosheet membranes; Mg^{2+}/Li^{+} separation; salt lake brine; permeability and selectivity

1. Introduction

Lithium, with its high specific heat capacity, allows it to efficiently absorb and release heat. Additionally, lithium's low density contributes to the development of lighter and more efficient batteries. Furthermore, lithium exhibits high electrochemical activity and good ductility [1], making it widely applicable in aerospace, new energy, and pharmaceutical industries [2,3]. The development and utilization of lithium resources are crucial to the global energy transition. Salt Lake brine contains 80% of China's lithium resources and is characterized by a relatively high mass ratio of Mg^{2+}/Li^{+} [4–6]. The application of new materials and adsorbents makes the lithium extraction process more efficient while reducing the environmental impact. Global trends indicate that with the increasing demand for electric vehicle batteries, brine lithium extraction technologies are rapidly evolving towards greater efficiency and environmental sustainability. Belonging to the diagonal elements, the physical and chemical properties of lithium and magnesium are quite similar [7]. Thus, the separation of Mg^{2+}/Li^+ in salt lake brine poses significant challenges for lithium extraction [8]. Common lithium extraction technologies include extraction, electrodialysis, adsorption, and precipitation [9–12]. However, these methods are hindered by severe equipment corrosion, high consumption of chemical reagents, and significant power requirements [13]. Therefore, developing a saline lithium extraction technology with high Mg^{2+}/Li^+ separation efficiency, energy efficiency, and environmental sustainability is essential.

In recent years, membrane separation technology has become a popular technology in desalination, food processing, and wastewater purification due to its high separation performance, low chemical reagent consumption, and no phase change [14,15]. The 2D membranes constructed with 2D materials have attracted increasing research interest owing to their unique interlayer permeation channels and high separation efficiency [16]. Commonly used 2D materials include graphene oxide (GO), transition metal carbides/nitrides (MXene), graphitic carbon nitride, and transition metal sulfides [17,18]. Among them, GO has the advantages of excellent hydrophilicity, abundant active functional groups (carboxyl, hydroxyl, carbonyl), and a large specific surface area [19–21]. Zhao et al. [22] used vacuum filtration to combine 0D GO quantum dots (QDs) with 2D GO nanosheets, successfully fabricating GO QDs/GO membranes with high permeability. The incorporation of GO QDs effectively enhanced the hydrophilicity and interlayer spacing of the membranes. Experimental results demonstrated that the membranes achieved excellent dye and macromolecular protein removal efficiencies, consistently maintaining removal rates above 99%. Moreover, the membranes exhibited permeation performance that was two to four times higher than that of pristine GO membranes. However, the study still did not solve the problem of poor anti-fouling ability of the membranes. Zhan et al. [23] precisely regulated the interlayer channels of GO membranes, overcoming the trade-off effect and achieving outstanding desalination ability. POSS@GO hybrid membranes were fabricated by compositing GO nanosheets with aminopropyl isobutyl polyhedral oligomeric silsesquioxane (NH_2-POSS). Glutaraldehyde was used as the crosslinking agent. In order to improve the permeability and desalination performance of the composite membranes (water flux = 112.7 kg/($m^2 \cdot h$), rejection of NaCl > 99%), the polyhedral structure of POSS could effectively increase the interlayer spacing of layered GO, providing a suitable spatial hindrance effect. However, due to its excellent hydrophilicity, GO membranes readily form hydrogen bonds with water molecules in aqueous solutions, leading to swelling phenomena.

MXene is a family of 2D transition metal carbides, nitrides, or carbonitrides that can be synthesized by etching the precursor $M_{n+1}AX_n$ phase [24,25]. M is the element of the transition metal, A is usually the element Al, and X is the element of carbon or nitrogen. The structural formula of MXene is generally written as $M_{n+1}X_nT_x$, where T_x represents the −O-O, −OH, and −F groups formed during etching [26,27]. These functional groups endow MXene nanosheets with good chemical modification and hydrophilicity [28]. Wang et al. [29] intercalated MXene as a guest material into rGO membranes, significantly improving their hydrophilicity. The 2D/2D rGO/MXene permeation channels were constructed, and the composite membrane demonstrated outstanding performance, with a pure water flux of 62.1 $L \cdot m^{-2} \cdot h^{-1} \cdot bar^{-1}$ and 91.4% methyl orange rejection. However, the study did not address the long-term stability of the membrane, a critical parameter for assessing membrane performance. Mg^{2+} and Li^+ have similar ionic radii (4.3 A and 3.8 A, respectively), making selective separation challenging when relying solely on the sieving effect of layer spacing [30]. The surface charge properties of 2D membranes significantly influence the electrostatic repulsion of Mg^{2+} and Li^+ ions, as explained by the Donnan effect [31,32]. Positively charged membrane surfaces have different electrostatic repulsive forces for Mg^{2+} and Li^+. This results in differing mass transfer rates, thereby enabling the selective separation of Mg^{2+} and Li^+ ions [33,34].

In this study, a simple pressure-assisted filtration and surface coating method to prepare PEI-GO/MXene composite membranes with a 2D/2D structure and positive charge.

Firstly, high-quality MXene nanosheets were synthesized through etching with LiF and HCl. Subsequently, GO and MXene were homogeneously dispersed and pump-filtered onto polyethersulfone (PES) substrate membranes. The surface of the composite membrane was treated with a polyethyleneimine (PEI) coating via an electrostatic attraction process. This process altered the surface charge characteristics of the composite membrane. Furthermore, the effects of different PEI coating conditions on membrane separation performance were investigated. The composite membrane's Mg^{2+}/Li^+ separation performance was also evaluated. In this study, the intercalation of MXene into the GO layer exhibited enhanced stability, which was conducive to the long-term use of the membrane. Compared with most membranes used for magnesium-lithium separation, PEI-GO/MXene membranes significantly reduced Li^+ retention due to the Donnan effect.

2. Results and Discussion
2.1. Characterization of Materials and Composite Membranes

The microscopic structures of GO and MXene nanomaterials were examined using scanning electron microscopy (SEM). GO nanosheets exhibited a typical 2D layered structure, as depicted in Figure S1a. The interlayer bonding was relatively tight, and the lateral size of the GO nanosheets was comparatively large. The MAX phase displayed a dense, blocky structure and was characterized by a tightly packed arrangement with minimal gaps between layers (Figure S1b). In contrast, MXene nanomaterials exhibited increased interlayer spacing following etching and ultrasonic stripping (Figure S1c). The interlayer spacing increased, and the lateral dimensions of the sheets were about a few micrometers. This loose interlayer structure makes MXene highly suitable for two-dimensional membrane construction.

As illustrated in Figure 1, SEM images of the M0 and M3 membranes are presented. A distinctive folded structure was observed on the surface of the M0 composite membrane (Figure 1a), which is a characteristic morphology of GO-based double-layered 2D membranes [35]. Furthermore, the observed flakes were larger in size and identified as GO-related. However, the overlay of larger GO nanosheets on smaller MXene nanosheets obscured these features, making them less apparent on the composite membrane's surface. In contrast, the surface folds of the M3 membrane were significantly reduced, resulting in a markedly smoother appearance (Figure 1b). This phenomenon was attributed to hydrogen bonding and electrostatic attractions between the amine groups of PEI and the oxygen-containing functional groups of GO and MXene. These interactions likely contribute to the formation of a dense organic layer on the membrane surface [36]. The dense organic layer enhances Mg^{2+} retention by the membrane. Additionally, Figure 1c illustrates that the cross-section of the M0 composite membrane exhibited a typical lamellar structure with 2D nano-channels capable of intercepting larger substances and facilitating the transport of small targets and water molecules [37]. After PEI surface coating (Figure 1d), the cross-section of the composite membrane showed no obvious changes, and the thickness of the laminate structure remained nearly unchanged. This suggests that the PEI surface coating had minimal impact on the membrane's cross-sectional structure.

Elemental distribution in the composite membrane was characterized using EDS mapping on M3, as shown in Figure 2. C, N, and O elements were uniformly distributed throughout the PES base membrane and its laminar structure. S element was primarily distributed in the PES membrane, while the aggregation of the Ti element was more obvious at the top of the membrane. This distribution pattern is attributed to the uniform dispersion of MXene nanosheets within the separated layers of the composite membrane, which prevents agglomerate formation.

Atomic force microscopy (AFM) analysis was performed on two composite membranes, as shown in Figure 3. Before modification, the surface roughness of the composite membrane was 38.2 ± 2.3 nm, which slightly decreased to 37.9 ± 1.4 nm after modification. The coating did not significantly impact the surface roughness of the membrane. The considerable surface roughness prior to modification contributed to the enhanced permeability of the composite membrane [38].

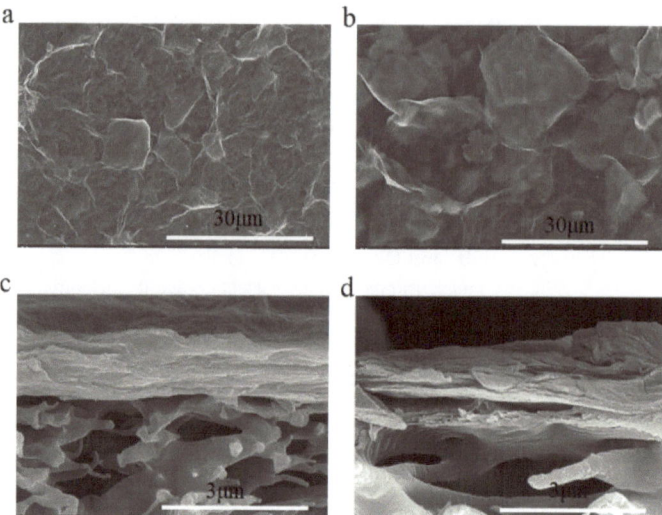

Figure 1. (**a**,**b**) SEM images showing the surfaces of M0 and M3; (**c**,**d**) SEM images of the cross-sections of M0 and M3.

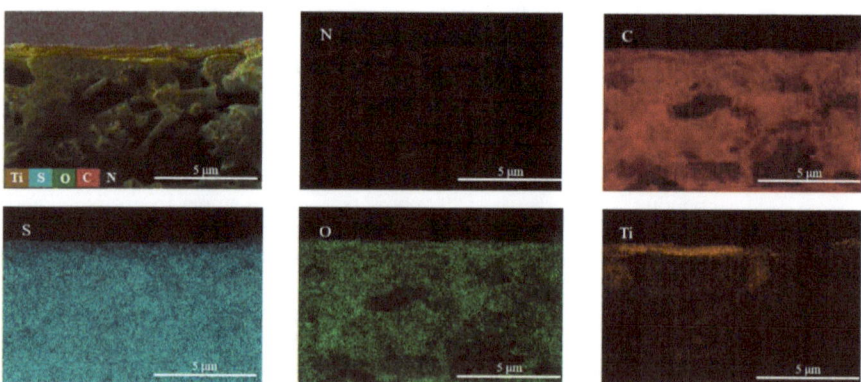

Figure 2. The EDS-mapping of M3.

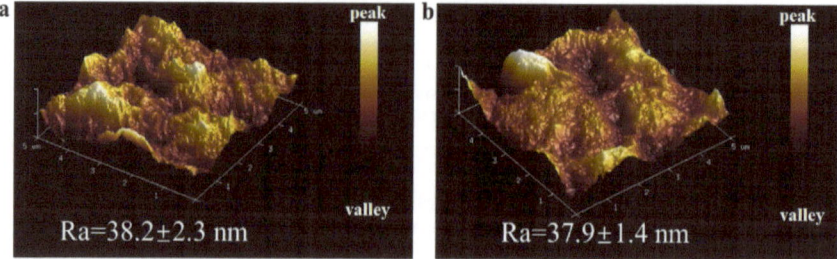

Figure 3. (**a**,**b**) Images of the M0 and M3 samples obtained by AFM.

Figure 4 presents the XPS energy spectrum of the M3 composite membrane, with Table 1 providing the elemental composition. Significant characteristic peaks at 284.8 eV (C 1s), 399.3 eV (N 1s), 531.5 eV (O 1s), and 457 eV (Ti 2p) were observed in the full spectrum (Figure 4a) [39–41]. The presence of numerous amine groups in PEI produced a more pronounced N 1s peak in the full spectrum of the composite membrane, confirming successful PEI incorporation. The O 1s peaks originated from oxygen-containing groups on the surface of GO and MXene. C 1s peaks indicated the presence of carbon, mainly from GO, MXene, and PEI. Ti was derived from MXene, confirming its successful intercalation into GO. The C 1s spectrum (Figure 4b) revealed the presence of C–C/C=C, C–O, and C=O/COOH functional groups in GO nanosheets, with peaks at 284.8 eV, 286.9 eV, and 288.5 eV, respectively. In addition, peaks at 401 eV (C–NH$_2$) and 399 eV (C–N) were detected in the N1s spectrum (Figure 4c). The O 1s spectrum (Figure 4d) showed a fitted peak corresponding to N–C=O. This peak resulted from nitrogen in the amine group of the PEI molecule [36]. Peaks at 455.8 eV (C–Ti–(O, OH)), 458.7 eV (Ti$_x$–O$_y$), 460.2 eV (TiO$_2$), 461.6 eV (Ti–C$_x$), and 464.4 eV (C–Ti–F) were observed, as shown in Figure 4e. These results confirm the successful preparation of PEI-GO/MXene composite membranes. According to the data in Figure 4f, the zeta potential of the PEI unmodified composite membrane was −41.9 mV. This negative potential is attributed to oxygen-containing functional groups on the surfaces of GO and MXene nanosheets, such as carboxyl (–COOH), hydroxyl (–OH), and oxygen (–O). PEI amine groups interacted and adhered to the membrane surface upon modification. Protonation of amine groups in water generated a positive potential (+11.8 mV) on the M3 membrane surface [28]. According to the Donnan effect, positively charged membrane surfaces exhibit greater electrostatic repulsion towards cations. Consequently, the introduction of PEI improved Mg^{2+} rejection in the PEI-GO/MXene membrane, enhancing its Mg^{2+}/Li$^+$ separation capacity.

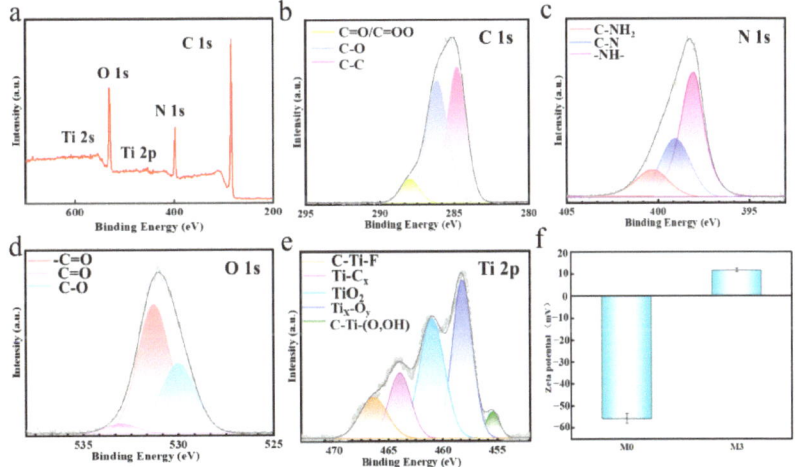

Figure 4. (a) Full spectra of XPS for M3; (b–e) The high-resolution spectra of the carbon 1s, nitrogen 1s, oxygen 1s, and titanium 2p electrons in M3; (f) Zeta potential of M0 and M3.

Table 1. Elemental composition of M0 and M3 based on XPS analysis.

Type of Membrane	C (%)	O (%)	Ti (%)	N (%)
M3	70.81	15.87	0.15	13.16

Figure 5a illustrates the Fourier-transform infrared spectroscopic (FTIR) results for the composite membrane. A notable peak at 1630 cm^{-1} corresponds to the stretching vibration of the carbonyl group, specifically the carboxyl group (C=O). This peak is attributed to the widespread presence of the oxygen-containing functional group –COOH in GO nanosheets [42]. The distinctive peak at 1410 cm^{-1} indicates the carbon-carbon out-of-plane stretching vibration on the sp^2 carbon skeleton. The peak at 1070 cm^{-1} is attributed to the stretching vibration of the C–O bond, originating from GO or MXene nanosheets. In the PEI-GO/MXene membrane, peaks at 2920 cm^{-1} and 2840 cm^{-1} correspond to the stretching vibrations of the amine group. This is due to abundant amine functional groups in PEI. These results suggest the successful incorporation of PEI into the composite membranes [36].

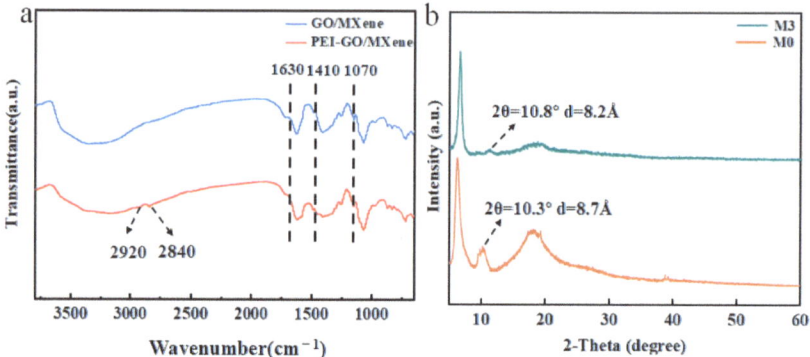

Figure 5. The FTIR (**a**) and XRD (**b**) patterns of M0 and M3.

Figure 5b displays the X-ray diffractometry (XRD) patterns of the M0 and M3 composite membranes. The GO characteristic peak in M0 corresponds to 2θ = 10.3, and according to the Bragg equation, the interlayer spacing of the M0 composite membrane was calculated to be 8.7 Å [43]. According to the literature [44], the diameter of hydrated Mg^{2+} (~8.56 Å) was slightly smaller than the interlayer spacing of M0. Furthermore, the Donnan repulsion of cations on the negatively charged surface of M0 is weak. Therefore, M0 could not achieve effective Mg^{2+} rejection through the combination of size sieving and the Donnan effect. For M3, the characteristic peak shifted to 10.8°, corresponding to an interlayer spacing of 8.2 Å. The formation of a polymer layer on the PEI membrane surface reduced the interlayer spacing of M3. As a consequence of the reduced layer spacing, the permeability coefficient of water molecules increases, while the overall permeability performance of the composite membrane decreases [28]. However, this reduction enhances the ion rejection capability of M3. As the PEI concentration in the composite membrane increases, the water contact angle gradually decreases (Figure 6a). With increasing PEI concentration, the number of amine groups on the membrane surface increases. This increase leads to the formation of additional hydrogen bonds, strengthening the interaction between the membrane surface and water molecules [37]. The dense PEI network structure reduces membrane flux and increases the Mg^{2+} retention rate.

2.2. Performance Testing of Composite Membranes

Water flux variation at 2 bar pressure for different composite membranes is shown in Figure 6b. The water flux of the M0 composite membrane reached 6.9 L·m^{-2}·h^{-1}. However, as the PEI concentration increased, the water flux decreased to 1.98 L·m^{-2}·h^{-1} at a PEI concentration of 1.5 wt%. Generally, PEI modification made the composite membrane more hydrophilic. PEI enhances the adhesion of water molecules to the surface, improving water flow through the material [45]. Nevertheless, PEI interacts with oxygen-containing functional groups on the surface of GO/MXene composite membranes through hydrogen bonding and electrostatic interactions, resulting in the formation of a dense membrane and reduced water flux [31].

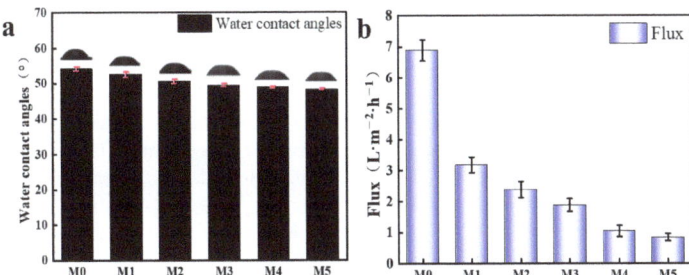

Figure 6. (**a**) The water contact angles and (**b**) the pure water flux of various composite membranes.

Additionally, the effect of coating time on composite membrane performance was evaluated (T = 5 min, 15 min, 30 min, 60 min), and the separation efficiency was investigated using a 1 g/L $MgCl_2$ solution. The composite membrane flux gradually decreased with increasing coating time (Figure 7a). This was due to the formation of thicker polymer layers over time, leading to higher resistance to water molecule permeation and reduced permeability performance. The rejection rate of the composite membrane for $MgCl_2$ gradually increased with longer PEI coating times. The final $MgCl_2$ rejection was 86.2% at the time of T4 (Figure 7b). This represents an improvement of over 20% compared to the 65.2% $MgCl_2$ rejection of the GO/MXene membrane. Therefore, the optimal PEI concentration was determined to be 1 wt%, with a coating time of 30 min.

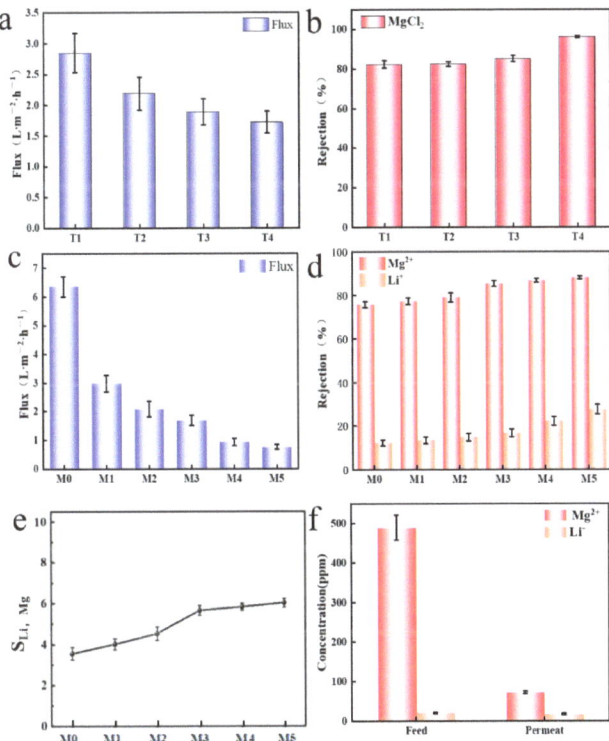

Figure 7. (**a**,**b**) Flux and Mg^{2+} rejection of composite membranes at different coating times; (**c**) Flux and (**d**) Mg^{2+} and Li^+ rejection for simulated salt lake brine; (**e**) $S_{Li,Mg}$ for different composite membranes; (**f**) Separation capacity of M3 for Mg^{2+} and Li^+.

Simulated salt lake brine (1866 ppm $MgCl_2$ and 134 ppm LiCl) was used to evaluate the Mg^{2+}/Li^+ separation performance of various composite membranes. The flux gradually decreased with increasing PEI concentration, which was similar to the decreasing trend of water flux (Figure 7c). Furthermore, the Mg^{2+} rejection rates of the composite membranes were 70.68%, 75.76%, 85.27%, 86.66%, and 87.99%, respectively (Figure 7d). $MgCl_2$ rejection increased rapidly before the PEI concentration of 1 wt%. When PEI concentrations were greater than 1 wt%, the increase in $MgCl_2$ rejection was not significant. The increased membrane surface charge at higher PEI concentrations enhanced electrostatic repulsion, improving $S_{Li,Mg}$ (Figure 7e). After further increasing the PEI concentration, there was no significant improvement in the Mg^{2+}/Li^+ selectivity of the composite membrane, so M3 was considered the best membrane ($S_{Li,Mg}$ = 5.7). The brine contained 489.4 ppm Mg^{2+} and 20.3 ppm Li^+, as illustrated in Figure 7f. Following treatment with M3, the permeate contained 58.8 ppm Mg^{2+} and 14.7 ppm Li^+. Combining interlayer sieving with electrostatic repulsion [46], M3 achieved 85.27% rejection of $MgCl_2$ and only 16.68% rejection of LiCl. The composite membrane exhibited a strong affinity for Mg^{2+}, resulting in significant retention. It demonstrated low Li^+ rejection, resulting in effective Mg^{2+}/Li^+ separation [47–49].

Due to the significant differences between the compositions of actual and simulated salt lake brines, this study further investigated the Mg^{2+}/Li^+ separation efficiency of composite membranes on real saltwater brines. Actual brine samples were sourced from a region of Tibet, China, and underwent flotation, filtration, and other pre-treatments, with specific compositions detailed in Table S2. The salt lake brine was diluted tenfold before use. The feed solution contained 377 ppm of Mg^{2+} and 84 ppm of Li^+, resulting in a Mg^{2+}/Li^+ ratio of 4.5. The concentrations of Mg^{2+} and Li^+ in the permeation solution after treatment with the composite membrane were 132 and 59 ppm, respectively, with Mg^{2+}/Li^+ = 2.2 (Figure 8a). These results demonstrated that the M3 significantly reduced the concentration of Mg^{2+} in the salt lake brine, thereby achieving the desired reduction in the Mg^{2+}/Li^+ mass ratio. As shown in Figure 8b, the Mg^{2+} rejection rate of M3 was 64.8%, and the Li^+ rejection rate was 29.8%, with $S_{Li,Mg}$ = 2. Compared to the simulated brine, the composite membrane's Mg^{2+}/Li^+ separation efficiency was decreased. First, the total salinity of the salt lake brine remained high even after dilution, intensifying the effect of concentration polarization. In addition, high salinity shielded some of the membrane charges and attenuated the Donnan effect, leading to a reduction in Mg^{2+} rejection. The Donnan effect imparted selectivity to the membrane for ions with different charges, which was particularly significant in this study. The Donnan effect also influenced the distribution of ions on both sides of the membrane, contributing to ion balance. Since the membrane carried a fixed internal charge, ion distribution on both sides had to maintain electrical neutrality, leading to a lower concentration of Mg^{2+} relative to Li^+ inside the membrane and enhancing the selective transmission of Li^+. PEI modification increased the membrane's positive charge, making it more selective for Li^+ passage while rejecting Mg^{2+}, thus achieving effective Mg^{2+}/Li^+ separation [50]. Additionally, the salt lake brine contained monovalent Na^+ and K^+, which competed with Li^+ for permeation and potentially hindered Mg^{2+}/Li^+ separation [51].

As illustrated in Figure 8c,d, the separation performance of the composite membrane was evaluated using 1 g/L of $CaCl_2$, $MgSO_4$, Na_2SO_4, and NaCl solutions. The flux order was NaCl > Na_2SO_4 > $CaCl_2$ > $MgSO_4$, while the salt rejection order was $MgSO_4$ (94.8%) > $CaCl_2$ (84.1%) > Na_2SO_4 (81.4%) > NaCl (36.9%). The PEI coating created a positively charged membrane surface, resulting in stronger electrostatic repulsion for divalent ions Mg^{2+} and Ca^{2+}. This decreased the permeability of the salt solutions but increased rejection efficiency. Due to the smaller charge and radius of Na^+ and the smaller hydration radius of Cl^- compared to SO_4^{2-}, NaCl rejection was lower than that of Na_2SO_4 [52,53].

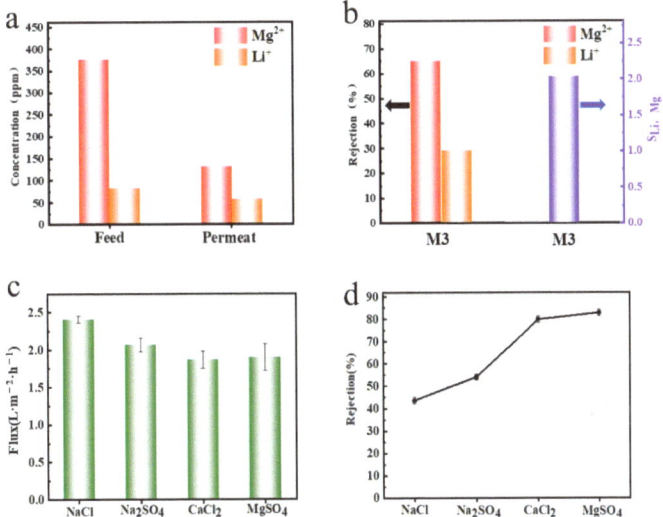

Figure 8. (**a**) Mg^{2+} and Li^+ concentration before and after filtration; (**b**) Effectiveness of M3 in selectively separating Mg^{2+} and Li^+ from diluted brine solutions; (**c**,**d**) Flux and rejection of NaCl, Na_2SO_4, $CaCl_2$ and $MgSO_4$ solutions by M3 composite membrane.

To test its swelling resistance, M3 was immersed in a beaker containing 50 mL of salt lake brine. The immersion times of the composite membrane were 24 h, 72 h, and 168 h. As shown in Figure S2a–c, the surface of M3 did not exhibit any peeling or color change after different immersion times. This indicates that the composite membrane demonstrated excellent resistance to swelling in salt lake brine, maintaining its stability and integrity throughout the immersion period. Furthermore, even when bent to 90° (Figure S2d), M3 remained intact without peeling.

3. Experiments

3.1. The Synthesis of MXene Nanosheets

The synthesis of MXene nanosheets was achieved through a two-step process involving the etching of LiF + HCl mixed solutions and ultrasound-assisted stripping [54,55]. The detailed preparation procedure is described in Supplementary Methods S2. Additionally, Supplementary Methods S1 lists the materials used, and Supplementary Methods S3 outlines the characterization conditions.

3.2. The Fabrication of a PEI-GO/MXene Composite Membrane

Initially, a specific quantity of GO and MXene powder was weighed and added to a glass vial containing deionized water, respectively, and ultrasonically dispersed for 2 h at a concentration of 0.5 mg/mL in an ice bath. The GO and MXene dispersions were then transferred into two beakers containing deionized water and ultrasonically dispersed for 15 min. The two dispersions were combined and sonicated for 30 min to create a homogeneous GO/MXene dispersion. The dispersion was filtered through a dead-end filtration device (Figure S3) with a PES-based membrane, followed by soaking in 20 mL of PEI solution at varying concentrations for a certain time. After coating, the PEI solution was removed, and the membrane surface was rinsed with deionized water. The membrane was then vacuum-dried in an oven at 40 °C for 1 h, resulting in the formation of a PEI-GO/MXene composite membrane, as illustrated in Figure 9. The ratios and numbers of different composite membranes are presented in Table S1.

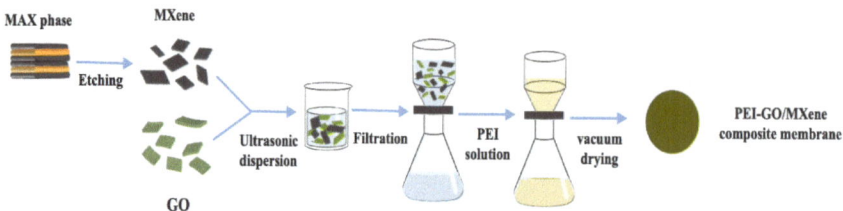

Figure 9. The schematic diagram depicts the methodology employed in the fabrication of the PEI-GO/MXene composite membrane.

4. Conclusions

This study employed PEI for surface coating modification to fabricate positively charged PEI-GO/MXene composite membranes. The preparation, characterization, and Mg^{2+}/Li^+ separation performance of the composite membranes were analyzed to elucidate the membrane-forming mechanism and the constitutive relationships of these new composite membranes. The findings indicated that $MgCl_2$ rejection by the composite membrane progressively improved with increasing PEI concentrations, while the water flux decreased correspondingly. When the coating time T3 was 30 min and PEI concentration was 1 wt%, the M3 composite membrane achieved a $MgCl_2$ rejection rate of 85.27%, a LiCl rejection rate of 16.68%, and the $S_{Li,Mg}$ value as high as 5.7, realizing efficient Mg^{2+}/Li^+ separation. In addition, the composite membrane demonstrated good Mg^{2+}/Li^+ selectivity and resistance to swelling when tested with actual salt lake brine. The experimental results show that strengthening the Donnan effect is a very effective starting point for magnesia-lithium separation, which will enhance the magnesia-lithium separation performance of the membrane. This study is expected to provide experimental guidance for constructing high-performance 2D composite membrane materials based on GO and offer innovative approaches for developing novel membrane materials for lithium extraction from salt lake brine. Although the membrane exhibits good magnesia-lithium separation performance, the narrow channels of the GO membrane limit its water flux. In future research, increasing the flux of the membrane is an urgent problem to be solved.

Supplementary Materials: The following supporting information can be downloaded at: https://www.mdpi.com/article/10.3390/molecules29184326/s1, Figure S1: (a–c) SEM images of GO nanosheets, MAX phase, and MXene nanosheets; Figure S2: Digital images of M3 after different immersion times in salt lake brine: (a) 24 h, (b) 72 h, (c) 168 h; (d) bending picture of M3; Figure S3: The dead-end filtration device; Table S1: Composition of different composite membranes. Table S2: Actual salt lake brine ion concentration in Tibet, China.

Author Contributions: Conceptualization, J.L.; methodology, Q.N.; software, Y.Z.; validation, A.W.; formal analysis, P.L.; investigation, C.L.; data curation, C.L.; writing—original draft preparation, J.W.; writing—review and editing, H.W.; visualization, X.Z.; supervision, A.W.; project administration, G.Z.; funding acquisition, J.W. and G.Z. All authors have read and agreed to the published version of the manuscript.

Funding: This research was supported by Sichuan Science and Technology Program (22NSFSC2972), Panzhihua municipal guiding science and technology plan project (2023ZD-G-3; 2023ZD-G-6); Panzhihua College general research project (20226501); Innovation and entrepreneurship training program for college students (S202311360058; S202411360050) and the TianfuYongxing Laboratory Organized Research Project Funding (NO. 2023KJGG13).

Institutional Review Board Statement: Not applicable.

Informed Consent Statement: Not applicable.

Data Availability Statement: The data that supports the findings of this study is available.

Acknowledgments: We expressed our gratitude to Arijit Sengupta (Homi Bhabha National Institute) for improving the English level of this manuscript and Taijun Chen (Sichuan GCL Lithium Power Technology Co., Ltd.) for providing some resources.

Conflicts of Interest: The authors declare no conflicts of interest.

References

1. Xu, P.; Hong, J.; Qian, X.; Xu, Z.; Xia, H.; Tao, X.; Xu, Z.; Ni, Q.-Q. Materials for lithium recovery from salt lake brine. *J. Mater. Sci.* **2021**, *56*, 16–23. [CrossRef]
2. Lebedeva, N.P.; Boon-Brett, L. Considerations on the Chemical Toxicity of Contemporary Li-Ion Battery Electrolytes and Their Components. *J. Electrochem. Soc.* **2016**, *163*, A821–A830. [CrossRef]
3. Meshram, P.; Pandey, B.D.; Mankhand, T.R. Extraction of lithium from primary and secondary sources by pre-treatment, leaching and separation: A comprehensive review. *Hydrometallurgy* **2014**, *150*, 192–208. [CrossRef]
4. Sun, Y.; Wang, Q.; Wang, Y.; Yun, R.; Xiang, X. Recent advances in magnesium/lithium separation and lithium extraction technologies from salt lake brine. *Sep. Purif. Technol.* **2021**, *256*, 117807. [CrossRef]
5. Swain, B. Recovery and recycling of lithium: A review. *Sep. Purif. Technol.* **2017**, *172*, 388–403. [CrossRef]
6. Xu, S.; Song, J.; Bi, Q.; Chen, Q.; Zhang, W.-M.; Qian, Z.; Zhang, L.; Xu, S.; Tang, N.; He, T. Extraction of lithium from Chinese salt-lake brines by membranes: Design and practice. *J. Membr. Sci.* **2021**, *635*, 119441. [CrossRef]
7. Lu, Z.; Wu, Y.; Ding, L.; Wei, Y.; Wang, H. A Lamellar MXene ($Ti_3C_2T_x$)/PSS Composite Membrane for Fast and Selective Lithium-Ion Separation. *Angew. Chem. Int. Ed.* **2021**, *60*, 22265–22269. [CrossRef]
8. Zhang, Y.; Wang, L.; Sun, W.; Hu, Y.; Tang, H. Membrane technologies for Li^+/Mg^{2+} separation from salt-lake brines and seawater: A comprehensive review. *J. Ind. Eng. Chem.* **2020**, *81*, 7–23. [CrossRef]
9. Foo, Z.H.; Thomas, J.B.; Heath, S.M.; Garcia, J.A.; Lienhard, J.H. Sustainable Lithium Recovery from Hypersaline Salt-Lakes by Selective Electrodialysis: Transport and Thermodynamics. *Environ. Sci. Technol.* **2023**, *57*, 14747–14759. [CrossRef]
10. Guo, Z.-Y.; Ji, Z.-Y.; Chen, Q.-B.; Liu, J.; Zhao, Y.-Y.; Li, F.; Liu, Z.-Y.; Yuan, J.-S. Prefractionation of LiCl from concentrated seawater/salt lake brines by electrodialysis with monovalent selective ion exchange membranes. *J. Clean. Prod.* **2018**, *193*, 338–350. [CrossRef]
11. Ren, P.; Yin, Z.; Wang, G.; Zhao, H.; Ji, P. The sustainable supply of lithium resources from the Qinghai-Tibet plateau salt lakes group: The selection of extraction methods and the assessment of adsorbent application prospects. *Desalination* **2024**, *583*, 117659. [CrossRef]
12. Yu, J.; Zhu, J.; Luo, G.; Chen, L.; Li, X.; Cui, P.; Wu, P.; Chao, Y.; Zhu, W.; Liu, Z. 3D-printed titanium-based ionic sieve monolithic adsorbent for selective lithium recovery from salt lakes. *Desalination* **2023**, *560*, 116651. [CrossRef]
13. Vera, M.L.; Torres, W.R.; Galli, C.I.; Chagnes, A.; Flexer, V. Environmental impact of direct lithium extraction from brines, Nature Reviews Earth & Environment. *Nat. Rev. Earth Environ.* **2023**, *4*, 149–165. [CrossRef]
14. Shahzad, A.; Oh, J.-M.; Azam, M.; Iqbal, J.; Hussain, S.; Miran, W.; Rasool, K. Advances in the Synthesis and Application of Anti-Fouling Membranes Using Two-Dimensional Nanomaterials. *Membranes* **2021**, *11*, 605. [CrossRef] [PubMed]
15. Werber, J.R.; Osuji, C.O.; Elimelech, M. Materials for next-generation desalination and water purification membranes. *Nat. Rev. Mater.* **2016**, *1*, 16018. [CrossRef]
16. Lu, Z.; Wei, Y.; Deng, J.; Ding, L.; Li, Z.-K.; Wang, H. Self-Crosslinked MXene ($Ti_3C_2T_x$) Membranes with Good Antiswelling Property for Monovalent Metal Ion Exclusion. *Angew. Chem. Int. Ed.* **2019**, *13*, 10535–10544. [CrossRef]
17. Ang, E.Y.M.; Ng, T.Y.; Yeo, J.; Lin, R.; Liu, Z.; Geethalakshmi, K.R. Investigations on different two-dimensional materials as slit membranes for enhanced desalination. *J. Membr. Sci.* **2020**, *598*, 117653. [CrossRef]
18. Shen, J.; Wu, J.; Wang, M.; Dong, P.; Xu, J.; Li, X.; Zhang, X.; Yuan, J.; Wang, X.; Ye, M.; et al. Surface Tension Components Based Selection of Cosolvents for Efficient Liquid Phase Exfoliation of 2D Materials. *Small* **2016**, *12*, 2741–2749. [CrossRef] [PubMed]
19. Chen, Z.-H.; Liu, Z.; Hu, J.-Q.; Cai, Q.-W.; Li, X.-Y.; Wang, W.; Faraj, Y.; Ju, X.-J.; Xie, R.; Chu, L.-Y. β-Cyclodextrin-modified graphene oxide membranes with large adsorption capacity and high flux for efficient removal of bisphenol A from water. *J. Membr. Sci.* **2020**, *595*, 117510. [CrossRef]
20. Compton, O.C.; Cranford, S.W.; Putz, K.W.; An, Z.; Brinson, L.C.; Buehler, M.J.; Nguyen, S.T. Tuning the Mechanical Properties of Graphene Oxide Paper and Its Associated Polymer Nanocomposites by Controlling Cooperative Intersheet Hydrogen Bonding. *Angew. Chem. Int. Ed.* **2012**, *6*, 2008–2019. [CrossRef]
21. Putz, K.W.; Compton, O.C.; Segar, C.; An, Z.; Nguyen, S.T.; Brinson, L.C. Evolution of Order During Vacuum-Assisted Self-Assembly of Graphene Oxide Paper and Associated Polymer Nanocomposites. *Angew. Chem. Int. Ed.* **2011**, *5*, 6601–6609. [CrossRef] [PubMed]
22. Zhao, G.; Hu, R.; Zhao, X.; He, Y.; Zhu, H. High flux nanofiltration membranes prepared with a graphene oxide homo-structure. *J. Membr. Sci.* **2019**, *585*, 29–37. [CrossRef]
23. Zhan, X.; Gao, Z.; Ge, R.; Lu, J.; Li, J.; Wan, X. Rigid POSS intercalated graphene oxide membranes with hydrophilic/hydrophobic heterostructure for efficient pervaporation desalination. *Desalination* **2022**, *543*, 116106. [CrossRef]
24. Ihsanullah, I. Potential of MXenes in Water Desalination: Current Status and Perspectives. *Nano-Micro Lett.* **2020**, *12*, 72. [CrossRef]

25. Karahan, H.E.; Goh, K.; Zhang, C.J.; Yang, E.; Yildirim, C.; Chuah, C.Y.; Ahunbay, M.G.; Lee, J.; Tantekin-Ersolmaz, S.B.; Chen, Y.; et al. MXene Materials for Designing Advanced Separation Membranes. *Adv. Mater.* **2020**, *32*, 1906697. [CrossRef]
26. Ding, L.; Li, L.; Liu, Y.; Wu, Y.; Lu, Z.; Deng, J.; Wei, Y.; Caro, J.; Wang, H. Effective ion sieving with $Ti_3C_2T_x$ MXene membranes for production of drinking water from seawater. *Nat. Sustain.* **2020**, *3*, 296–302. [CrossRef]
27. Ren, C.E.; Hatzell, K.B.; Alhabeb, M.; Ling, Z.; Mahmoud, K.A.; Gogotsi, Y. Charge- and Size-Selective Ion Sieving through $Ti_3C_2T_x$ MXene Membranes. *J. Phys. Chem. Lett.* **2015**, *6*, 4026–4031. [CrossRef]
28. Ding, M.; Xu, H.; Chen, W.; Yang, G.; Kong, Q.; Ng, D.; Lin, T.; Xie, Z. 2D laminar maleic acid-crosslinked MXene membrane with tunable nanochannels for efficient and stable pervaporation desalination. *J. Membr. Sci.* **2020**, *600*, 117871. [CrossRef]
29. Wang, X.; Zhang, H.; Wang, X.; Chen, S.; Yu, H.; Quan, X. Electroconductive RGO-MXene membranes with wettability-regulated channels: Improved water permeability and electro-enhanced rejection performance. *Front. Environ. Sci. Eng.* **2022**, *17*, 1. [CrossRef]
30. Yang, G.; Shi, H.; Liu, W.; Xing, W.; Xu, N. Investigation of Mg^{2+}/Li^+ Separation by Nanofiltration. *Chin. J. Chem. Eng.* **2011**, *19*, 586–591. [CrossRef]
31. Meng, B.; Liu, G.; Mao, Y.; Liang, F.; Liu, G.; Jin, W. Fabrication of surface-charged MXene membrane and its application for water desalination. *J. Membr. Sci.* **2021**, *623*, 119076. [CrossRef]
32. Chen, K.; Zhao, S.; Lan, H.; Xie, T.; Wang, H.; Chen, Y.; Li, P.; Sun, H.; Niu, Q.J.; Yang, C. Dual-electric layer nanofiltration membranes based on polyphenol/PEI interlayer for highly efficient Mg^{2+}/Li^+ separation. *J. Membr. Sci.* **2022**, *660*, 120860. [CrossRef]
33. Guo, C.; Qian, X.; Tian, F.; Li, N.; Wang, W.; Xu, Z.; Zhang, S. Amino-rich carbon quantum dots ultrathin nanofiltration membranes by double "one-step" methods: Breaking through trade-off among separation, permeation and stability. *Chem. Eng. J.* **2021**, *404*, 127144. [CrossRef]
34. He, J.-H.; Elgazery, N.; Elagamy, K.; Abd Elazem, N. Efficacy of a Modulated Viscosity-dependent Temperature/nanoparticles Concentration Parameter on a Nonlinear Radiative Electromagneto-nanofluid Flow along an Elongated Stretching Sheet. *J. Appl. Comput. Mech.* **2023**, *9*, 848–860. [CrossRef]
35. Zhang, P.; Gong, J.-L.; Zeng, G.-M.; Song, B.; Cao, W.; Liu, H.-Y.; Huan, S.-Y.; Peng, P. Novel "loose" GO/MoS_2 composites membranes with enhanced permeability for effective salts and dyes rejection at low pressure. *J. Membr. Sci.* **2019**, *574*, 112–123. [CrossRef]
36. Zhao, X.; Che, Y.; Mo, Y.; Huang, W.; Wang, C. Fabrication of PEI modified GO/MXene composite membrane and its application in removing metal cations from water. *J. Membr. Sci.* **2021**, *640*, 119847. [CrossRef]
37. Zhang, M.; Guan, K.; Ji, Y.; Liu, G.; Jin, W.; Xu, N. Controllable ion transport by surface-charged graphene oxide membrane. *Nat. Commun.* **2019**, *10*, 1253. [CrossRef]
38. Yadav, S.; Ibrar, I.; Altaee, A.; Samal, A.K.; Ghobadi, R.; Zhou, J. Feasibility of brackish water and landfill leachate treatment by GO/MoS_2-PVA composite membranes. *Sci. Total Environ.* **2020**, *745*, 141088. [CrossRef]
39. Cheng, P.; Chen, Y.; Gu, Y.-H.; Yan, X.; Lang, W.-Z. Hybrid 2D WS_2/GO nanofiltration membranes for finely molecular sieving. *J. Membr. Sci.* **2019**, *591*, 117308. [CrossRef]
40. Han, S.; Li, W.; Xi, H.; Yuan, R.; Long, J.; Xu, C. Plasma-assisted in-situ preparation of graphene-Ag nanofiltration membranes for efficient removal of heavy metal ions. *J. Hazard. Mater.* **2022**, *423*, 127012. [CrossRef]
41. Liu, T.; Liu, X.; Graham, N.; Yu, W.; Sun, K. Two-dimensional MXene incorporated graphene oxide composite membrane with enhanced water purification performance. *J. Membr. Sci.* **2020**, *593*, 117431. [CrossRef]
42. Yan, M.; Huang, W.; Li, Z. Chitosan cross-linked graphene oxide/lignosulfonate composite aerogel for enhanced adsorption of methylene blue in water. *Int. J. Biol. Macromol.* **2019**, *136*, 927–935. [CrossRef]
43. Ma, J.; Tang, X.; He, Y.; Fan, Y.; Chen, J.; Yu, H. Robust stable MoS_2/GO filtration membrane for effective removal of dyes and salts from water with enhanced permeability. *Desalination* **2020**, *480*, 114328. [CrossRef]
44. Liu, X.; Feng, Y.; Ni, Y.; Peng, H.; Li, S.; Zhao, Q. High-permeance Mg^{2+}/Li^+ separation nanofiltration intensified by quadruple imidazolium salts. *J. Membr. Sci.* **2023**, *667*, 121178. [CrossRef]
45. Ding, J.; Zhao, H.; Xu, B.; Yu, H. Biomimetic Sustainable Graphene Ultrafast-Selective Nanofiltration Membranes. *ACS Sustain. Chem. Eng.* **2020**, *8*, 8986–8993. [CrossRef]
46. Wang, H.; Zeng, G.; Yang, Z.; Chen, X.; Wang, L.; Xiang, Y.; Zeng, X.; Feng, Z.; Tang, B.; Yu, X.; et al. Nanofiltration membrane based on a dual-reinforcement strategy of support and selective layers for efficient Mg^{2+}/Li^+ separation. *Sep. Purif. Technol.* **2024**, *330*, 125254. [CrossRef]
47. Dixit, F.; Zimmermann, K.; Dutta, R.; Prakash, N.J.; Barbeau, B.; Mohseni, M.; Kandasubramanian, B. Application of MXenes for water treatment and energy-efficient desalination: A review. *J. Hazard. Mater.* **2022**, *423*, 127050. [CrossRef]
48. Mozafari, M.; Shamsabadi, A.A.; Rahimpour, A.; Soroush, M. Ion-Selective MXene-Based Membranes: Current Status and Prospects. *Adv. Mater. Technol.* **2021**, *6*, 2001189. [CrossRef]
49. Peng, H.; Zhao, Q. A Nano-Heterogeneous Membrane for Efficient Separation of Lithium from High Magnesium/Lithium Ratio Brine. *Adv. Funct. Mater.* **2021**, *31*, 2009430. [CrossRef]
50. Sun, S.-Y.; Cai, L.-J.; Nie, X.-Y.; Song, X.; Yu, J.-G. Separation of magnesium and lithium from brine using a Desal nanofiltration membrane. *J. Water Process Eng.* **2015**, *7*, 10–217. [CrossRef]

51. Wen, X.; Ma, P.; Zhu, C.; He, Q.; Deng, X. Preliminary study on recovering lithium chloride from lithium-containing waters by nanofiltration. *Sep. Purif. Technol.* **2006**, *49*, 230–236. [CrossRef]
52. Gao, J.; Sun, S.-P.; Zhu, W.-P.; Chung, T.-S. Polyethyleneimine (PEI) cross-linked P84 nanofiltration (NF) hollow fiber membranes for Pb^{2+} removal. *J. Membr. Sci.* **2014**, *452*, 300–310. [CrossRef]
53. Gumbi, N.N.; Li, J.; Mamba, B.B.; Nxumalo, E.N. Relating the performance of sulfonated thin-film composite nanofiltration membranes to structural properties of macrovoid-free polyethersulfone/sulfonated polysulfone/O-MWCNT supports. *Desalination* **2020**, *474*, 11476. [CrossRef]
54. Zeng, G.; Liu, Y.; Lin, Q.; Pu, S.; Zheng, S.; Ang, M.B.M.Y.; Chiao, Y.-H. Constructing composite membranes from functionalized metal organic frameworks integrated MXene intended for ultrafast oil/water emulsion separation. *Sep. Purif. Technol.* **2022**, *293*, 121052. [CrossRef]
55. Yang, Z.; Lin, Q.; Zeng, G.; Zhao, S.; Yan, G.; Ang, M.B.M.Y.; Chiao, Y.-H.; Pu, S. Ternary hetero-structured BiOBr/Bi_2MoO_6@MXene composite membrane: Construction and enhanced removal of antibiotics and dyes from water. *J. Membr. Sci.* **2023**, *669*, 121329. [CrossRef]

Disclaimer/Publisher's Note: The statements, opinions and data contained in all publications are solely those of the individual author(s) and contributor(s) and not of MDPI and/or the editor(s). MDPI and/or the editor(s) disclaim responsibility for any injury to people or property resulting from any ideas, methods, instructions or products referred to in the content.

Article

Remote Sulfonylation of Anilines with Sodium Sulfinates Using Biomass-Derived Copper Catalyst

Xiaoping Yan [1], Jinguo Wang [1], Chao Chen [1,*], Kai Zheng [1], Pengfei Zhang [2] and Chao Shen [1,*]

[1] Key Laboratory of Pollution Exposure and Health Intervention of Zhejiang Province, College of Biology and Environmental Engineering, Zhejiang Shuren University, Hangzhou 310015, China
[2] Key Laboratory of Organosilicon Chemistry and Material Technology, College of Material, Chemistry and Chemical Engineering, Ministry of Education, Hangzhou Normal University, Hangzhou 311121, China
* Correspondence: chencc@zjsru.edu.cn (C.C.); shenchaozju@zjsru.edu.cn (C.S.)

Abstract: A biomass-based catalyst, Cu_xO_y@CS-400, was employed as an excellent recyclable heterogeneous catalyst to realize the sulfonylation reaction of aniline derivatives with sodium sulfinates. Various substrates were compatible, giving the desired products moderate to good yields at room temperature. In addition, this heterogeneous copper catalyst was also easy to recover and was recyclable up to five times without considerably deteriorating in catalytic efficiency. Importantly, these sulfonylation products were readily converted to the corresponding 4-sulfonyl anilines via a hydrolysis step. The method offers a unique strategy for synthesizing arylsulfones and has the potential to create new possibilities for developing heterogeneous copper-catalyzed C-H functionalizations.

Keywords: biomass derivation; sulfonylation; heterogeneous; C-H activation

1. Introduction

Sulfone is widely utilized in diverse fields such as agricultural chemistry, medicine, or advanced materials due to its notable chemical, pharmaceutical, and biological activities [1–3]. Some biologically active aryl sulfones are shown in Figure 1. Therefore, a significant amount of research has been focused on the synthesis of aryl sulfones through a variety of pathways, including sulfide oxidation, sulfinate alkylation, arene sulfonylation of the Friedel–Crafts type, and electrophilic substitution of aromatics with sulfonyl halides or sulfonic acids [4–7].

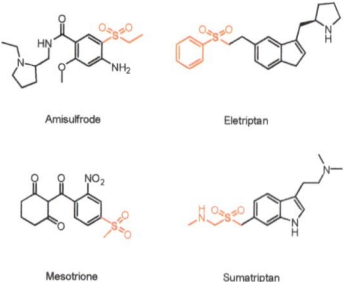

Figure 1. Examples of aryl sulfone-based bioactive molecules.

In recent years, the formation of transition metal-catalyzed C-H functional syntheses of sulfur-containing compounds by direct C-S bonding has proven to be one of the most effective protocols, because this strategy makes the C-S cross-coupling simpler and more efficient [8]. Traditional aromatic sulfonation reactions usually require metal catalysts and chemical oxidants [9–13]. For example, Manolikakes's group and Wu's group have independently mentioned the remote para-selective sulfonylation of 1-naphthylamides using Cu(OAc)$_2$-catalyzed sodium sulfinates (Scheme 1a,b) [14,15]. Xiong's group developed the use of CuBr$_2$ as a catalyst for highly selective functionalization of the C4-H position of 1-naphthylamides using sodium sulfinates (Scheme 1c) [16]. However, most copper catalysts that are used are homogeneous and not environmentally friendly. Consequently, creating a green reaction pathway to satisfy this requirement for sulfonylation remains significant and difficult.

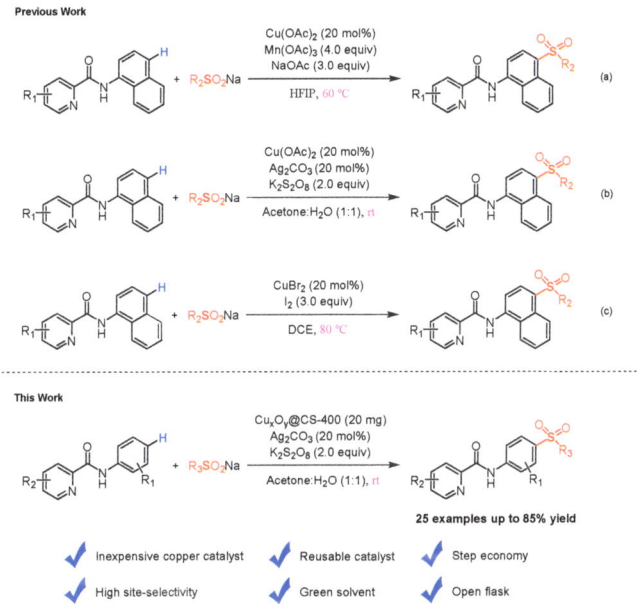

Scheme 1. Direct sulfonylation of 1-naphthylamides or anilines at the C4 site.

Compared to homogeneous catalysts [17–25], heterogeneous catalysts provide the advantage of being able to be retrieved from reaction mixtures and reused, resulting in reduced economic and environmental expenses [26–30]. Biomass-derived heterogeneous catalysts are garnering an increasing amount of attention [31–34]. Various organic reactions for heterogeneous biomass-based catalysts have been reported by our group in recent years [35–39]. The biomass resources used for these catalyst materials have the advantages of leading to high-efficiency catalysts, being low-cost materials, and requiring a simple preparation process. Herein, we report the synthesis of a biomass-based copper catalyst and its successful application in the sulfonylation of aniline derivatives with sodium sulfinates, producing sulfonylation products in acceptable to good yields (Scheme 1c).

2. Results and Discussion

According to our previous synthesis method [36], the Cu$_x$O$_y$@CS-T catalyst was synthesized using a sequential approach outlined in Scheme 2, where T represents the pyrolysis temperature. Firstly, a straightforward and practical method was used to create the catalyst precursor, which was a metal–chitosan complex obtained by combining Cu(OAc)$_2$ with commercial chitosan in water at 50 °C. Then, the catalyst precursor underwent drying at 60 °C and pyrolysis at high temperatures (300 °C, 400 °C, and 500 °C), which formed the Cu$_x$O$_y$@CS-T catalyst. Since the catalyst is stable to air and moisture both in the solid form and in a solution, it was kept at room temperature in a screw-capped vial without any additional air protection.

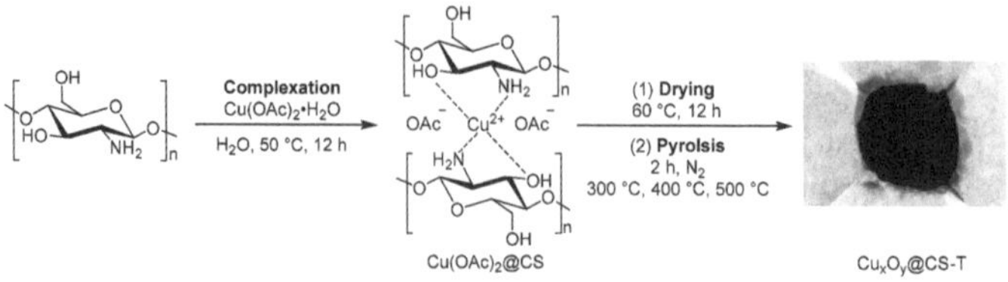

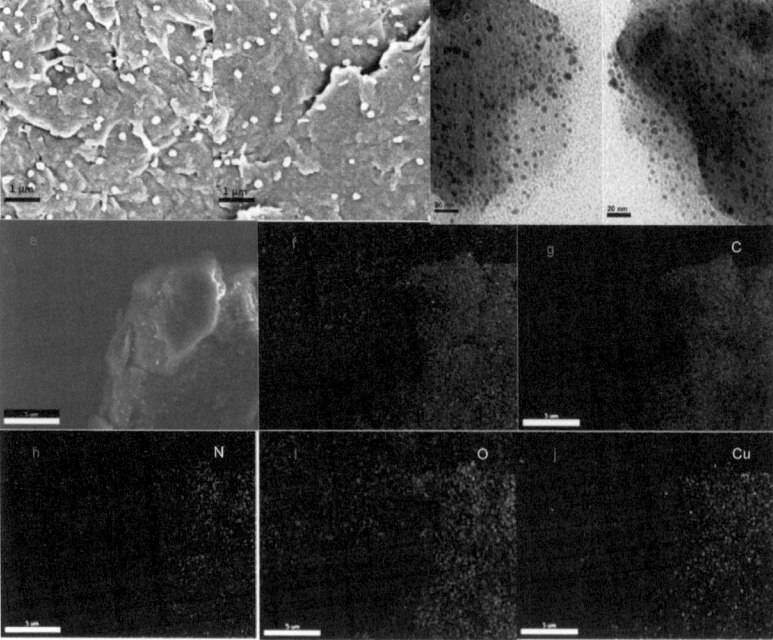

Scheme 2. Synthesis of Cu_xO_y@CS-T catalyst.

A scanning electron microscope (SEM) (Figure 2a,b) and a transmission electron microscope (TEM) (Figure 2c,d) were used to examine the Cu_xO_y@CS-400 catalyst. As shown in Figure 2, the copper particles were evenly distributed on the catalyst carrier surface, and a size of approximately 0.3–0.5 um was obtained. The structural characterization showed that the copper was evenly distributed on the biochar without obvious agglomeration. To confirm the element composition of the Cu_xO_y@CS-400, energy-dispersive spectroscope (EDS) element mapping was conducted (Figure 2e,j). The element mapping demonstrated the uniform dispersion of carbon, nitrogen, and oxygen in conjunction with copper components. In addition, the catalyst was characterized using X-ray diffraction (XRD) and photoelectron spectroscopy (XPS). Our previous studies have shown that the catalyst features three components, Cu, Cu_2O, and CuO (Figure S1) [36].

Figure 2. SEM images of Cu_xO_y@CS-400 (**a,b**). TEM images of Cu_xO_y@CS-400 (**c,d**). Elemental mapping of Cu_xO_y@CS-400 (**e–j**).

With the characterized catalysts in hand [31], we started the investigation by examining the response of N-(o-tolyl)picolinamide (**1a**) with sodium benzenesulfinates (**2a**) as a model reaction to adjust the reaction conditions (Table 1). We were pleased to discover that by using Cu_xO_y@CS-400 as a catalyst at room temperature in a mixed solvent (acetone/H_2O = 1:1), the expected sulfonylation yield of 82% could be achieved (Table 1, entry 1). Encouraged by this result, the reaction conditions for sulfonylation were modified by adjusting the metal catalysts, silver cocatalysts, oxidants, solvents, and temperatures.

An investigation was conducted to analyze the impact of several copper catalysts on the model reaction (Table 1, entries 1–5). It is important to highlight that the reaction did not occur in the absence of a copper catalyst, indicating the unique catalytic function of copper in this process (Table 1, entry 6). The combination of Ag-peroxydisulfate has shown significant efficacy in several sulfonylated reactions [3,15,36]. Therefore, screening of several silver salts showed that the yield of **3a** could not be improved by using either $AgNO_3$ or AgOAc instead of Ag_2CO_3, whereas no reaction occurred in the presence of $AgSbF_6$ (Table 1, entries 7–9). Subsequently, other oxidants such as $Na_2S_2O_8$, $(NH_4)_2S_2O_8$, TBHP, and H_2O_2 were investigated (Table 1, entries 10–13), and it was found that $K_2S_2O_8$ was the most efficient. When Ag_2CO_3 or $K_2S_2O_8$ was not present, the reaction did not produce any product (Table 1, entries 14 and 15). These findings suggest that organic transformation is determined by the Ag_2CO_3 and $K_2S_2O_8$ synergistic action. Then, other solvents (acetone, H_2O, EtOH, DMSO) were also evaluated (Table 1, entries 16–19), and a mixed solvent of acetone and H_2O turned out to enhance the yield. In order to further increase the yield of the reaction, we explored the impact of temperature on the reaction outcomes. Changing the temperature did not improve the yield of the sulfonylation product (Table 1, entry 20).

Table 1. Optimization of reaction conditions [a].

Entry	Variation from Standard Conditions	Yield (%) [b]
1	none	82
2	$Cu(OAc)_2$@CS as catalyst	35
3	CuO as catalyst	28
4	Cu_xO_y@CS-300 as catalyst	69
5	Cu_xO_y@CS-500 as catalyst	65
6	Without Cu_xO_y@CS-400	0
7	$AgNO_3$ as cocatalyst	69
8	AgOAc as cocatalyst	73
9	$AgSbF_6$ as cocatalyst	0
10	$Na_2S_2O_8$ as oxidant	76
11	$(NH_4)_2S_2O_8$ as oxidant	74
12	TBHP as oxidant	22
13	H_2O_2 as oxidant	11
14	Without Ag_2CO_3	0
15	Without $K_2S_2O_8$	0
16	Acetone as solvent	28
17	H_2O as solvent	54
18	EtOH as solvent	23
19	DMSO as solvent	25
20	60 °C as reaction temperature	72

[a] Reaction conditions: **1a** (0.2 mmol), **2a** (2.0 equiv.), Cu_xO_y@CS-400 (20 mg), Ag_2CO_3 (20 mol%), $K_2S_2O_8$ (2.0 equiv.), acetone/H_2O =1:1 (3 mL), stirred at rt, under air, 12 h. [b] Isolated yields.

After obtaining the most suitable reaction conditions, we proceeded to investigate the various anilines that could be utilized for sulfonylation (Scheme 3). Substrates bearing an electron-donating substituent at the C2 and C3 positions of the aniline ring were well tolerated (**3a–3c**). Changing substituents to unsubstituted aniline could also smoothly facilitate the reaction to obtain **3d** and **3e** in 84% and 68% yields, respectively. Notably, when the substrates containing electron-withdrawing groups (-Cl, -I) were used, the reaction failed to occur (**3f** and **3g**). In addition, 1-naphthylamide and quinolin-5-amine could also provide the required product in a high yield (**3h** and **3i**). The further application of this method focused on the substituted picolinamide moiety. When the picolinamide moiety was varied, products **3j–3m** were smoothly obtained. Subsequently, we conducted further research on the wide range of sodium sulfinates. Our catalytic system was highly compatible with functionalized sodium sulfinates, resulting in sulfonylation products with yields ranging from moderate to good. In particular, comparable compounds (**3n–3w**) were produced in 60–83% yields by benzene sodium sulfinates with either electron-donating or electron-withdrawing groups at the C2, C3, or C4 sites. Reactions of *N*-(2-ethylphenyl)picolinamide with both naphthalene, heterocyclic, and aliphatic sodium sulfinates proceeded successfully as well, resulting in the formation of corresponding products (**3x–3aa**) in 45–82% yields. Unfortunately, sodium trifluoromethanesulfinate could not be converted into **3ab** smoothly.

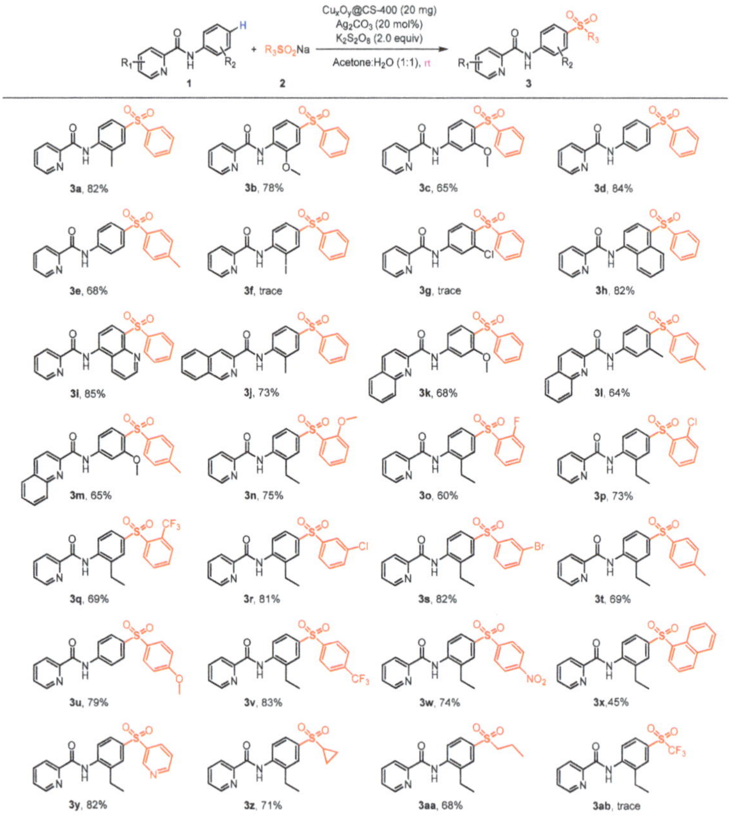

Scheme 3. Substrate scope of anilines and sodium sulfonates. Reaction conditions: anilines (0.2 mmol), sodium sulfonates (2.0 equiv.), Cu$_x$O$_y$@CS-400 (20 mg.), Ag$_2$CO$_3$ (20 mol%), K$_2$S$_2$O$_8$ (2.0 equiv.), acetone/H$_2$O (1:1) (3 mL), stirred at rt, under air, 12 h. Isolated yields.

With the goal of confirming the effectiveness and applicability of this approach, we conducted a large-scale reaction (Scheme 4a), resulting in the production of C4 sulfonylation product **3a** with a yield of 78% using the standard conditions. After that, the directing group could be removed by hydrolysis to give 2-methyl-4-(phenylsulfonyl)aniline and picolinic acid in 92% and 82% yields, respectively. In addition, we conducted a few radical trapping studies to further explain the reaction mechanism (Scheme 4b). The addition of the radical inhibitors TEMPO ((2,2,6,6-tetramethylpiperidin-1-yl)oxyl), BHT (butylated hydroxytoluene), or DPE (1,1-diphenylethene) somewhat hindered the sulfonylation process. The reaction was hindered by radical scavengers, indicating the participation of a free radical pathway in the reaction mechanism.

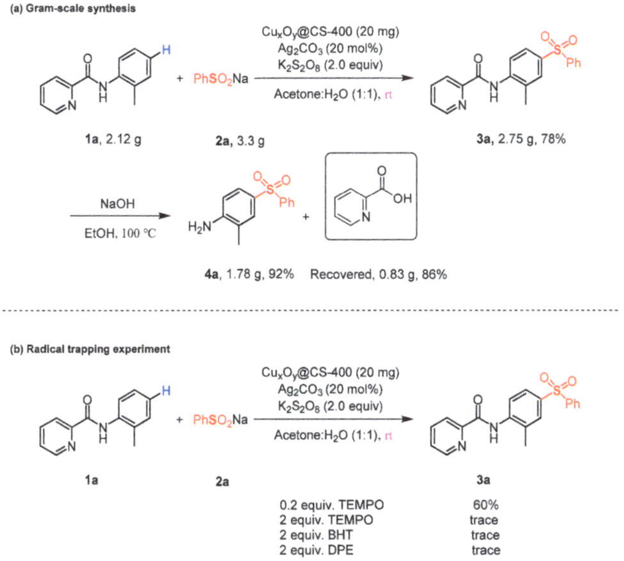

Scheme 4. Synthetic applications and control experiment. (**a**) Gram-scale synthesis; (**b**) Radical trapping experiment.

On the basis of the abovementioned investigations, we suggest that the sulfonylation of anilines with sodium sulfinate occurs by the mechanism described in Scheme 5 [14–16,40]. Firstly, the Cu$_x$O$_y$@CS-400 catalyst and *N*-(o-toly)picolinamide (**1a**) create an aninicimidaate–copper complex **A**. The complex **A** took place as an intermolecular single electron transfer (SET) between the o-toluidine and K$_2$S$_2$O$_8$, furnishing the radical complex **B**. Meanwhile, the benzene sulfonyl radical was produced by reacting sodium benzenesulfinate (**2a**) with silver salts and oxidants. Next, the formation of complex **C** occurred by a radical coupling reaction between radical complex B and the benzenesulfonyl radical. Ultimately, complex **C** undergoes a proton transfer process to form complex **D**, whose dissociation yields the intended product **3a** and involves catalyst regeneration to complete the catalytic cycle.

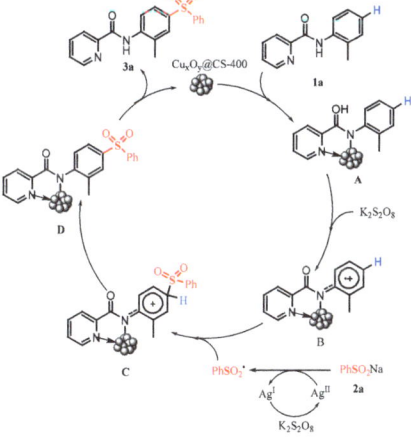

Scheme 5. Proposed mechanism.

The appealing elements of heterogeneous catalysis include the use of milder reaction conditions and the option of reusing catalysts. In order to assess the effectiveness and demonstrate the benefits of this approach, the viability of reusing the catalyst was further investigated. After the reaction, the Cu_xO_y@CS-400 catalyst was simply recovered by filtration and washed with H_2O and EtOH. It was reused in the sulfonylation reaction of N-(o-tolyl)picolinamide and benzenesulfinate, and the results are shown in Figure 3. For the recycling experiment, the Cu_xO_y@CS-400 catalyst that had been separated was recharged with a new substrate to be used again in the following run while keeping the same reaction conditions. It was observed that the catalyst maintained its catalytic activity even after being reused five times.

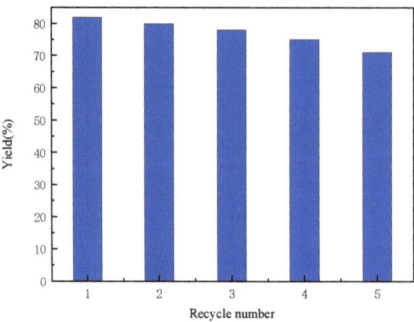

Figure 3. Catalyst reutilization studies.

3. Experimental Section

3.1. Chemicals and Materials

Chitosan powder (MW: 10,000–50,000, deacetylation degree 95%, purchased from Aladdin reagent (Shanghai) Co., Ltd., Shanghai, China) was used without further purification. Acetophenone and sulfinic acid salts were purchased from Alfa Aesar. Other chemicals were obtained commercially and used without any prior purification. ^1H NMR spectra were recorded on a Bruker AvanceII 400 spectrometer using TMS as the internal standard. All products were isolated by short chromatography on a silica gel (200–300 mesh) column using petroleum ether (60–90 °C) unless otherwise noted. All compounds were characterized by ^1H NMR, ^{13}C NMR.

3.2. General Procedure for Synthesis of Biomass-Derived Copper Catalysts

We dissolved anhydrous copper acetate (90.83 mg, 0.500 mmol) in H_2O (40 mL) in a 100 mL round-bottom flask equipped with a reflux condenser and magnetic stir bar. Then, chitosan (690 mg) was added to obtain a suspension, which was stirred at 50 °C for 3 h. After the solution was cooled to room temperature, H_2O was slowly removed under reduced pressure. The light-blue solid obtained was dried under vacuum at 60 °C for 12 h. We transferred the dried sample to a porcelain boat and placed it in the oven. The oven was evacuated and purged with nitrogen for 30 min. Then, we heated the oven to the appropriate temperature (e.g., 300 °C, 400 °C, 500 °C) with a temperature gradient of 2 °C/min and maintained the same temperature under a nitrogen atmosphere for 2 h. Then, we let the oven cool to room temperature. Throughout the process, the furnace was continuously purged with nitrogen. The prepared catalysts were stored in screw-cap vials at room temperature without special air protection.

3.3. General Procedure for C–H Sulfonylation of Anilines Derivatives

We used a 25 mL Schlenk tube equipped with a magnetic stir bar and added N-phenylpicolinamide derivative **1** (0.2 mmol), sodium sulfinate **2** (0.4 mmol), Cu_xO_y@CS-400 (20 mg), Ag_2CO_3 (20 mol%), and $K_2S_2O_8$ (2.0 equiv.) in acetone/H_2O (1:1) (3.0 mL). The resulting mixture was stirred at room temperature for 3 h in air. After completion, the mixture was added to H_2O (20 mL) and extracted three times with ethyl acetate (10 mL). The combined organic layer was dried over anhydrous Na_2SO_4 and filtered. After the solvent was evaporated in a vacuum, the residue was purified by silica gel column chromatography using petroleum ether/ethyl acetate as a detergent to obtain pure product 3.

4. Conclusions

In conclusion, the remote C–H sulfonylation of anilines with sodium sulfinate at room temperature was facilitated by a biomass-based Cu_xO_y@CS-400 catalyst, which could be recycled and produced the required products in moderate to good yields. The technique exhibited many advantages, including straightforward and gentle reaction conditions, a small impact on the environment, reduced energy consumption, and excellent tolerance towards functional groups. In addition, the control experiments demonstrated the involvement of a radical route in the reaction. More importantly, the catalyst was reutilized five times without a considerable deterioration in catalytic efficiency.

Supplementary Materials: The following supporting information can be downloaded at: https://www.mdpi.com/article/10.3390/molecules29204815/s1. Figure S1: XPS spectra of Cu_xO_y@CS-400: (a) Survey spectrum. (b) Cu 2p. (c) O 1s. (d) Cu LMM. (e) XRD spectra of Cu_xO_y@CS-400. Reference [41] is cited in the supplementary materials.

Author Contributions: X.Y.: Data curation, Investigation, Writing—original draft. J.W.: Methodology and Writing—review and editing. C.C.: Revision and Writing—review and editing. K.Z.: Revision and Writing—review and editing. P.Z.: Funding acquisition, Methodology, and Writing—review and editing. C.S.: Funding acquisition, Methodology, and Writing—review and editing. All authors have read and agreed to the published version of the manuscript.

Funding: This work was supported by the Key Research & Development Project of Science Technology Department of Zhejiang Province (No. 2024C01203), the Zhejiang Shuren University Basic Scientific Research Special Funds.

Institutional Review Board Statement: Not applicable.

Informed Consent Statement: Not applicable.

Data Availability Statement: The original contributions presented in this study are included in the article/Supplementary Materials. Further inquiries can be directed to the corresponding author/s.

Conflicts of Interest: The authors declare that they have no known competing financial interests or personal relationships that could have appeared to influence the work reported in this paper.

References

1. Liang, S.; Shaaban, S.; Liu, N.; Hofman, K.; Manolikakes, G. Recent advances in the synthesis of C-S bonds via metal-catalyzed or -mediated functionalization of C-H bonds. *Adv. Organonet. Chem.* **2018**, *69*, 135–207. [CrossRef]
2. Feng, M.; Tang, B.; Liang, H.S.; Jiang, X. Sulfur containing scaffolds in drugs: Synthesis and application in medicinal chemistry. *Curr. Top. Med. Chem.* **2016**, *16*, 1200–1216. [CrossRef] [PubMed]
3. Xu, J.; Shen, C.; Qin, X.; Wu, J.; Zhang, P.; Liu, X. Oxidative sulfonylation of hydrazones enabled by synergistic Copper/Silver catalysis. *J. Org. Chem.* **2021**, *86*, 3706–3720. [CrossRef]
4. Rezaeifard, A.; Jafarpour, M.; Naeimi, A.; Haddad, R. Aqueous heterogeneous oxygenation of hydrocarbons and sulfides catalyzed by recoverable magnetite nanoparticles coated with copper(II) phthalocyanine. *Green Chem.* **2012**, *14*, 3386–3394. [CrossRef]
5. Amarnath Reddy, M.; Surendra Reddy, P.; Sreedhar, B. Iron(III) chloride-catalyzed direct sulfonylation of alcohols with sodium arenesulfinates. *Adv. Synth. Catal.* **2010**, *352*, 1861–1869. [CrossRef]
6. Shen, C.; Zhang, P.; Sun, Q.; Bai, S.; Andy Hor, T.S.; Liu, X. Recent advances in C–S bond formation via C–H bond functionalization and decarboxylation. *Chem. Soc. Rev.* **2015**, *44*, 291–314. [CrossRef]
7. Shaaban, S.; Liang, S.; Liu, N.; Manolikakes, G. Synthesis of sulfones via selective C–H-functionalization. *Org. Biomol. Chem.* **2017**, *15*, 1947–1955. [CrossRef] [PubMed]
8. Xu, J.; Shen, C.; Zhu, X.; Zhang, P.; Ajitha, M.J.; Huang, K.; An, Z.; Liu, X. Remote C–H activation of quinolines through copper-catalyzed radical cross-coupling. *Chem. Asian J.* **2016**, *11*, 882–892. [CrossRef]

9. Lu, F.; Li, J.; Wang, T.; Li, Z.; Jiang, M.; Hu, X.; Pei, H.; Yuan, F.; Lu, L.; Lei, A. Electrochemical oxidative C-H sulfonylation of anilines. *Asian J. Org. Chem.* **2019**, *8*, 1838–1841. [CrossRef]
10. Johnson, T.C.; Elbert, B.L.; Farley, A.T.M.; Gorman, T.W.; Genicot, C.; Lallemand, B.; Pasau, P.; Flasz, J.; Schofield, C.J.; Smith, M.D.; et al. Direct sulfonylation of anilines mediated by visible light. *Chem. Sci.* **2017**, *55*, 12212–12215. [CrossRef]
11. Sherman, E.S.; Chemler, S.R.; Tan, T.B.; Gerlits, O. Copper(II) acetate promoted oxidative cyclization of arylsulfonyl-o-allylanilines. *Org. Lett.* **2004**, *6*, 1573–1575. [CrossRef]
12. Alizadeh, A.; Khodaei, M.M.; Nazari, E. Rapid and mild sulfonylation of aromatic compounds with sulfonic acids via mixed anhydrides using Tf$_2$O. *Tetrahedron Lett.* **2019**, *48*, 6805–6808. [CrossRef]
13. Sarkar, S.; Sahoo, T.; Sen, C.; Ghosh, S.C. Copper(II) mediated ortho C-H alkoxylation of aromatic amines using organic peroxides: Efficient synthesis of hindered ethers. *Chem. Commun.* **2021**, *51*, 8949–8952. [CrossRef]
14. Liang, S.; Bolte, M.; Manolikakes, G. Copper-catalyzed remote para-C-H functionalization of anilines with sodium and lithium sulfinates. *Chem.—Eur. J.* **2017**, *23*, 96–100. [CrossRef] [PubMed]
15. Bai, P.; Sun, S.; Li, Z.; Qiao, H.; Su, X.; Yang, F.; Wu, Y.; Wu, Y. Ru/Cu photoredox or Cu/Ag catalyzed C4–H sulfonylation of 1-naphthylamides at room temperature. *J. Org. Chem.* **2017**, *82*, 12119–12127. [CrossRef] [PubMed]
16. Zhou, X.; Yu, R.; Wang, J.; Liao, X.; Xiong, Y. Copper-catalyzed remote sulfonylation of 1-naphthylamides with sodium-sulfinates. *Chin. J. Org. Chem.* **2021**, *41*, 4370–4377. [CrossRef]
17. Zhu, J.; Hong, Y.; Wang, Y.; Guo, Y.; Zhang, Y.; Ni, Z.; Li, W.; Xu, J. Synthesis of 1-(halo)alkyl-3-heteroaryl bicyclo[1.1.1]pentanes enabled by a photocatalytic minisci-type multicomponent reaction. *ACS Catal.* **2024**, *14*, 6247–6258. [CrossRef]
18. Guo, Y.; Zhu, J.; Wang, Y.; Li, Y.; Hu, H.; Zhang, P.; Xu, J.; Li, W. General and modular route to (halo)alkyl BCP-heteroaryls enabled by α-aminoalkyl radical-mediated halogen-atom transfer. *ACS Catal.* **2024**, *14*, 619–627. [CrossRef]
19. Huang, L.; Xu, J.; He, L.; Liang, C.; Ouyang, Y.; Yu, Y.; Li, W.; Zhang, P. Rapid Alkenylation of quinoxalin-2(1*H*)-ones enabled by the sequential mannich-type reaction and solar photocatalysis. *Chin. Chem. Lett.* **2021**, *32*, 3627–3631. [CrossRef]
20. Xu, J.; Liang, C.; Shen, J.; Chen, Q.; Li, W.; Zhang, P. Photoinduced, metal- and photosensitizer-free decarboxylative C-H (amino)alkylation of heteroarenes in a sustainable solvent. *Green Chem.* **2023**, *25*, 1975–1981. [CrossRef]
21. Zhu, J.; Guo, Y.; Zhang, Y.; Li, W.; Zhang, P.; Xu, J. Visible-light-induced direct perfluoroalkylation/heteroarylation of [1.1.1]propellane to diverse bicyclo[1.1.1]pentanes (BCPs) under metal and photocatalyst-free conditions. *Green Chem.* **2023**, *25*, 986–992. [CrossRef]
22. Zhang, L.; Wang, Y.; Shen, J.; Xu, H.; Shen, C. Platform for 3-fluoro-3-hydroxyoxindoles: Photocatalytic C-N cross-coupling and deaminative oxidation-fluorohydroxylation. *Org. Chem. Front.* **2024**, *11*, 2727–2732. [CrossRef]
23. Zhang, L.; Zheng, K.; Zhang, P.; Jiang, M.; Shen, J.; Chen, C.; Shen, C. Visible-light-enabled multicomponent synthesis of trifluoromethylated 3-indolequinoxalin- 2(1*H*)-ones without external photocatalysis. *Green Syn. Catal.* **2024**, *5*, 51–56. [CrossRef]
24. Lu, W.; Mao, J.; Xing, J.; Tang, H.; Liao, J.; Quan, Y.; Lu, Z.; Yang, Z.; Shen, C. Palladium-catalyzed synthesis of indanone via C–H annulation reaction of aldehydes with norbornenes. *J. Org. Chem.* **2024**, *89784*–*89792*. [CrossRef]
25. Zheng, K.; Liang, K.; Zhu, J.; Chen, H.; Zhang, P.; Shen, C.; Cao, J. Self-catalytic photochemical three-component reaction for the synthesis of multifunctional 3,3-disubstituted oxindoles. *Mol. Catal.* **2024**, *565*, 114379. [CrossRef]
26. Zhu, D.; Zheng, K.; Qiao, J.; Xu, H.; Chen, C.; Zhang, P.; Shen, C. One-step synthesis of PdCu@Ti$_3$C$_2$ with high catalytic activity in the Suzuki-Miyaura coupling reaction. *Nanoscale Adv.* **2022**, *4*, 3362–3369. [CrossRef]
27. Zhou, E.; Jin, J.; Zheng, K.; Zhang, L.; Xu, H.; Shen, C. Novel recyclable Pd/H-MOR catalyst for Suzuki-Miyaura coupling and application in the synthesis of crizotinib. *Catalysts* **2021**, *11*, 1213. [CrossRef]
28. Wang, Z.; Dai, L.; Yao, J.; Guo, T.; Hrynsphan, D.; Tatsiana, S.; Chen, J. Enhanced adsorption and reduction performance of nitrate by Fe-Pd-Fe$_3$O$_4$ embedded multi-walled carbon nanotubes. *Chemosphere* **2021**, *281*, 130718. [CrossRef]
29. Wang, Z.; Fu, W.; Hu, L.; Zhao, M.; Guo, T.; Hrynsphan, D.; Tatsiana, S.; Chen, J. Improvement of electron transfer efficiency during denitrification process by Fe-Pd/multi-walled carbon nanotubes: Possessed redox characteristics and secreted endogenous electron mediator. *Sci. Total Environ.* **2021**, *781*, 146686. [CrossRef]
30. Lv, S.; Zheng, F.; Wang, Z.; Dai, L.; Liu, H.; Hrynsphan, D.; Tatsiana, S.; Chen, J. Effects of bamboo-charcoal modified by bimetallic Fe/Pd nanoparticles on n-hexane biodegradation by bacteria Pseudomonas mendocina NX-1. *Chemosphere* **2023**, *318*, 137897. [CrossRef]
31. Shen, C.; Qiao, J.; Zhao, L.; Zheng, K.; Jin, J.; Zhang, P. An efficient silica supported chitosan@vanadium catalyst for asymmetric sulfoxidation and its application in the synthesis of esomeprazole. *Catal. Commun.* **2017**, *92*, 114–118. [CrossRef]
32. Shen, C.; Xu, J.; Ying, B.; Zhang, P. Heterogeneous chitosan@copper(II)-catalyzed remote trifluoromethylation of aminoquinolines with the Langlois reagent by radical cross-coupling. *ChemCatChem* **2016**, *8*, 3560–3564. [CrossRef]
33. Shen, C.; Shen, H.; Yang, M.; Xia, C.; Zhang, P. Novel D-glucosamine-derived pyridyl-triazole@palladium catalyst for solvent-free Mizoroki-Heck reactions and its application in the synthesis of axitinib. *Green Chem.* **2015**, *17*, 225–230. [CrossRef]
34. Shen, C.; Xu, J.; Yu, W.; Zhang, P. A highly active and easily recoverable chitosan@copper catalyst for the C-S coupling and its application in the synthesis of Zolimidine. *Green Chem.* **2014**, *16*, 3007–3012. [CrossRef]
35. Li, S.; Wang, J.; Jin, J.; Tong, J.; Shen, C. Recyclable cellulose-derived Fe$_3$O$_4$/Pd NPs for Highly Selective C-S formation by heterogeneously C-H sulfenylation of indoles. *Catal. Lett.* **2020**, *150*, 2409–2414. [CrossRef]
36. Qiao, J.; Wang, T.; Zheng, K.; Zhou, E.; Shen, C.; Jia, A.; Zhang, Q. Magnetically reusable Fe$_3$O$_4$@NC@Pt catalyst for selective reduction of nitroarenes. *Catalysts* **2021**, *11*, 1219. [CrossRef]

37. Zheng, K.; Zhou, E.; Zhang, L.; Zhang, L.; Yu, W.; Xu, H.; Shen, C. Catalyst controlled remote C–H activation of 8-aminoquinolines with NFSI for C–N versus C–F coupling. *Catal. Commun.* **2021**, *158*, 106336–106344. [CrossRef]
38. Lin, Z.; Jin, J.; Qiao, J.; Tong, J.; Shen, C. Facile fabrication of glycosylpyridyl-triazole@Nickel nanoparticles as recyclable nanocatalyst for acylation of amines in Water. *Catalysts* **2020**, *10*, 230. [CrossRef]
39. Zhao, L.; Zheng, K.; Tong, J.; Jin, J.; Shen, C. Novel biomass derived Fe_3O_4@Pd NPs as efficient and sustainable nanocatalyst for nitroarene reduction in aqueous media. *Catal. Lett.* **2019**, *149*, 2607–2613. [CrossRef]
40. Li, J.; Wang, Y.; Yu, Y.; Wu, R.; Weng, J.; Llu, G. Copper-catalyzed remote C–H functionalizations of naphthylamides through a coordinating activation strategy and single-electron-transfer (SET) mechanism. *ACS Catal.* **2017**, *7*, 2661–2667. [CrossRef]
41. Hangzhou Vocational & Technical College. Sulfonyl Pyridine Amide Derivatives and Its Preparation Method. CN112142656 A, 29 December 2020.

Disclaimer/Publisher's Note: The statements, opinions and data contained in all publications are solely those of the individual author(s) and contributor(s) and not of MDPI and/or the editor(s). MDPI and/or the editor(s) disclaim responsibility for any injury to people or property resulting from any ideas, methods, instructions or products referred to in the content.

MDPI AG
Grosspeteranlage 5
4052 Basel
Switzerland
Tel.: +41 61 683 77 34

Molecules Editorial Office
E-mail: molecules@mdpi.com
www.mdpi.com/journal/molecules

Disclaimer/Publisher's Note: The title and front matter of this reprint are at the discretion of the Guest Editors. The publisher is not responsible for their content or any associated concerns. The statements, opinions and data contained in all individual articles are solely those of the individual Editors and contributors and not of MDPI. MDPI disclaims responsibility for any injury to people or property resulting from any ideas, methods, instructions or products referred to in the content.

www.ingramcontent.com/pod-product-compliance
Lightning Source LLC
LaVergne TN
LVHW072327090526
838202LV00019B/2366